A Course in Computational Number Theory

A Course in Computational Number Theory

David Bressoud
Macalester College

Stan Wagon
Macalester College

CD-ROM
Included

Key College Publishing
Innovators in Higher Education

www.keycollege.com

in cooperation with

Springer

David Bressoud
Macalester College
St. Paul, MN 55105
bressoud@macalester.edu

Stan Wagon
Macalester College
St. Paul, MN 55105
wagon@macalester.edu

About the Cover: The disk on the cover shows the prime numbers in the Gaussian integers ($m + n\sqrt{-1}$, where m and n are integers), colored according to their reachability from the one closest to the origin. The yellow primes are reachable using steps of size at most $\sqrt{2}$; the blue primes are reachable using steps of size at most 2; and the red primes are the network based on size $\sqrt{8}$. The background is a close-up of the first quadrant. A famous unsolved problem (see Section 9.2) asks whether such a reachability set can ever be infinite.

Key College Publishing was founded in 1999 as a division of Key Curriculum Press in cooperation with Springer-Verlag New York, Inc. It publishes innovative curriculum materials for undergraduate courses in mathematics, statistics, and other physical sciences.

Key College Publishing
1150 65th Street
Emeryville, CA 94608
www.keycollege.com
(510) 595-7000

Library of Congress Cataloging-in-Publication Data
Bressoud, David M., 1950–
    A course in computational number theory / David Bressoud, Stan
  Wagon.
      p. cm.
    Includes bibliographical references and index.
    ISBN 1-930190-10-7 (alk. paper)
    1. Number theory. 2. Algorithms. I. Wagon, S. II. Title.
QA241.B788 1999
512'.7—dc21   99-16037

Editorial Director: Jeremiah J. Lyons
Production Editor: Steven Pisano
Manufacturing Manager: Erica Bresler
Text and Cover Design: Wanda Kossak
Composition: Louis J. D'Andria (using the authors' *Mathematica* files)
Printer/Binder: Hamilton Printing Co., Rensselaer, NY

Printed in the United States of America.

9 8 7 6 5 4 3 2 1   04 03 02 01 00

ISBN 1-930190-10-7 Key College Publishing SPIN 10764541

Preface

This book is a successor to *Factorization and Primality Testing* (Bressoud, Springer-Verlag, 1989), but it is a totally new creation and should not be considered a new edition. The two books are both introductions to number theory, structured so that the motivation arises from interesting computational problems, but that is almost all they have in common.

In 1989, when *Factorization and Primality Testing* appeared, the two computational problems of the title were still sufficiently undeveloped that much of what was known could be presented in a first course in number theory, providing the motivation for that course. Today, that is no longer true. The field of factorization and primality testing has advanced so far in the past decade that no slim book could both introduce number theory to undergraduates and do justice to current techniques of factorization and primality testing.

While this book retains a discussion of the basic methods of factorization and primality testing, the emphasis has shifted to a broader range of computational issues. More is done with continued fractions, pseudoprime tests, Gaussian primes, Pell's equation, and computer explorations in general. And there are presentations of several applications that have not appeared in texts at this level before: the Madelung constant for salt, the Yao millionaire problem as an extension of the ideas of RSA encryption, and the use of check digits both to catch errors and to correct errors. The entire topic of elliptic curves has been avoided. There is not space to do it justice. And the detailed discussion of the quadratic sieve has been replaced by a detailed discussion of the continued fraction algorithm (CFRAC). While CFRAC is no longer one of the most powerful factorization algorithms, neither is the simplistic version of the multiple polynomial quadratic sieve that was presented in *Factorization and Primality Testing*. The advantage of CFRAC is that it illustrates the fundamental processes common to all of the Kraitchik family of factorization algorithms, including the quadratic sieves and the number field sieves, while having a relatively simple implementation.

But the greatest difference between these books lies in their style of presentation. The relatively infrequent programs written in pseudocode have been replaced by a rich assortment of *Mathematica* programs that are included on the accompanying CD-ROM. The student using this book does not need to be able to program and should be able to get on the computer quickly to begin doing his or her own experiments with the patterns of the integers. Our presentation of the theoretical structure of number theory is tightly integrated with these explorations, arising out of them to confirm, explain, or deny what is observed experimentally, and feeding back into them to enable ever deeper and more sophisticated investigations. For those who want to program, there is a rich assortment of problems, challenging the student to turn the theory into practical algorithms.

This is designed to be a one-semester introduction to number theory, though there is more material than can be covered in a single semester. The book is more than a text, and we hope that the reader will continue to return to it for ideas and challenges.

We are grateful to the students of our number theory classes at Macalester who have always been enthusiastic about the use of computers in their explorations. In particular, the work of Celine Liu, Einar Mykletun, Tamas Nemeth, Craig Ortner, Bill Owens, and Scott Turnquest has led to improvements in the package and exposition. We thank Daniel Bleichenbacher and Ilan Vardi for sharing their expertise, both theoretical and computational, on several topics covered in the book. The comments of Ellen Gethner and Patrick Mitchell, who used a preliminary version in their classes, were much appreciated. The careful production work of Steven Pisano (Springer-Verlag) and Louis J. D'Andria (Wolfram Research, Inc.) was superb. We thank them. We also thank our editor Jerry Lyons for his confidence in this project and his feelings that the incorporation of computational ideas in traditional courses is the right direction for contemporary education.

The CNT package that accompanies this book (*Mathematica* is required for its use) has a lot of powerful techniques and ideas encoded in it. Readers who feel it would benefit from the inclusion of additional functions should feel free to communicate with the authors. We have prepared a file of solutions to selected exercises and will provide that to instructors on request.

David Bressoud
bressoud@macalester.edu

Stan Wagon
wagon@macalester.edu

Contents

Preface   **v**

Notation   **xi**

Chapter 1. Fundamentals   **1**

1.0 Introduction.. 1

1.1 A Famous Sequence of Numbers... 2

1.2 The Euclidean Algorithm.. 6

      The Oldest Algorithm
      Reversing the Euclidean Algorithm
      The Extended GCD Algorithm
      The Fundamental Theorem of Arithmetic
      Two Applications

1.3 Modular Arithmetic.. 25

1.4 Fast Powers.. 30

      A Fast Algorithm for Exponentiation
      Powers of Matrices, Big-O Notation

Chapter 2. Congruences, Equations, and Powers   **41**

2.0 Introduction.. 41

2.1 Solving Linear Congruences.. 41

      Linear Diophantine Equations in Two Variables
      Linear Equations in Several Variables
      Linear Congruences
      The Conductor
      An Important Quadratic Congruence

2.2 The Chinese Remainder Theorem....................................... 49

2.3 PowerMod Patterns.. 55

    Fermat's Little Theorem
    More Patterns in Powers

2.4 Pseudoprimes... 59

    Using the Pseudoprime Test

Chapter 3. Euler's ϕ Function   **65**

3.0 Introduction... 65

3.1 Euler's ϕ Function.. 65

3.2 Perfect Numbers and Their Relatives 72

    The Sum of Divisors Function
    Perfect Numbers
    Amicable, Abundant, and Deficient Numbers

3.3 Euler's Theorem... 81

3.4 Primitive Roots for Primes.. 84

    The Order of an Integer
    Primes Have Primitive Roots
    Repeating Decimals

3.5 Primitive Roots for Composites... 90

3.6 The Universal Exponent.. 93

    Universal Exponents
    Power Towers
    The Form of Carmichael Numbers

Chapter 4. Prime Numbers   **99**

4.0 Introduction... 99

4.1 The Number of Primes.. 100

    We'll Never Run Out of Primes
    The Sieve of Eratosthenes
    Chebyshev's Theorem and Bertrand's Postulate

4.2 Prime Testing and Certification.. 114

    Strong Pseudoprimes
    Industrial-Grade Primes
    Prime Certification Via Primitive Roots
    An Improvement
    Pratt Certificates

4.3 Refinements and Other Directions.. 131

    Other Primality Tests
    Strong Liars Are Scarce
    Finding the nth Prime

4.4 A Dozen Prime Mysteries.. 141

Chapter 5. Some Applications   **145**

5.0 Introduction... 145

5.1 Coding Secrets.. 145

    Tossing a Coin into a Well
    The RSA Cryptosystem
    Digital Signatures

5.2 The Yao Millionaire Problem................................. 155

5.3 Check Digits... 158

    Basic Check Digit Schemes
    A Perfect Check Digit Method
    Beyond Perfection: Correcting Errors

5.4 Factoring Algorithms... 167

    Trial Division
    Fermat's Algorithm
    Pollard Rho
    Pollard $p - 1$
    The Current Scene

Chapter 6. Quadratic Residues   **179**

6.0 Introduction... 179

6.1 Pépin's Test... 179

    Quadratic Residues
    Pépin's Test
    Primes Congruent to 1 (Mod 4)

6.2 Proof of Quadratic Reciprocity............................. 185

    Gauss's Lemma
    Proof of Quadratic Reciprocity
    Jacobi's Extension
    An Application to Factoring

6.3 Quadratic Equations.. 194

Chapter 7. Continued Fractions   **201**

7.0 Introduction... 201

7.1 Finite Continued Fractions..................................... 202

7.2 Infinite Continued Fractions................................... 207

7.3 Periodic Continued Fractions................................. 213

7.4 Pell's Equation... 227

7.5 Archimedes and the Sun God's Cattle................... 232

    Wurm's Version: Using Rectangular Bulls
    The Real Cattle Problem

7.6 Factoring via Continued Fractions......................... 238

Chapter 8. Prime Testing with Lucas Sequences   **247**

8.0 Introduction.. 247

8.1 Divisibility Properties of Lucas Sequences.. 248

8.2 Prime Tests Using Lucas Sequences.. 259

      Lucas Certification
      The Lucas–Lehmer Algorithm Explained
      Lucas Pseudoprimes
      Strong Quadratic Pseudoprimes
      Primality Testing's Holy Grail

Chapter 9. Prime Imaginaries and Imaginary Primes   **279**

9.0 Introduction.. 279

9.1 Sums of Two Squares.. 279

      Primes
      The General Problem
      How Many Ways
      Number Theory and Salt

9.2 The Gaussian Integers... 302

      Complex Number Theory
      Gaussian Primes
      The Moat Problem
      The Gaussian Zoo

9.3 Higher Reciprocity... 325

Appendix A. *Mathematica* Basics   **333**

A.0 Introduction... 333

A.1 Plotting... 335

A.2 Typesetting.. 338

      Sending Files By E-Mail

A.3 Types of Functions... 341

A.4 Lists... 343

A.5 Programs... 345

A.6 Solving Equations.. 347

A.7 Symbolic Algebra.. 349

Appendix B. Lucas Certificates Exist   **351**

References   **355**

Index of *Mathematica* Objects   **359**

Subject Index   **363**

Notation

Mathematics Notation

τ	The golden ratio, $\left(1+\sqrt{5}\right)/2$
ω	The rank of the Lucas sequence
$\overline{\alpha}$	The quadratic conjugate of the quadratic irrational α
Λ	Lucas lambda function
$\Phi_{(P,Q)}$	The analog for the $\{P,\ Q\}$-Lucas sequence of the Euler ϕ function
ϕ	The Euler phi function: the number of integers in $\{1,\ ...,\ n\}$ that are relatively prime to n
F_n	The nth Fibonacci number; also the nth Fermat number, 2^{2^n}
$\pi(x)$	The number of primes less than or equal to x
$\mathbb{Z}$	The integers, $\{...,\ -3,\ -2,\ -1,\ 0,\ 1,\ 2,\ 3,\ ...\}$
$\mathbb{Z}_n$	The integers modulo n
$\mathbb{Z}_n^*$	The integers modulo n that are relatively prime to n
$\mathbb{Z}[i]$	The Gaussian integers: $\{a+bi:a,\ b\in\mathbb{Z}\}$
$[a_0;a_1,\ a_2,\ ...,\ a_n]$	The finite continued fraction $a_0+\cfrac{1}{a_1+\cfrac{1}{a_2+\cfrac{}{\ddots+\cfrac{1}{a_{n-1}+\cfrac{1}{a_n}}}}}$
$[a_0;a_1,\ a_2,\ ...]$	The infinite continued fraction $a_0+\cfrac{1}{a_1+\cfrac{1}{a_2+...}}$
$[a_0;a_1,\ a_2,\ ...,\ a_s,\ \overline{b_1,\ b_2,\ ...,\ b_n}]$	The periodic continued fraction $[a_0;a_1,\ a_2,\ ...,\ a_s,\ b_1,\ b_2,\ ...,\ b_n,\ b_1,\ b_2,\ ...,\ b_n,\ b_1,\ ...]$
$\mathbb{N}$	The natural numbers $\{0,\ 1,\ 2,\ 3,\ ...\}$
λ	Carmichael's lambda function: the least universal exponent
$\text{li}(x)$	The logarithmic integral function, $\int_0^x \frac{1}{t}dt$
$\log(x)$	The natural logarithm of x
$\sigma(n)$	The sum of the divisors of n
$\sigma_k(n)$	The sum of the kth powers of the divisors of n
$\sigma^-(n)$	The sum of the divisors of n, excluding n
$\sigma(n,\ 1)$	The sum of the divisors of n that are congruent to 1 (mod 4)
$\tau(n)$	The number of divisors of n
$\phi(n)$	The Euler phi function: the number of integers in $\{1,\ ...,\ n\}$ that are relatively prime to n
$\gcd(a,\ b)$	The greatest common divisor

$\text{lcm}(a, b)$	The least common multiple
$\log^*(a, s)$	The largest integer n such that $a \uparrow n \le s$
$O(f(n))$	Big-O: a function that is bounded by a constant times $f(n)$
$Q[\{q_1, \ldots, q_k\}]$	The continuant function
$a \mid n$	a divides n
$a \nmid n$	a does not divide n
$a \equiv b \pmod{n}$	a is congruent to b modulo n
$a \not\equiv b \pmod{n}$	a is not congruent to b modulo n
$N(\alpha)$	Norm of a Gaussian integer α
$V_\phi(n)$	The ϕ-valence of n: the number of times n occurs as $\phi(m)$
$\left(\frac{a}{p}\right)$	The Legendre symbol (and the Jacobi symbol) of a with respect to p
$\left(\frac{\alpha}{\pi}\right)_4$	The quartic residue symbol, where π is a prime in the Gaussian integers
$\text{ord}_m(a)$	The order of a modulo m
$a \uparrow n$	The tower of powers with n terms: $a^{a^{a^{\cdot^{\cdot^{\cdot}}}}}$

Mathematica Notation

#, #1, #2	The generic variables in pure functions
/@	Map
&	A delimiter for the end of a pure function
&&	And
||	Or
@@	Apply
//	Notation allowing a function to be applied after its argument (postfix notation)
/.	ReplaceAll
/;	Conditional
_	Blank: stands for any pattern
__	Double blank: stands for a nonempty sequence of objects
___	Triple blank: stands for a sequence of objects, possibly empty
:=	Delayed assignment
->, →	A rule
:>, :→	A delayed rule
==	Is numerically equal to
===	Is identical too
!	Not
<<	Get (the file)

CHAPTER 1

Fundamentals

1.0 Introduction

Consider the following sequence of rational numbers:

$$\frac{1}{1}, \frac{3}{2}, \frac{7}{5}, \frac{17}{12}, \frac{41}{29}, \frac{99}{70}, \dots . \tag{1}$$

After starting with $^{1}/_{1}$, the next fraction is obtained by adding numerator and denominator to get the new denominator, and then adding the old and new denominators to get the next numerator. Compare these values to the square root of 2. These are not only good approximations, they are fantastically good. The more familiar approximation $^{141}/_{100}$ is in error by no more than $^{1}/_{200}$. But the error for a fraction in (1) is less than the reciprocal of twice the *square* of the denominator; for example, $^{99}/_{70} - \sqrt{2} = 0.000072\dots$, while $1/(2 \cdot 70^2) = 0.00010\dots.$

This is an ancient algorithm, known to the Greeks of the classical era and rediscovered in other times and places. A more modern algorithm, one that is less than 150 years old, is the following test for whether or not $m = 2^n - 1$ is prime when n is an odd integer greater than 2. Start with $S = 4$ and iterate the following sequence of commands $n - 2$ times: square S, subtract 2, divide the resulting number by m, and assign to S the value of the remainder that you get from the division. After $n - 2$ iterations, look at the value of S. If it is 0, then m is prime; if it is not 0, then m is not prime. For example, when n is 5 and $m = 2^5 - 1 = 31$ we get the following sequence of values for S:

$$\mathbf{4}, \; 4^2 - 2 = \mathbf{14}, \; 14^2 - 2 = 194 = 6 \cdot 31 + \mathbf{8}, \; 8^2 - 2 = 62 = 2 \cdot 31 + \mathbf{0} .$$

Given that this algorithm is valid, 31 must be prime. In 1999, this algorithm was used to prove that $2^{6972593} - 1$ is prime, setting the record for the world's largest known prime.

These are two of the algorithms of number theory. We shall see an explanation of the first algorithm as well as an exploration of its extension to other square roots and eventually to other irrational numbers in Chapter 7. The second algorithm will be explained and justified in Chapter 8.

Number theory, the study of properties of the integers, has long been inseparable from **algorithms**, precisely specified procedures that produce desired output. To find efficient numerical algorithms, we must understand the structure of the integers. To explore this structure, we are aided by the algorithms that we already possess. The advent of the electronic computer has strengthened this symbiosis. Raw computer power is seldom sufficient for really interesting problems.

Factorization of large integers gives us an illustration of the need to understand the structure of the integers. Integers 100 digits long can now be factored routinely. It might seem that this is a problem for which a naive approach and a lot of computer power would work well: you generate a list of prime numbers and just run trial division until you find a prime divisor. As we shall see, generating the possible prime divisors is easy. The problem is that there are too many of them. If the smallest prime divisor of our integer has 50 digits (and these days, that is not considered to be especially large), then the Prime Number Theorem (Chapter 4) tells us that there are about $8.7 \cdot 10^{47}$ primes that we have to test before we get to this divisor. That is a lot of trial divisions!

If we imagine an ideal processor that does a trillion (10^{12}) trial divisions per second, and set up a million of these processors in parallel, each processor taking a different subset of the possible divisors, we would be able to test 10^{18} different primes each second. That sounds like a lot, but it is not. There are about $3.2 \cdot 10^7$ seconds in a year, so it would take over 10^{22} years to complete the factorization. Our universe is less than $2 \cdot 10^{10}$ years old.

In fact, there is no number theory problem that stretches our understanding of integers so much as the problem of factorization. In recent years, it has seemed that whatever new insight we gain into the structure of the integers, it is directly applicable to the factorization problem or to its twin, the problem of recognition of prime numbers. Algorithms for factorization and primality testing will appear throughout this book.

1.1 A Famous Sequence of Numbers

The following sequence of numbers plays a central role in elementary number theory: 0, 1, 1, 2, 3, 5, 8, 13, ..., where each number is obtained by adding the preceding two. These are the **Fibonacci numbers**, with the indices starting at 0; $F_0 = 0$, $F_1 = 1$, $F_2 = 1$, and so on, with $F_n = F_{n-1} + F_{n-2}$. One could use a recursive method to generate these numbers, but that is very inefficient (see Exercise 1.5). It is much, much better to use an iterative approach, as follows.

```
Fib[0] = 0;
Fib[n_] := Module[{a = 0, b = 1},
  Do[ {a, b} = {b, a + b}, {n - 1}]; b]
```

```
Map[Fib, {0, 1, 2, 3, 4, 5, 6, 7, 100}]
```

{0, 1, 1, 2, 3, 5, 8, 13, 354224848179261915075}

The parallel definition of a and b makes it clear that a matrix formulation will work well, since $\{a, b\}$ is updated to $\begin{pmatrix} 0 & 1 \\ 1 & 1 \end{pmatrix}\begin{pmatrix} a \\ b \end{pmatrix}$. This means that the nth Fibonacci number is the number in the (1, 2) position of $\begin{pmatrix} 0 & 1 \\ 1 & 1 \end{pmatrix}^n$.

```
MatrixPower[{{0, 1}, {1, 1}}, 5] // MatrixForm
```

$\begin{pmatrix} 3 & 5 \\ 5 & 8 \end{pmatrix}$

```
Table[MatrixPower[( 0  1
                     1  1 ), n][[1, 2]], {n, 0, 10}]
```

{0, 1, 1, 2, 3, 5, 8, 13, 21, 34, 55}

In fact, the Fibonacci numbers are built into *Mathematica* as follows.

```
Fibonacci[{10, 100}]
```

{55, 354224848179261915075}

An important question concerns the growth rate of F_n. Figure 1.1(a) shows the raw Fibonacci numbers; not too much is visible except that the growth seems to follow a steady pattern. But plotting the logarithm of F_n will cause the underlying exponential growth rule to jump out, since $F = \alpha^n$ becomes $\log F = n(\log \alpha)$, a linear function of n.

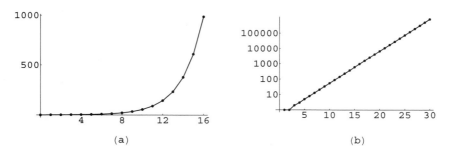

(a) (b)

Figure 1.1. (a) The Fibonacci numbers grow at a regular rate. (b) Plotting $\log F_n$ against n makes the exponential nature of the Fibonacci growth rate jump out.

A beautiful and important fact about the Fibonacci numbers is due to Binet. Let τ denote the **golden ratio**, which is $(1 + \sqrt{5})/2$, or 1.61803...; this famous number, the larger root of the quadratic equation $x^2 = x + 1$, is built into *Mathematica* as GoldenRatio. Note that the other root, which we will denote by σ, is $-1/\tau$. Binet's formula is the surprising result that a certain combination of nth powers of τ and σ equals F_n. The formula is given in Proposition 1.1. Here is a computation in support, done using floating-point arithmetic (caused by the decimal point after the "1").

$$\texttt{Table}\Big[\frac{1.}{\sqrt{5}}\Big(\texttt{GoldenRatio}^{n} - \Big(-\frac{1}{\texttt{GoldenRatio}}\Big)^{n}\Big), \{n, 0, 10\}\Big]$$

$$\{0, 1., 1., 2., 3., 5., 8., 13., 21., 34., 55.\}$$

Proposition 1.1. Binet's Formula.

$$F_n = \frac{\tau^n - \sigma^n}{\sqrt{5}}$$

Proof. It is easy to verify the formula when $n = 0$ or 1. For the latter observe that $1/\tau = (-1 + \sqrt{5})/2$. Thus, it suffices to show that the right side obeys the same recursive rule as the Fibonacci numbers; that is, we need

$$\frac{\tau^n - \sigma^n}{\sqrt{5}} + \frac{\tau^{n+1} - \sigma^{n+1}}{\sqrt{5}} = \frac{\tau^{n+2} - \sigma^{n+2}}{\sqrt{5}} .$$

But this follows immediately from the fact that $\tau^n + \tau^{n+1} - \tau^{n+2} = \tau^n(1 + \tau - \tau^2) = 0$, and the identical relationship that holds for σ. □

Because $|\sigma| < 1$, powers of σ approach 0. This means that Binet's formula has the following corollary.

Corollary 1.2. F_n equals the nearest integer to $\tau^n / \sqrt{5}$.

Proof. Because $|\sigma| < 1$ and $\sqrt{5} > 2$, the $\sigma^n / \sqrt{5}$ part of Binet's formula is, in absolute value, less than $1/2$; this suffices. □

Exercises for Section 1.1

1.1. Prove that the gcd of two consecutive Fibonacci numbers is 1.

1.2. Investigate the ratio F_{n+1}/F_n of consecutive Fibonacci numbers and try to identify the limit.

1.3. Investigate the sum $F_0 + F_1 + \cdots + F_n$. Find a formula for this sum and prove it by induction.

1.4. Investigate the numbers $F_{n-1} \cdot F_{n+1}$. Find a formula for this product and prove it by induction.

1.5. The following code is a recursive definition of the Fibonacci numbers, with a counter added to count the number of additions performed. The output following the code shows that, when `Fib[6]` is called, the resulting value of `count` is 12. Explain where 12 comes from. Do the same for values other than 6. Find a formula for `sumCount` when `Fib[n]` is called. If you do this correctly you will see why this method, simple as it is, is useless for generating Fibonacci numbers.

```
Clear[Fib];
Fib[0] = 0;
Fib[1] = 1;
Fib[n_] := (sumCount ++; Fib[n - 1] + Fib[n - 2])

sumCount = 0;
Fib[6];
sumCount
```

12

1.6. (a) Find all n between 1 and 100 for which F_n is prime. These numbers are large, so use `PrimeQ` or an equivalent function in other software. Such functions are almost always correct, as will be discussed in detail in Chapters 4 and 8. Do you see any patterns?

(b) Prove that F_n divides F_{kn}, and show how this fact helps in the search for prime Fibonacci numbers. *Hint*: Use the fact that $F_m = A^m$ where A is a 2×2 matrix. Show that if a 2×2 matrix B has the property that its two off-diagonal elements are divisible by b, then that property is valid for any power of b.

1.7. (E. Lucas, 1876; see [Jon]) Prove that the pairs of positive integers x, y that satisfy $y^2 - yx - x^2 = \pm 1$ are precisely the pairs of consecutive Fibonacci numbers. Figure 1.2 shows how the pairs of Fibonacci numbers sit on these two hyperbolas. The relationship of the Fibonacci numbers to the theory of Diophantine equations and computers is a deep one; see [Mat, Jon] for the connections with Hilbert's tenth problem. In particular, a central role is played by the fact that there is a Diophantine definition of the set of pairs (n, F_n). This is somewhat more difficult than the result of this exercise, which gives a Diophantine definition of the set of pairs (F_n, F_{n+1}).

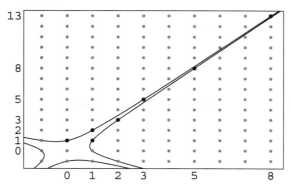

Figure 1.2. Pairs of consecutive Fibonacci numbers live on the two hyperbolas given by $y^2 - yx - x^2 = \pm 1$.

1.2 The Euclidean Algorithm

◼ The Oldest Algorithm

One of the oldest and most important algorithms of number theory is the one that Euclid described at the beginning of Book VII of *The Elements*, written around 300 BCE. The algorithm finds the greatest common divisor of two integers and is essentially unchanged after 2300 years. It appears as a subroutine in many algorithms we will study. Most modern factorization techniques rely on finding integers that are likely to share a nontrivial factor with n, the integer that is to be factored. They then accomplish the factorization by applying the Euclidean algorithm to find this nontrivial factor of n. On a more basic level, computer algebra systems such as *Mathematica* typically use the Euclidean algorithm whenever a rational number is encountered, so that it can remove common factors and express the fraction in lowest terms. An example:

$$\frac{8888888}{31111108}$$

$$\frac{2}{7}$$

Before we explain the Euclidean algorithm, we must pin down our terminology. We say that an integer n is ***divisible*** by the integer d, or that d ***divides*** n, if there is an integer m such that $n = d \cdot m$. If d divides n then it is a ***divisor*** of n. While this definition may look obvious, it will be important for discovering and understanding proofs. An integer larger than 1 that is not divisible by any other integer larger than 1 is called a ***prime*** or a ***prime number***. An integer larger than 1 that is not prime is called ***composite***; thus, a composite integer has a divisor larger than 1 and less than itself. The ***prime factorization*** of a positive integer is a representation of n as a product of primes. Thus, the prime factorization of 620 is $2^2 \cdot 5 \cdot 31$.

```
FactorInteger[620]
```

```
{{2, 2}, {5, 1}, {31, 1}}
```

The CNT package that accompanies this book contains a function, FactorForm, that puts factorizations in familiar form. The //fcn notation is a way of applying a function to an expression without typing brackets at both ends. First we load the package.

```
Needs["CNT`"]
```

```
FactorInteger[30!] // FactorForm
```

$2^{26} \, 3^{14} \, 5^7 \, 7^4 \, 11^2 \, 13^2 \, 17 \, 19 \, 23 \, 29$

The number 1 is neither prime nor composite. It is an example of a *unit*, an integer that divides 1. When working with ordinary integers, there are only two units, the other one being -1. For purposes of factorization, we shall ignore negative integers. The significance of the primes is that they are the building blocks under multiplication for all integers greater than 1.

We assume that we have a division algorithm: given any two integers a and b ($b \neq 0$), we can find the unique integers r and q such that $a = qb + r$, with $0 \leq r < |b|$. The integer r represents the remainder when a is divided by b. *Mathematica* and many other computer languages use Mod for this function: Mod[a, b] is the remainder when a is divided by b. So b divides a if and only if Mod[a, b] is 0.

Mod[25, 7]

4

Mod[10^{100}, 7919]

4806

The quotient — q in the preceding paragraph — can also be obtained quickly. This can be useful in programs because it is faster than Floor[a / b]. (Reason: Quotient avoids the invocation of the Euclidean algorithm to reduce to lowest terms.)

Quotient[25, 7]

3

Because the division algorithm fails in other domains, it is worthwhile to see why it is valid in the integers. The main point is that every set of nonnegative integers has a least element. Thus, the remainder r can be defined as the smallest element of $\{a - qb : q \in \mathbb{Z}, a - qb \geq 0\}$. From this, it is not hard to prove (Exercise 1.8) that $0 \leq r < |b|$ and that r and the choice of q are unique.

The **greatest common divisor**, or **gcd**, of two integers is the largest integer that divides both; it is denoted by $\gcd(a, b)$. Two integers are called **relatively prime** if their gcd is 1. The **least common multiple**, or **lcm**, of two integers is the smallest positive integer that is divisible by both. We leave it as an exercise to prove that for any two positive integers, a and b, $\gcd(a, b) \cdot \text{lcm}(a, b) = ab$ (and so the lcm is easily computed from the gcd). The Euclidean algorithm for calculating the gcd is based on the following fact.

Proposition 1.3. If $a - b$ is divisible by n, then $\gcd(a, n) = \gcd(b, n)$.

Proof. The hypothesis tells us that there is an integer t such that $a - b = nt$. Let d be any common divisor of b and n, say $b = dy$ and $n = dz$. Then $a = b + nt = dy + dzt = d(y + zt)$, so d is also a divisor of a and therefore a common divisor of a and n. And if c is any common divisor of a and n, say $a = cu$ and $n = cv$, then $b = a - nt = cu - cvt = c(u - vt)$. So c divides b and is a common divisor of b and n.

Because the set of common divisors of b and n is identical to the set of common divisors of a and n, the largest elements of these sets must be the same. □

Because a and Mod[a, b] differ by a multiple of b, Proposition 1.1 tells us that gcd(a, b) = gcd(Mod[a, b], b). This is the basis for a recursive definition of the Euclidean algorithm.

Algorithm 1.1. The Euclidean Algorithm

```
gcd[a_, 0] := a;
gcd[a_, b_] := gcd[b, Mod[a, b]]
```

From the definition of Mod[a, b], after each recursive step the second argument is strictly smaller than the first (unless $b > a$, but then they switch roles after the first step). Therefore the algorithm must terminate. The question of the number of steps required is addressed in Theorem 1.6 and Exercise 1.14.

```
gcd[97, 18]
```

1

Alternatively, the algorithm can be defined using a single If statement. Here we add a Print statement so we can watch the algorithm at work.

```
gcd1[a_, b_] := (Print[{a, b}]; If[b == 0, a, gcd1[b, Mod[a, b]]])
```

```
gcd1[97, 18]
```
```
{97, 18}
{18, 7}
{7, 4}
{4, 3}
{3, 1}
{1, 0}
```

1

The values produced by the Euclidean algorithm starting with a, b are called the **remainders**, and we will refer to them as $r_0 = a$, $r_1 = b$, ..., $r_n = d$, $r_{n+1} = 0$, where d = gcd(a, b). The **quotients** are denoted q_1, ..., q_n. For the detailed study of the algorithm later in this chapter it is important to be clear and consistent in the use of subscripts. The FullGCD function shows the progress of the algorithm.

FullGCD`[97, 18, Labels → True]`

Remainders	Quotients
r_0 = a = 97	
r_1 = b = 18	q_1 = 5
r_2 = 7	q_2 = 2
r_3 = 4	q_3 = 1
r_4 = 3	q_4 = 1
r_5 = 1	q_5 = 3
r_6 = 0	

Of course, gcd is built into *Mathematica* (as is lcm, as LCM`[a, b]`).

`GCD[97, 18]`

1

One can appreciate the speed of this algorithm by using very large integers. Here we use two random 1000-digit integers; the computation is almost instantaneous.

`{a, b} = Table[Random[Integer, {`10^{999}`, `10^{1000}` - 1}], {2}];`
`Timing[GCD[a, b]]`

`{0.0166667 Second, 1}`

In a moment we will find the values of a and b below any given bound that cause the Euclidean algorithm to take the largest number of steps. Exercise 1.14 contains some code that one could use to search for such integers.

The Euclidean algorithm is blindingly fast, widely useful, and implementable in a single line of code, all of which combine to make it one of the most beautiful algorithms of number theory.

▮ Reversing the Euclidean Algorithm

It often happens in mathematics that reversing an algorithm leads to new insights. For example, the ability to multiply two integers is something all numerate humans can do. Yet the reversal of this operation — going from an integer n that has the form $p \cdot q$, where p and q are primes, to the individual numbers p, q — is extremely difficult; this observation is the key to modern secure encryption systems.

For several reasons — deeper understanding of the algorithm, its connection with continued fractions (Chapter 7), its use for writing an integer as a sum of squares (Chapter 9) — it is useful to consider how to reverse the Euclidean algorithm. Given the quotients, we try to find the remainders. Suppose the quotients are q_1, ..., q_n, where q_1 is the quotient formed by a/b, and q_n, the last quotient, is r_{n-1}/d. For simplicity assume that a and b are relatively prime, so d, the gcd of a and b, is 1. Can one reconstruct the sequence of remainders — the r-values, including a and b —

from the q-values? The answer is YES, for it is fairly clear that the quotients can be used, bottom-up, to determine the remainders. Keep in mind the fundamental relationship of these numbers:

$$r_{k-1} = q_k r_k + r_{k+1}. \tag{2}$$

Thus $r_n = d = 1$, $r_{n-1} = q_n$, $r_{n-2} = q_{n-1}r_{n-1} + r_n = q_{n-1}q_n + 1$, and so on. It is instructive to work out the value of r_{n-3} in terms of the q's; then try r_{n-4}. Can you see a pattern in the definition of the r-values in terms of the q-values? A further exercise: Suppose the quotient sequence is {5, 2, 1, 1, 3}. Determine a and b.

Let us use Q to represent the function that takes a list of quotients and produces the topmost remainder of the corresponding Euclidean algorithm remainder sequence. This is called the **continuant** function; and the procedure we are about to describe was known to Euler. Keeping with our assumption that $\gcd(a, b) = 1$, we can say that $Q[\{\}] = 1$ and, as observed a moment ago, $Q[\{q_1\}]$ is simply q_1. Equation (2) leads to the following recursive definition of Q for lists of length 2 or more:

$$Q[\{q_1, ..., q_k\}] = q_1 \, Q[\{q_2, ..., q_k\}] + Q[\{q_3, ..., q_k\}]. \tag{3}$$

We can use this to define Q in *Mathematica* and investigate some examples. It is good practice to use descriptive names, so we use ContinuantQ instead of just Q.

```
ContinuantQ[{}] = 1;
ContinuantQ[{q_}] := q
ContinuantQ[q_] :=
  q[[1]] ContinuantQ[Rest[q]] + ContinuantQ[Drop[q, 2]]
```

We can now recover our a-value, 97.

```
qVals = {5, 2, 1, 1, 3};
ContinuantQ[qVals]
```

```
97
```

And we can recover the entire r-sequence by applying Q to all the tails of the q-sequence. Here are the tails.

```
Table[Take[qVals, i], {i, -5, 0}]
```

```
{{5, 2, 1, 1, 3}, {2, 1, 1, 3}, {1, 1, 3}, {1, 3}, {3}, {}}
```

```
Table[ContinuantQ[Take[qVals, -i]], {i, -5, 0}]
```

```
{97, 27, 16, 11, 5, 1}
```

Now comes a surprising fact: Q is invariant under the reversal of the quotient list.

`ContinuantQ[{3, 1, 1, 2, 5}]`

97

To understand why this is so, we use `ContinuantQ` to examine the remainders as abstract functions of the quotients. Such an approach oftens leads to valuable symbolic information. The version of `ContinuantQ` in the CNT package expands the output and formats it in a more natural sorted order than the default, so we load the package now and use that version.

`Get["CNT`"]`

`ContinuantQ[{q₁, q₂, q₃, q₄, q₅}]`

$q_1 + q_3 + q_5 + q_1 q_2 q_3 + q_1 q_2 q_5 + q_1 q_4 q_5 + q_3 q_4 q_5 + q_1 q_2 q_3 q_4 q_5$

`ContinuantQ[Table[qᵢ, {i, 6}]]`

$1 + q_1 q_2 + q_1 q_4 + q_1 q_6 + q_3 q_4 + q_3 q_6 + q_5 q_6 + q_1 q_2 q_3 q_4 +$
$q_1 q_2 q_3 q_6 + q_1 q_2 q_5 q_6 + q_1 q_4 q_5 q_6 + q_3 q_4 q_5 q_6 + q_1 q_2 q_3 q_4 q_5 q_6$

Can you detect the pattern in the results? There is indeed a simple rule at work, which we express in Proposition 1.4. Note that the product of no numbers is taken to be 1 (just like the sum of the empty set is 0). Before reading the proof, make sure you see the connection with the sums just obtained: starting with, say, $abcde$, one can delete a single pair to get abc, abe, ade, or cde, or one can delete a double pair to get a, c, or e (but not b or d). In the 6-case, deleting all pairs yields the empty set, whose product is 1.

Proposition 1.4. $Q[\{q_1, ..., q_k\}]$ is the sum of all products obtained by starting with $q_1 q_2 \cdots q_k$ and deleting adjacent pairs in all possible ways.

Proof. Let R denote the pair-deletion function. Then R and Q agree on the empty set and singletons. Thus it is sufficient to show that R obeys the same recursive law that defines Q. So we want to show that $R[\{q_1, ..., q_k\}] = q_1 R[\{q_2, ..., q_k\}] + R[\{q_3, ..., q_k\}]$. But this is easy, since every summand on the left side arises by the deletion of pairs that do not involve q_1, in which case the summand shows up in the first term on the right side, or by the deletion of a pair that does involve q_1, in which case q_2 must have been deleted too, so it shows up in the second term on the right. A similar argument shows that every term on the right side appears in the sum on the left side. □

This immediately explains the reversal invariance, because the sum of products is invariant under reversal. This fact will play a key role in Chapter 9, where it is used in a very fast algorithm for writing an integer as a sum of two squares.

Corollary 1.5. Continuant Reversal $Q[\{q_1, ..., q_k\}] = Q[\{q_k, ..., q_1\}]$

Proposition 1.4 sheds a strong light on the issue of worst-case running times for the Euclidean algorithm. Suppose we seek the smallest number a for which some Euclidean algorithm sequence starting with a has six steps. What would the quotient sequence look like? It cannot end in a 1 for that would mean that $r_n = r_{n-1}$, which cannot happen. So the smallest last quotient is 2. But all the other quotients can be 1. So the quotient sequence can look like $\{1, 1, 1, 1, 1, 2\}$. Proposition 1.4 tells us that any other q-sequence of length six would lead to a larger value of a, since a is a simple algebraic sum of quotients that can only increase when a quotient increases. This means that the n-element sequence $\{1, 1, 1, ..., 1, 2\}$ is the quotient sequence that leads to the smallest a-value for which a Euclidean algorithm sequence has length n. Moreover, the corresponding b-value comes from the $(n-1)$-term sequence of the same form (just delete the first 1) and so it is the worst a-value for length $n-1$.

Identifying these numbers is easy.

```
Table[ContinuantQ[Append[Table[1, {n - 1}], 2]], {n, 1, 10}]
```

```
{2, 3, 5, 8, 13, 21, 34, 55, 89, 144}
```

These are the Fibonacci numbers (defined in Section 1.1). It follows immediately from equation (3) that $Q[\{1, 1, 1, ..., 1, 2\}] = F_{n+2}$, where there are n terms in the argument to Q. These observations provide a proof of the following result, commonly known as Lamé's theorem.

Theorem 1.6. Lamé's Theorem.   The smallest positive integer a for which there is a smaller number b so that the Euclidean algorithm applied to a and b stops in n steps is the $(n+2)$nd Fibonacci number. And the corresponding b is the $(n+1)$st Fibonacci number.

Here is an example showing the length-6 behavior for the seventh and eighth Fibonacci numbers.

```
FullGCD[Fibonacci[8], Fibonacci[7], Labels → Short]
```

Remainders	Quotients
$r_0 = a = 21$	
$r_1 = b = 13$	$q_1 = 1$
8	1
5	1
3	1
$r_5 = 2$	$q_5 = 1$
$r_6 = 1$	$q_6 = 2$
$r_7 = 0$	

Binet's formula (Corollary 1.2) tells us that the $(n+2)$nd Fibonacci numbers is the closest integer to $\tau^{n+2}/\sqrt{5}$, where τ is the golden ratio. Thus, we get the following result on the worst-case running time of the Euclidean algorithm.

Corollary 1.7.   If $1 \le b < a$, then the number of steps in the Euclidean algorithm on the pair $\{a, b\}$ is at most $4.8\log_{10} a$.

Proof.   Lamé's theorem states that the number of steps, n, is at most $k-2$, where F_k is the largest Fibonacci number under a. But then, by the corollary to Binet's formula, $\tau^k/\sqrt{5} \le a + 0.5$, so $k \le \log_\tau \sqrt{5} + \log_\tau(a+0.5) = 1.68 + 4.79\log_{10}(a+0.5)$. Therefore, $n \le k - 2 \le -0.32 + 4.79\log_{10}(a+0.5)$. If $a \ge 3$ this last expression is bounded by $4.8\log_{10} a$. The single case where $a = 2$ and $b = 1$ stops in 1 step, and $1 \le 4.8\log_{10} 2$.   □

From a practical perspective, average-case behavior can be more important than worst-case behavior. The average behavior of the Euclidean algorithm is discussed in Exercise 1.14.

▣ The Extended GCD Algorithm

With just a little more work we can get a lot more information out of this algorithm. As Euclid realized, given integers a and b, a small modification of this algorithm can be used to find integers s and t such that

$$sa + tb = \gcd(a, b).$$

Moreover, the s and t value are tremendously important for a variety of reasons, one of which is their central role in the proof that numbers can be factored uniquely into primes (Theorem 1.11).

The proof that s and t exist is straightforward. Recall the central relationship (3) among the remainders in the Euclidean algorithm, which we rephrase as $r_{k+1} = r_{k-1} - q_k r_k$. This tells us that any remainder is an integer linear combination of the previous two remainders. Now, if r_n is the gcd of a and b, then r_n is an integer combination of r_{n-1} and r_{n-2}. But r_{n-1} is an integer combination of r_{n-2} and r_{n-3}, so we now have a combination of r_{n-2} and r_{n-3}. We can work all the way back up to a and b in this way, getting the desired representation. An example will clarify how this would work.

Remainders	Quotients	{s, t}
97		{1, 0}
18	5	{0, 1}
$7 = 97 - 5 \cdot 18$	2	{1, −5}
$4 = 18 - 2 \cdot 7$	1	{−2, 11}
$3 = 7 - 4$	1	{3, −16}
$1 = 4 - 3$	3	{−5, 27}

Working backwards up the list of remainders, we see that

$$1 = 4 - (7 - 4) = 2*4 - 7 = 2*(18 - 2*7) - 7 =$$
$$2*18 - 5*7 = 2*18 - 5*(97 - 5*18) = 27*18 - 5*97.$$

This tells us that $1 = 27 \cdot 18 - 5 \cdot 97$. But a much better way to organize this process is to start at the top, rather than the bottom. It is clear that each of a and b is an integer combination of a and b: just use $1 \cdot a + 0 \cdot b$ and $0 \cdot a + 1 \cdot b$. But whenever each of r_{k-1} and r_{k-2} is such a combination, say $r_{k-2} = a s_{k-2} + b t_{k-2}$ and $r_{k-1} = a s_{k-1} + b t_{k-1}$, one can use the fundamental relationship (2) to get the next remainder in such a form as well: $r_k = r_{k-2} - q_k r_{k-1} = a(s_{k-2} - q_k s_{k-1}) + b(t_{k-2} - q_k t_{k-1})$.

This tells us that the new coefficients, s_k and t_k, are $s_{k-2} - q_k s_{k-1}$ and $t_{k-2} - q_k t_{k-1}$, respectively. Algorithm 1.2 implements this recursion in an iterative manner, making use of the fact that the definition of the new s- and t-values can be defined by multiplying the old values by a 2×2 matrix. *Mathematica* uses lists of lists to represent matrices, with a dot for matrix multiplication.

```
{{0, 1}, {1, -q}}.{{s_{k-2}, t_{k-2}}, {s_{k-1}, t_{k-1}}}
```

```
{{s_{-1+k}, t_{-1+k}}, {s_{-2+k} - q s_{-1+k}, t_{-2+k} - q t_{-1+k}}}
```

However, by using `TraditionalForm` for output one can gets results that are much easier to read.

$$\begin{pmatrix} 0 & 1 \\ 1 & -q \end{pmatrix} \cdot \begin{pmatrix} s_{k-2} & t_{k-2} \\ s_{k-1} & t_{k-1} \end{pmatrix} \text{ // TraditionalForm}$$

$$\begin{pmatrix} s_{k-1} & t_{k-1} \\ s_{k-2} - q s_{k-1} & t_{k-2} - q t_{k-1} \end{pmatrix}$$

In fact, equation (3), which generates the new remainder from q and the previous remainders, is identical to the s- and t-equations, so the remainder computation can be included in the matrix operation as well:

$$\begin{pmatrix} 0 & 1 \\ 1 & -q \end{pmatrix} \cdot \begin{pmatrix} r_{k-2} & s_{k-2} & t_{k-2} \\ r_{k-1} & s_{k-1} & t_{k-1} \end{pmatrix} \text{ // TraditionalForm}$$

$$\begin{pmatrix} r_{k-1} & s_{k-1} & t_{k-1} \\ r_{k-2} - q r_{k-1} & s_{k-2} - q s_{k-1} & t_{k-2} - q t_{k-1} \end{pmatrix}$$

Algorithm 1.2. The Extended Euclidean Algorithm

Given integers a and b, this algorithm calculates their gcd, d, and outputs $\{d, \{s, t\}\}$ where $d = s \cdot a + t \cdot b$. The three columns of the 2×3 matrix `rst` represent the two latest r-values, s-values, and t-values, respectively.

```
extendedgcd[a_, b_] :=
  Module[{rst = (a 1 0
                 b 0 1)}, While[rst[[2, 1]] ≠ 0,

    q = Quotient[rst[[1, 1]], rst[[2, 1]]]; rst = (0  1
                                                   1 -q).rst];

   {rst[[1, 1]], rst[[1, {2, 3}]]}]
```

```
extendedgcd[97, 18]
```

```
{1, {-5, 27}}
```

This function is built into *Mathematica* as `ExtendedGCD`. Here's an example using the hundredth and thousandth prime.

```
ExtendedGCD[541, 7919]
```

```
{1, {-1010, 69}}
```

It never hurts to verify results, especially when the verification is as simple as it is here.

```
-1010 * 541 + 69 * 7919
```

```
1
```

We can see the algorithm at work by use the `ExtendedGCDValues` option to `FullGCD`. Having the full list of s- and t-values illustrates several important points:

1. The equation $r_k = s_k a + t_k b$ holds at every step.
2. The signs of the s- and t-sequences alternate.
3. If $\gcd(a, b) = 1$ and if one carries the iteration an extra step, then the final s- and t-values agree with b and a (except for sign).
4. The complete s- and t- sequences, viewed in reverse and ignoring the minus signs, are Euclidean algorithm remainder sequences starting from b and a, respectively.
5. $s_k t_{k+1} - t_k s_{k+1} = (-1)^k$.

```
FullGCD[97, 18, ExtendedGCDValues → True, Labels → Short]
```

Remainders	Quotients	s	t
r_0 = a = 97		s_0 = 1	t_0 = 0
r_1 = b = 18	q_1 = 5	s_1 = 0	t_1 = 1
7	2	1	-5
4	1	-2	11
r_4 = 3	q_4 = 1	s_4 = 3	t_4 = -16
r_5 = 1	q_5 = 3	s_5 = -5	t_5 = 27
r_6 = 0		s_6 = 18	t_6 = -97

Proofs are left as an exercise (Exercise 1.18). Note also that if one wants only the final values s_n and t_n, then one could skip the t-computation entirely and get t_n as simply $(d - s_n a)/b$. But in fact it is useful to have the full sequence, and for that the iterative method of Algorithm 1.2 is just as efficient as getting each s_k from t_k.

Theorem 1.8. Let a and b be integers with greatest common divisor d. There exist integers s and t such that $d = sa + tb$.

One corollary of Theorem 1.8 is that any integer that divides both a and b must divide their greatest common divisor. In other words, the gcd is not merely the largest integer in the set of common divisors. It is actually divisible by every element of the set of common divisors.

A second corollary that is so important that we shall call it a proposition is the following.

Proposition 1.9. If d divides mn and d is relatively prime to m, then d must divide n.

Proof. Because d divides mn, we can write $mn = du$ for some integer u. And because d and m are relatively prime, we can find integers s and t with $1 = sd + tm$. We multiply each side of this equation by n and then replace mn by du:

$$n = nsd + ntm = nsd + dut = d(ns + ut).$$

Because $ns + ut$ is an integer, this proves that d divides n. □

An important special case is given in the following result.

Corollary 1.10. If a prime number divides the product of two integers, then it must divide at least one of those integers.

Proof. If a prime p divides mn but does not divide m, then it is relatively prime to m (because p and 1 are the only divisors of p). Proposition 1.9 then tells us that p divides n. □

▨ The Fundamental Theorem of Arithmetic

We are now in a position to prove the fundamental theorem of arithmetic, the theorem that asserts that not only do we have a prime factorization for each positive integer, but that it must be unique. First, note that it is very easy to prove that every integer is a product of primes. Use induction: either n is prime or it equals $a \cdot b$ where $1 < a, b < n$. By induction a and b are each a product of primes. Multiplying all the primes together shows that the same is true for ab. But this is far short of showing that the prime factorization is unique; that is more subtle and requires the gcd machinery we have just built.

Theorem 1.11.   The Fundamental Theorem of Arithmetic.   Every positive integer may be factored into primes, and the factorization is unique up to order.

This says that even though there may be several ways of ordering the primes that go into a factorization ($30 = 2 \cdot 3 \cdot 5 \cdot 5 = 5 \cdot 3 \cdot 5 \cdot 2$), we cannot change the primes that appear. As obvious as this theorem may appear, it actually says something that is significant, a fact that was driven home in the nineteenth century when mathematicians began to use **extended integers** such as numbers of the form $m + n\sqrt{10}$, where m and n are ordinary integers. Integers such as these underlie the theory that explains the algorithm (mentioned in Section 1.0) for testing whether or not $2^n - 1$ is prime.

Like ordinary integers, these extended integers form a set that is closed under addition, subtraction, and multiplication. We can also apply the notion of divisibility: $12 + 3\sqrt{10}$ is divisble by $2 + \sqrt{10}$ because there is an extended integer that we can multiply $2 + \sqrt{10}$ by to get $12 + 3\sqrt{10}$:

$$\left(2 + \sqrt{10}\right)\left(1 + \sqrt{10}\right) = 12 + 3\sqrt{10}.$$

In Exercise 1.22 you will be asked to prove that 2, 3, $4 + \sqrt{10}$, and $4 - \sqrt{10}$ cannot be written as products of smaller extended integers — thus, they are analogous to primes in this system. But factorization is not unique! Here are two factorizations of 6:

$$6 = 2 \cdot 3 = \left(4 + \sqrt{10}\right)\left(4 - \sqrt{10}\right).$$

Proof of Theorem 1.11. We use induction. The base case is $n = 2$; but 2 is prime and so can be factored in only one way: 2. Suppose every integer strictly between 1 and n has unique factorization. To prove it for n, suppose it fails. So

$$n = p_1 p_2 \cdots p_r = q_1 \cdots q_s$$

where the p_i and q_j are primes that are not necessarily distinct, but where the second factorization is not simply a reordering of the first. If p_1 were equal to any of the q_j, then we could divide each product by p_1 to get an integer smaller than n with nonunique factorization. So p_1 is relatively prime to each q_j. We now invoke Proposition 1.10. We know that p_1 divides n and so it must divide $q_1 q_2 \cdots q_s$. But p_1 is relatively prime to q_1, so it must divide $q_2 \cdots q_s$. But p_1 is also relatively prime to q_2, so it divides $q_3 \cdots q_s$. Continuing to the end we see that p_1 divides q_s, a contradiction. $\square$

The most common use of the fundamental theorem is to represent an integer in the form $p_1^{a_1} p_2^{a_2} \cdots p_r^{a_r}$ where the p_i are prime and $a_i \geq 1$. *Mathematica*'s internal factoring algorithm finds such factorizations for modestly sized integers, or large integers that have only small factors.

```
FactorInteger[35113151674694180853418027S] // FactorForm
```

$$3^{11} \ 5^2 \ 47^2 \ 541^2 \ 7919 \ 15485863$$

▦ Two Applications

Here are two more applications of the ideas of gcd and the Fundamental Theorem of Arithmetic.

Irrational Numbers

A real number that can be expressed as a ratio of integers, such as $3/5$ or $22/7$, is called a **rational** number. Nonrational numbers are **irrational**. The first irrational number, in the sense that it was the first number to be proved irrational, is $\sqrt{2}$. This discovery, by the Pythagoreans in 500 BCE, is extremely important, for it showed the world that irrational numbers do exist.

Theorem 1.12. The square root of 2 is irrational.

Proof. We will need the fact that if s is a square, then all primes in s occur to an even power, which is an easy consequence of the fundamental theorem because $(p^a)^2 = p^{2a}$. Now, assume that $\sqrt{2}$ can be written as m/n where m and n are integers. Then $2n^2 = m^2$. But the power of 2 in m^2 is even, while the power of 2 in $2n^2$ is odd. Contradiction. □

In fact, $\sqrt{n}$ is irrational whenever n is a nonsquare (Exercise 1.25).

Pythagorean Triples

The next application involves **Pythagorean triples**, triples of integers that satisfy $x^2 + y^2 = z^2$, and a truly ancient algorithm for finding them. Without loss of generality, we can assume that x, y, and z are positive and share no common factor. Such a triple is called a **primitive triple**.

The most famous Pythagorean triples are (3, 4, 5) and (5, 12, 13). The triple (6, 8, 10) is Pythagorean, but is not a primitive triple, because it is obtained by multiplying (3, 4, 5) by 2. If we want to find all Pythagorean triples, it is sufficient to first find all primitive triples and then take multiples of them. The following proposition shows how to generate all primitive triples.

Proposition 1.13. For any pair of relatively prime integers, (m, n), such that one of them is odd, the other is even, and $m > n > 0$, the three numbers $(m^2 - n^2, 2mn, m^2 + n^2)$ form a primitive triple. Furthermore, every primitive triple is of this form.

Note that (3, 4, 5) corresponds to $m = 2$, $n = 1$, while (5, 12, 13) comes from $m = 3$, $n = 2$.

Proof. Simple algebra verifies that the triple in question is Pythagorean. You should verify for yourself that $(2mn)^2 + (m^2 - n^2)^2 = (m^2 + n^2)^2$. We leave the proof that the triple has no factor in common as an exercise. The more interesting part of this result is that there are no other primitive triples.

Let (x, y, z) be a primitive triple. Because x, y, and z are not all even, at most one of them is even. If x and y are both odd, then x^2 and y^2 are each one more than a multiple of 4 (because $(2k + 1)^2 = 4k^2 + 4k + 1$) and so $x^2 + y^2$ must be 2 more than a multiple of 4. But that says that z^2 is divisible by 2 and not by 4, a contradiction to Proposition 1.10. Thus one of x or y must be even. By symmetry in x and y, we can assume that it is y that is even, say $y = 2k$, and x and z are odd.

We can solve for k^2.

$$k^2 = \frac{z^2 - x^2}{4} = \frac{z - x}{2} \frac{z + x}{2}$$

Because x and z are odd, $(z - x)/2$ and $(z + x)/2$ are integers. Furthermore, they must be relatively prime because any common divisor would have to divide both their sum (which is z) and their difference (which is x), and we assumed that x and z have no common divisors.

Each prime p that divides $(z - x)/2$ also divides k^2, and so p^2 divides k^2. Because p does not divide $(z + x)/2$, p^2 must divide $(z - x)/2$. Thus the prime factorization of $(z - x)/2$ will have only even exponents, which is another way of saying that $(z - x)/2$ is a perfect square. Similarly, $(z + x)/2$ must be a perfect square. Let us write

$$\frac{z + x}{2} = m^2 \quad \text{and} \quad \frac{z - x}{2} = n^2.$$

As we have just shown, m and n are relatively prime. Because $m^2 + n^2 = z$ is odd, one of m or n is odd and the other even. Also, $m > n$. We now solve for x, y, and z.

$$z = m^2 + n^2$$
$$y = 2mn$$
$$x = m^2 - n^2 \quad \square$$

The characterization is easy to program. Indeed, we can immediately get a giant triple as follows.

```
{m² - n², 2 m n, m² + n²} /. {m → 10¹⁰ + 1, n → 10¹⁰}
```

```
{20000000001, 200000000020000000000, 200000000020000000001}
```

Going from the generators to the triples and vice versa are excellent programming exercises, and more specific instructions are given in Exercise 1.28. When one generates tables of triples, several patterns jump out, and a few are mentioned in Exercise 1.29.

A computational problem of a different sort comes from trying to visualize the triples. One approach is to view the generators as points in the m-n plane; such an image is given in Figure 1.3. The transformation from the generators to the actual triples is left to the viewer's imagination.

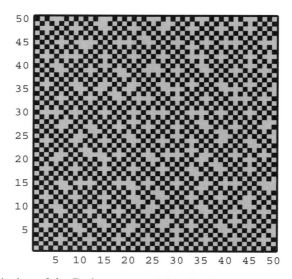

Figure 1.3. A view of the Pythagorean triples. Each dark square represents a pair (m, n) that is a generator of a primitive triple.

A much more fruitful way of viewing the triples comes by observing that the central relationship, $a^2 + b^2 = c^2$, can be viewed as $(a/c)^2 + (b/c)^2 = 1$. This can be viewed as a relationship between two rational numbers: $r^2 + s^2 = 1$. Thus a Pythagorean triple corresponds to a point on the unit circle, both of whose coordinates are rational; such a point is called a **rational point**. Note that a triple and any multiple of it correspond to the same rational point — the multiplier disappears in the fraction — so this point of view is somewhat purer than the integer context.

Here is some code to turn a triple into a rational point and then to look at the rational points corresponding to the first n triples. It uses the package function `PythagoreanTriples[n]`, which generates the first n primitive triples.

```
RationalPoints[n_] :=
  PythagoreanTriples[n, UseGridBox → False] /.
   {a_, b_, c_} :⟶ {a / c, b / c}
```

```
RationalPoints[10]
```

$$\left\{\left\{\frac{3}{5}, \frac{4}{5}\right\}, \left\{\frac{5}{13}, \frac{12}{13}\right\}, \left\{\frac{8}{17}, \frac{15}{17}\right\}, \left\{\frac{7}{25}, \frac{24}{25}\right\}, \left\{\frac{20}{29}, \frac{21}{29}\right\},\right.$$
$$\left.\left\{\frac{9}{41}, \frac{40}{41}\right\}, \left\{\frac{12}{37}, \frac{35}{37}\right\}, \left\{\frac{11}{61}, \frac{60}{61}\right\}, \left\{\frac{28}{53}, \frac{45}{53}\right\}, \left\{\frac{33}{65}, \frac{56}{65}\right\}\right\}$$

Because $a < b$ in our setup, these points will lie only on a 45° arc of the circle. Symmetry and/or negatives will fill up the circle. Figure 1.4 shows the first 15 triples and then 400 of them. The natural question is whether these points are dense. There are two regions where the rational points seem rare: near $(0, 1)$ and near $(^3/_5, {}^4/_5)$. Exercise 1.27 addresses these questions, and in a constructive way, so that, for example, a triple that is very "close" to $(3, 4, 5)$ can be found.

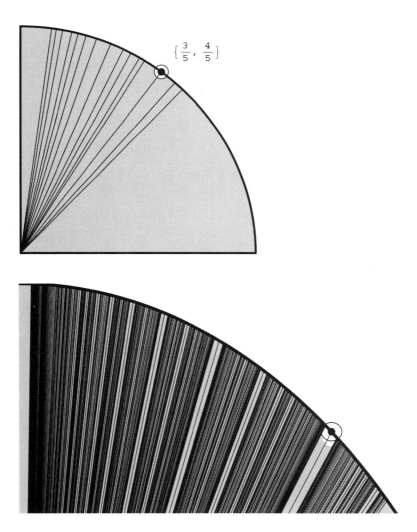

Figure 1.4. The Pythagorean triples viewed as rational points on the unit circle. The upper diagram shows the points corresponding to the first 15 triples, while the lower one is a close-up of the image of the first 400 points. There seem to be two gaps, one near $(0, 1)$ and the other near $(^3/_5, {}^4/_5)$, but these gaps disappear eventually because the set of rational points really is dense in the circle.

Exercises for Section 1.2

1.8. Complete the proof of the division algorithm by showing that, given integers a and b with $b \neq 0$, and then defining r to be the smallest element of $\{a - qb : q \in \mathbb{Z}, a - qb \geq 0\}$ leads to a pair, r and q, that satisfies the conclusion of the division algorithm. Moreover, there is a unique pair r, q that works.

1.9. Let $d = \gcd(a, b)$. Prove that a/d and b/d are relatively prime.

1.10. Prove that $\operatorname{lcm}(a, b) = ab/\gcd(a, b)$. This can be done by appealing to the fundamental theorem of arithmetic, but it is more elegant to avoid prime factorizations. Find such a proof as follows. Let $d = \gcd(a, b)$.

(a) Show that ab/d is an integer and is a multiple of both a and b.

(b) Let n be any common multiple of a and b, say $n = ar = bs$. Prove that a/d is relatively prime to b/d and therefore a/d divides s, say $s = (a/d)t$. Prove that ab/d divides n. It follows that ab/d is the least common multiple of a and b.

1.11. Write a program that accepts a list of integers and returns `True` if the entries are pairwise relatively prime, and `False` otherwise. Important: Try to find a way of doing this that avoids checking all pairs from the list and avoids factoring the inputs.

1.12. The recursive definition of `gcd` can run into problems, because *Mathematica* has built-in recursion limits. Thus, if the recursion involves too many steps an error will result. It is generally better to use traditional iterative programming, as exemplified by the following routine based on a `While` loop. The parentheses group the three sentences of the program into a single compound sentence.

```
gcdIterative[a_, b_] := ({aa, bb} = {a, b};
   While[bb ≠ 0, {aa, bb} = {bb, Mod[aa, bb]}]; aa)
```

Alternatively, it is a little faster to use the matrix formulation as described in Algorithm 1.2.

```
gcdIterative[a_, b_] := (r = {a, b}; While[r[[2]] ≠ 0,
   q = Apply[Quotient, r]; r = {{0, 1}, {1, -q}}.r];
   r[[1]])
```

```
gcdIterative[97, 18]
```

1

Compare the timing of the recursive and iterative approaches on pairs of large Fibonacci numbers. To get reasonable timings you may wish to repeat a calculation 100 or times, via `Do[…, {100}]`. Find such a pair for which the recursive method fails.

1.13. Investigate the question: What percentage of random pairs of integers are relatively prime? For example, look at a couple of thousand random pairs between 1 and 10^6 and compute the percentage that are relatively prime. The theoretical probability (defined as a limit as the bounds approach infinity, because the notion of a random integer between 1 and infinity does not make sense) turns out to be a very simple number involving π^2; use the results of your experiments to try to find this formula. Use `Random[Integer, {a, b}]` to get a random integer between a and b.

1.14. Here is a recursive routine to compute the length of the Euclidean algorithm sequence.

```
EALength[a_, b_] := If[b == 0, 0, 1 + EALength[b, Mod[a, b]]]
EALength[97, 18]
```

```
5
```

Turn this into an iterative routine, avoiding recursion. Investigate Euclidean algorithm lengths for a variety of inputs. Look at the average length of the sequence, over hundreds of samples. A result of H. Heilbronn [Knu, §4.5.3, Exercises 33, 34] says that there is a constant c such that the average length of this sequence is asymptotically equal to c times the length of the input. More precisely: the average length over all inputs less than M divided by $\log_{10}(M)$ approaches c as $M \to \infty$. Do some computations to make an estimate of c. For comparison, recall that the worst-case behavior is about $4.78 \log_{10}(M)$. Generate some graphics to illustrate the convergence to c. (*Warning*: The convergence is a little slow. It is better to try large random inputs than to look at all integers under some bound.)

1.15. Work out the proper definition and implementation of a continuant function when the starting point — the gcd of a and b — is greater than 1.

1.16. Write a function `FullQ` to turn a quotient sequence into the entire remainder sequence.

1.17. Algorithm 1.2 outputs only the final s- and t-values. Write a program that accepts a and b as input and generates the entire s- and t-sequences corresponding to the Euclidean algorithm for a and b. The output should give the sequences in the terms of two lists, one for the s-values, one for the t-values.

1.18. Let a, b, r_k, q_k, s_k, t_k be as in the discussion of the extended Euclidean algorithm, and let n be the number of steps, so that $r_n = \gcd(a, b)$.

(a) Prove that $r_k = s_k\, a + t_k\, b$. (Use induction and the definition of s_k and t_k.)

(b) Prove that the signs of the s- and t-sequences alternate. (Use induction and the definition of s_k and t_k.)

(c) The complete s- and t- sequences, when viewed in reverse and ignoring the minus signs, are Euclidean algorithm remainder sequences starting from b and a, respectively. (Use the defining equations of s and t. Then observe that the quotients are the same, and so the continuant reversal phenomenon (Theorem 1.5) can be used.) Note one small exception: If $b > a/2$, then the reversed t-sequence is a remainder sequence provided the initial 0 and 1 are ignored.)

(d)   Prove, under the assumption that $\gcd(a, b) = 1$, that $|s_{n+1}| = b$ and $|t_{n+1}| = a$. (Use (c).)

(e)   Prove that $s_k t_{k+1} - t_k s_{k+1} = (-1)^k$ (Induction.)

(f)   Prove that $\gcd(s_k, t_k) = 1$ if $k > 1$.

1.19.   Prove that in the system of extended integers of the form $m + n\sqrt{10}$, $3 + \sqrt{10}$ divides 1 and therefore it divides every integer. Such a number is called a ***unit***.

1.20.   Prove that in the system $\mathbb{Z}\left[\sqrt{10}\right]$ of extended integers of the form $m + n\sqrt{10}$, $a + b\sqrt{10}$ divides 1 if and only if $a^2 - 10b^2 = \pm 1$. It is useful to define the ***norm*** of one of these objects, $N\left(a + b\sqrt{10}\right)$, to be $(a^2 - 10b^2)$. Then this exercise can be rephrased as if $\alpha \in \mathbb{Z}\left[\sqrt{10}\right]$, then α divides 1 if and only if $N(\alpha) = \pm 1$.

1.21.   Prove that if $a + b\sqrt{10}$ divides $m + n\sqrt{10}$, then $a^2 - 10b^2$ divides $m^2 - 10n^2$.

1.22.   Prove that there are no integers a and b for which $a^2 - 10b^2 = 2, -2$, 3, or -3. *Hint*: First show that every perfect square is either a multiple of 5, 1 more than a multiple of 5, or 1 less than a multiple of 5.

1.23.   Prove that for $2 = 2 + 0\sqrt{10}$, $3 = 3 + 0\sqrt{10}$, $4 + \sqrt{10}$, and $4 - \sqrt{10}$, if we represent these as a product of two extended integers, then at least one of those extended integers must be a unit. In other words, we cannot factor the four given numbers any further.

1.24.   Explain why the following formula gives the power e of a given prime p in $n!$:

$$e = \sum_{i=1}^{\log_p n} \left\lfloor \frac{n}{p^i} \right\rfloor$$

Then write a program that, on input n and p, returns e. Explain how this can be used to predict the number of 0s at the right end of $n!$.

1.25.   Prove that $\sqrt{n}$ is an irrational number whenever n is not a perfect square. (*Hint*: Look at the prime factorization of n and use a prime whose exponent is odd.)

1.26.   Show that no integer greater than 1 divides each entry of the Pythagorean triple $(m^2 - n^2, 2mn, m^2 + n^2)$, where m and n are relatively prime.

1.27. Show that the set of rational points on the unit circle is **dense**, meaning that, for any point P on the circle and any $\varepsilon > 0$, there is a rational point within distance ε of P. Find a rational point within distance 10^{-6} of $(^3/_5, {}^4/_5)$. *Hint*: Let $P_{m,n}$ be the rational point corresponding to the primitive triple whose generators are (m, n). Work out the slope of the line from $P_{m,n}$ to $(-1, 0)$. The formula for this slope will be sufficient to allow you to construct a rational point near any given point.

1.28. Implement some functions to go from a primitive Pythagorean triple (a, b, c) to the generators (m, n) and vice versa. For example, a routine might accept max as input and return all the triples whose generators are less than max in each coordinate. Or it could return all the primitive triples such that $c \le$ max.

1.29. (a) Show that any Pythagorean triple has an entry that is divisible by 3.

(b) Show that any Pythagorean triple has an entry that is divisible by 4.

(c) Show that any Pythagorean triple has an entry that is divisible by 5.

1.3 Modular Arithmetic

Much of the insight into the structure of the integers we need comes from an understanding of modular arithmetic. Some readers may be familiar with its structure and so may wish to skip this section. This is intended as an introduction for those who are less familiar with it.

We begin with a postage stamp problem. Given positive integers a and b, which integers can and cannot be represented by $ma + nb$ where m and n are nonnegative integers? In other words, what postage amounts can you make if you are restricted to an unlimited supply of a¢ and b¢ stamps?

First consider the concrete example of 5¢ and 8¢ stamps. To find the largest amount that cannot be made, we begin by listing the nonnegative integers in five columns.

0	1	2	3	4
5	6	7	8	9
10	11	12	13	14
15	16	17	18	19
20	21	22	23	24
25	26	27	28	29
30	31	32	33	34
35	36	37	38	39
⋮	⋮	⋮	⋮	⋮

If we highlight (bold in the table) the amounts that can be made, there are several easy, and important, observations:

- When we can make any postage amount in any column, we can also make any larger amount in that column (just add 5¢ stamps).

- The first number in each column that is highlighted is a multiple of 8 (if it was not a multiple of 8, then we would need to use some 5¢ stamps to make it, and so we could take away one of those 5¢ stamps to make the number above it).

- Each of the multiples of 8 from $0 \cdot 8 = 0$ to $4 \cdot 8 = 32$ appears in a different column.

When we have a multiple of 8 in each column, we can make any larger postage amount. The largest amount that cannot be made is the number directly above the last highlighted multiple of 8: 27¢ is the largest amount that cannot be made. Moreover, there are exactly fourteen unmakeable values.

We can do the same thing with a¢ stamps and b¢ stamps. List all nonnegative integers in a columns that begin with 0 through $a - 1$. When we can make an amount in a given column, we can make any amount below it by adding a¢ stamps. And the smallest amount that we can make in any column must be a multiple of b. Do we always have a multiple of b in every column?

A little experimentation should make it clear that the answer is NO whenever a and b are not relatively prime. If $\gcd(a, b) > 1$, then any linear combination of a and b must be divisible by this greatest common divisor, and you can only make amounts in columns all of whose entries are divisible by this greatest common divisor.

Central Question: What if a and b are relatively prime? Do the first a multiples of b, $(0, b, 2b, 3b, ..., (a - 1)b)$, have to lie in different columns? You are encouraged to examine some examples.

To answer this question, it is very helpful to switch to the terminology of modular arithmetic. The integers in each column share an important property: They each differ from the number at the top of the column by a multiple of a. In modular arithmetic, we take this entire column of integers, call it a ***residue class***, and replace it by a single integer. Usually, this is the integer at the top of the column, but any integer in the column can be used to represent this residue class. When two integers lie in the same residue class, we say that they are ***congruent*** modulo a, and this is written using a modified equality symbol, as in: $27 \equiv 12 \pmod 5$.

The Mod command that we used at the beginning of Section 1.2 takes us from an integer to the top integer in its residue class. The first argument is the input integer and the second argument is the ***modulus*** or number of columns; for example, Mod[27, 5] is 2.

We can do. arithmetic on residue classes. For the modulus 5, if we take any integer in the residue class of 2 and add 7 to it, we always wind up in the residue class of 9, which is the same as the residue class of 4: $2 + 7 \equiv 4 \pmod 5$. The congruence sign, $\equiv$, means that the two sides differ by a multiple of the modulus. This also works for multiplication. If we take any integer in the residue class of 2 and multiply it by 9, we always wind up in the residue class of 3: $2 \cdot 9 \equiv 3 \pmod 5$.

We write the symbol for congruence to look like an equality because it acts very much like the equality symbol. We can add congruent amounts to each side of a congruence:

$$\left(\begin{array}{c} a \equiv b \pmod m \\ \text{and} \\ c \equiv d \pmod m \end{array} \right) \Rightarrow a + c \equiv b + d \pmod m.$$

We can multiply congruent amounts to each side of a congruence:

$$\left(\begin{array}{c} a \equiv b \pmod m \\ \text{and} \\ c \equiv d \pmod m \end{array} \right) \Rightarrow a \cdot c \equiv b \cdot d \pmod m.$$

We can also subtract congruent amounts from each side of a congruence.

Division, however, must be approached with great care. We have that $3 \equiv 15 \pmod{12}$ but dividing both sides by 3 is highly illegal, since $1 \not\equiv 5 \pmod{12}$. To pinpoint what goes wrong here, we need a precise definition of congruence: a is **congruent** to b modulo m if and only if m divides the difference between a and b. It is easy to check (Exercise 1.30) that if m divides $a - b$ and m divides $c - d$, then m divides both $(a + c) - (b + d)$ and $ac - bd$. On the other hand, if m divides $ac - bc$, we cannot be certain that m divides $a - b$.

Proposition 1.9 does come to our aid here with a partial solution. If m divides $ac - bc = (a - b)c$ and m is relatively prime to c, then m must divide $a - b$. In other words, we have yet another corollary of Proposition 1.9.

Corollary 1.14. If $ac \equiv bc \pmod m$ and $\gcd(c, m) = 1$, then $a \equiv b \pmod m$.

Cancellation *is* always possible, provided one is careful to change the modulus. Here is an important rule, whose proof is left as an exercise.

Proposition 1.15. If $ac \equiv bc \pmod m$, then $a \equiv b \pmod{m / \gcd(m, c)}$.

We also get some counterintuitive divisibility properties in modular arithmetic. For example, 3 divides 1 modulo 5 because $1 \equiv 3 \cdot 2 \pmod 5$. If we go back to the central question on page 26, we see that it is equivalent to asking if b must divide *everything* when b is relatively prime to the modulus.

Proposition 1.16. If $\gcd(b, m) = 1$ then b is a divisor of every residue class modulo m.

Proof. We shall prove that b divides 1. This suffices, because if b divides 1, b divides everything. So we must show that there is an integer a, called the ***multiplicative inverse*** of b (modulo m), such that $ab \equiv 1 \pmod{m}$.

Because b and m are relatively prime, we can use the extended Euclidean algorithm to find integers s and t such that $sb + tm = 1$. This implies that m divides $sb - 1$. The integer s is therefore a multiplicative inverse of b modulo m.   □

Notice that this is much more than a proof that a multiplicative inverse exists; the argument provides a method for finding it. The solution s might be negative, but a residue class is easily extended to cover negative integers. If we need a positive solution, we just add the appropriate multiple of m to s.

As an example, to find the multiplicative inverse of 17 (mod 73), we apply the extended gcd algorithm to 73 and 17.

    ExtendedGCD[73, 17]

    {1, {7, -30}}

This means that $7 \cdot 73 - 30 \cdot 17 = 1$. So we could use -30 as a multiplicative inverse of 17, or any other integer that is congruent to -30. The smallest positive integer that is a multiplicative inverse of 17 is $-30 + 73 = 43$.

    Mod[43 * 17, 73]

    1

And here is a simple result that allows us to paste together congruences modulo relatively prime factors to get a congruence modulo their product.

Proposition 1.17. If $x \equiv y \pmod{m_1}$ and $x \equiv y \pmod{m_2}$, where m_1 and m_2 are relatively prime, then $x \equiv y \pmod{m_1 \cdot m_2}$.

Exercises for Section 1.3

1.30. Prove that if $a \equiv b \pmod{m}$ and $c \equiv d \pmod{m}$, then $a + c \equiv b + d \pmod{m}$ and $ac \equiv bd \pmod{m}$.

1.31. Prove that if $a \equiv b \pmod{m}$ and r is any integer, then $a^r \equiv b^r \pmod{m}$. Find an example to show that r being congruent to s modulo m does not guarantee that $a^r \equiv a^s \pmod{m}$. Fixing m to be, say 11, experiment to find conditions on r and s that do guarantee that, for any a, $a^r \equiv a^s \pmod{m}$.

1.32. Prove Proposition 1.15.

1.33. Prove Proposition 1.17.

1.34. Find a version of Proposition 1.17 that is valid when the moduli fail to be relatively prime.

1.35. Binomial coefficients are critical to combinatorics, but also play an important role in number theory. The **binomial coefficient** $\binom{n}{k}$ is defined to be $\frac{n!}{k!(n-k)!}$. In *Mathematica* they are computed as `Binomial[n, k]`.

```
Table[Binomial[n, i], {n, 0, 10}, {i, 0, n}] // ColumnForm
```

```
{1}
{1, 1}
{1, 2, 1}
{1, 3, 3, 1}
{1, 4, 6, 4, 1}
{1, 5, 10, 10, 5, 1}
{1, 6, 15, 20, 15, 6, 1}
{1, 7, 21, 35, 35, 21, 7, 1}
{1, 8, 28, 56, 70, 56, 28, 8, 1}
{1, 9, 36, 84, 126, 126, 84, 36, 9, 1}
{1, 10, 45, 120, 210, 252, 210, 120, 45, 10, 1}
```

When arranged in a triangular array, the result is called **Pascal's triangle**.

PascalTriangle[14]

0								1								
1							1		1							
2						1		2		1						
3					1		3		3		1					
4				1		4		6		4		1				
5			1		5		10		10		5		1			
6		1		6		15		20		15		6		1		
7	1		7		21		35		35		21		7		1	
8	1		8		28		56		70		56		28		8		1
9	1	9		36		84		126		126		84		36		9	1
10	1	10	45		120		210		252		210		120		45	10	1
11	1	11	55	165		330		462		462		330		165	55	11	1
12	1	12	66	220	495		792		924		792		495	220	66	12	1
13	1	13	78	286	715	1287		1716		1716		1287	715	286	78	13	1
14	1 14	91	364	1001	2002	3003	3432	3003	2002	1001	364	91	14 1			

Show that if p is prime, then every entry (except the 1s) in the pth row of Pascal's triangle is divisible by p.

1.36. (Erdős and Szekeres, 1978) Prove that any two entries (excluding the 1s) in the same row of Pascal's triangle have a common factor. Amazingly, this fact was not proved until 1978. *Hint:* Prove the following formula, and then use it to derive the result in question:

$$\binom{n}{k}\binom{k}{r} = \binom{n}{r}\binom{n-r}{k-r}.$$

1.37. Suppose a and b are relatively prime, the Euclidean algorithm applied to a and b requires n steps, and the quotients are q_1, ..., q_n. Then the continuant reversal phenomenon (Corollary 1.5) tells us that if Q is applied to all initial segments of $\{q_n, ..., q_1\}$ the result will be a Euclidean algorithm remainder sequence of the form $\{a, b_1, ...\}$. What is the relationship between b_1 and b? That is, given b, find (and prove) a simple formula that will produce b_1. Then answer these same questions when $\gcd(a, b) = d > 1$. *Hint*: The s- and t-sequences of the extended gcd algorithm play an important role in generating the desired formulas. In particular, it will be useful to find a formula for t_n in terms of a modular inverse in the

1.38. (Emma Lehmer) If F_n is the nth Fibonacci number, there is an interesting divisibility relation that holds for primes. It is not true that p divides F_p, but something very similar is true. Can you discover what it is? Does your property characterize the prime numbers? *Hint*: Consider the mod-5 residue of p.

1.4 Fast Powers

■ A Fast Algorithm for Exponentiation

We hope that it is evident by now that the Euclidean algorithm and the extended Euclidean algorithm are two of the fundamental subroutines of algorithmic number theory. They are built into *Mathematica* and virtually every high-level language. There is one more basic algorithm that rivals these in importance: fast exponentiation.

The idea behind fast exponentiation is that if the exponent is a power of 2, then we can exponentiate by successively squaring:

$$b^8 = \left((b^2)^2\right)^2 \text{ and } b^{256} = \left(\left(\left(\left(\left(\left(\left(b^2\right)^2\right)^2\right)^2\right)^2\right)^2\right)^2\right)^2$$

If the exponent is not a power of 2, then we use its binary representation; for example, $291 = 2^8 + 2^5 + 2^1 + 2^0 = 256 + 32 + 2 + 1$, which gets us b^{291} as follows.

$$b^{291} = b^{256} b^{32} b^2 b$$

So we can calculate b^{291} with only 11 multiplications: 8 for the powers of 2 and 3 to combine them. In general, b^n can be calculated with at most $2\lfloor \log_2 n \rfloor$ multiplications.

No one cares to compute the giant integers that arise as such powers. Rather, it is their values modulo m that are of critical importance to prime testing, factoring, and many other parts of number theory. The binary method just presented allows one to compute such residues efficiently because one can reduce modulo m at each step, and so the numbers never get large. This is worth repeating: ***Never*** use raw exponentiation (^) to compute exponents. Some languages might use logarithms, which will result in round-off error that can make the computation meaningless. And the testing of a 100-digit integer r for primality might call for examining $2^n \pmod{p}$ where n also has 100 digits. Any attempt to evaluate the exact integer $2^{(10^{100})}$ will crash your computer system, for this number has more binary digits than there are particles in the universe.

The power-of-2 method described above can easily be turned into a precise algorithm, one that scans the binary digits of the exponent from right to left. It turns out that it is more efficient to scan the digits from left to right, using a slightly different method, and we will take that approach as our main exponentiation technique.

Most computers have numbers stored in base 2, so the extraction of the base-2 digits is very fast. And high-level languages typically have a built-in function to do this. Here is how to do it in *Mathematica*.

IntegerDigits[41, 2]

{1, 0, 1, 0, 0, 1}

So $41 = 101001_2$. To get b^{41}, start by setting ans to b and then scan the digits from the left, ignoring the leading 1. At each digit square ans, and at each 1 follow that by multiplying ans by b. Using base-2 for the exponents, the values in ans will be: b^1, b^{10}, b^{100}, b^{101}, b^{1010}, b^{10100}, b^{101000}, b^{101001}; the last term is b^{41}. If we are working modulo m, then each multiplication is followed by a modular reduction so that the numbers stay under m. The package function PowerAlgorithm allows one to watch this algorithm in action.

PowerAlgorithm[b, 41]

Binary Digits of 41	Answer
1	b
0	b^2
1	b^4
	b^5
0	b^{10}
0	b^{20}
1	b^{40}
	b^{41}

The sequence of exponents, which has the property that each one is a sum of two previous ones and the first term is 1, is called an ***addition chain***, and that too can be generated by a package function. Further properties of addition chains are discussed in Exercises 1.40 and 1.41.

AdditionChain[41, BinaryLeftToRight]

{1, 2, 4, 5, 10, 20, 40, 41}

Here is the algorithm in action modulo 7.

PowerAlgorithm[3, 41, 7]

Binary Digits of 41	Answer
1	3
0	2
1	4
	5
0	4
0	2
1	4
	5

And here's an example with larger numbers.

PowerAlgorithm[3, 7918, 7919]

Binary Digits of 7918	Answer
1	3
1	9
	27
1	729
	2187
1	7812
	7598
0	94
1	917
	2751
1	5356
	230
1	5386
	320
0	7372
1	6206
	2780
1	7375
	6287
1	2640
	1
0	1

Each nonzero bit ("bit" comes from "binary digit") causes two operations: a squaring and a multiply-by-3. It is very common in number theory to want $b^n \pmod{m}$, where n and m are large, but b is small, often 2 or 3. Thus, it is noteworthy that the multiplication step involves multiplication by b, which, for small b, will be much faster than the multiplication of two integers near m. The right-to-left method alluded to earlier involves multiplication of two large numbers at each multiply step, and that is why it is inferior to the left-to-right method (Exercise 1.39).

Algorithm 1.3. Fast Powers, Left to Right

```
powermodLR[a_, n_, m_] :=
 Module[{ans = a, dig = IntegerDigits[n, 2]},
  Do[ans = Mod[ans², m]; If[dig[[i]] == 1, ans = Mod[ans * a, m]],
   {i, 2, Length[dig]}];
  ans]
```

Setting up a Do-loop for operations on a list is not efficient in *Mathematica*. Rather, one should work on the list more abstractly using the concept of a pure function. An expression such as (#^2)& stands for the function that takes its argument and squares it. For functions of more than one argument, #1 and #2 are used for the variables.

```
Fold[f, 1, {3, 4, 5}]
```

```
f[f[f[1, 3], 4], 5]
```

```
FoldList[#1 + 3 #2 &, 1, {3, 4}]
```

```
{1, 10, 22}
```

This leads to the following alternative, faster, and shorter way of coding the exponentiation algorithm.

```
powermodLR1[a_, n_, m_] :=
 Fold[ Mod[Mod[#1 #1, m] * If[#2 == 1, a, 1], m] &,
  a, Rest[IntegerDigits[n, 2]]]
```

```
{powermodLR[3, 41, 7], powermodLR1[3, 41, 7]}
```

```
{5, 5}
```

This is built into *Mathematica* as PowerMod.

```
PowerMod[3, 41, 7]
```

```
5
```

And very large integers can be handled easily.

```
PowerMod[1000001, 10000000, 999999999999]
```

```
399680399680
```

A nifty bonus is that PowerMod accepts -1 as an exponent, in which case it returns the modular inverse, using the extended gcd method described in Section 1.3.

▣ Powers of Matrices

The ideas of fast exponentiation are especially important when the item being raised is a matrix. All the same ideas apply, but we must use matrix multiplication (a dot in *Mathematica*) instead of normal multiplication. The modification to the left-to-right algorithm is very simple.

Algorithm 1.4. Powers of Matrices

The following function is not built into *Mathematica*, but it is included in the CNT package.

```
MatrixPowerMod[a_, n_, m_] :=
  Module[{ans = a, dig = IntegerDigits[n, 2]},
    Do[ans = Mod[ans.ans, m]; If[dig[[i]] == 1, ans = Mod[ans.a, m]],
    {i, 2, Length[dig]}]; ans]
```

The powers of $\left(\begin{smallmatrix} 0 & 1 \\ 1 & 1 \end{smallmatrix}\right)$ are interesting because they are the Fibonacci numbers (proved in Section 1.1). Now, suppose we wish to know the rightmost 10 digits of the 10^{100}th Fibonacci number. The raw integer is way too large to compute easily, but the mod-10^{10} reductions make the computation almost instantaneous. The following output tells us that the googol'th Fibonacci number ends in ...9560546875 (a ***googol*** is 10^{100}).

```
MatrixPowerMod[{{0, 1}, {1, 1}}, 10^100, 10^10]
```

```
{{2900390626, 9560546875}, {9560546875, 2460937501}}
```

▣ Big-O Notation

Computers are fast enough that you can calculate 37^{291} by multiplying 37 by itself 290 times. They are not fast enough to be able to calculate $37^{(10^{50})}$ by multiplying 37 by itself $10^{50} - 1$ times. On the other hand, fast exponentiaton requires 11 multiplications in the first case, but only 223 in the second. The time required is growing much more slowly.

Big-O notation (the "O" stands for "order of growth") is a way of describing how fast the computation time grows. Running time is measured as a function of the length of the input. Let us take n for the input to our algorithm (though many algorithms have more than one input) and N for the physical length of n. Then N is typically $\log_2 n$ or $\log_{10} n$; the choice of base is irrelevant as far as the big picture is concerned. An algorithm is said to run in ***linear time*** if there is a constant c such that, on input n, the number of steps it takes is less than cN. For such an algorithm, if the input length is doubled, so is the running time. For a quadratic algorithm — that is, time bounded by cN^2— doubling the input length multiplies the running time by a factor of 4.

An algorithm is said to run in time $O(f(N))$ if there is some constant c so that the running time for input n is bounded by $c \cdot f(N)$. If f is a polynomial function of N, then the algorithm runs in **polynomial time**, and this is generally good. An algorithm whose running time exceeds all polynomials will typically be useless on very large integers (though it can still be very helpful for work on small integers). For example, factorization of n by trial division requires up to $\sqrt{n}$ trials (one needs to be able to check all prime divisors up to $\sqrt{n}$). Because $\sqrt{n} = 2^{(\log_2 n)(1/2)} = (\sqrt{2})^N$, this is not a polynomial-time algorithm. Number field sieve methods for factorization run in $O(\text{Exp}[K N^{1/3} (\ln N)^{2/3}])$ for some constant K. The reader should verify that for any value of the constant K, $\text{Exp}[K N^{1/3} (\log N)^{2/3}]$ grows more slowly than $(\sqrt{2})^N$. The rub is in that unknown constant c out front.

Note that "steps" really refers to steps involving individual bits. Thus, the multiplication of two N-digit numbers, using the algorithm we all learned in elementary school, has time complexity $O(N^2)$.

Here is an example of how to estimate running times from timing data: the squaring of an $n \times n$ matrix. This involves n^3 multiplications. We use 100-digit integers to slow down the multiplication process and bring out more clearly the time increase due to the increase in n. Because the integer-lengths are fixed, the input length is measured as a constant times the number of entries in the matrix, which is n^2. Thus, the true complexity of the algorithm should be $O(N^{3/2})$. For simplicity, we will work with functions of n, instead of functions of n^2, and so we will see whether an n^3 complexity shows up. The following function squares an $n \times n$ matrix and returns the time only. Figure 1.5(a) shows a plot of the data as n goes from 10 to 60 in steps of 5.

```
time[n_] :=
  (mat = Table[Random[Integer, {10^100, 10^101}], {n}, {n}];
   Timing[mat.mat][[1, 1]])

data = Table[{n, time[n]}, {n, 10, 60, 5}];
```

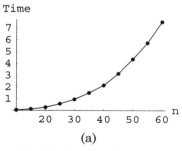
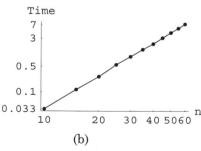

(a)   (b)

Figure 1.5. (a) One would expect that the time needed to square an $n \times n$ matrix would be proportional to n^3. The timing plot supports this. (b) A log–log plot (base 10) of the timing data yields an approximate straight line having slope very close to 3 $[(\log_{10} 60 - \log_{10} 10)/(\log_{10} 7 - \log_{10} 0.033) = 2.9898...]$.

A relationship of the form $y = x^s$ becomes, when logarithms of both x and y are taken, $\log y = s \log x$. Thus, when a power law is suspected, one should look at a log–log plot to see if a straight line results. And if so, the slope indicates the power. In this case (Figure 1.5(b)), the cubic relationship pops right out. In *Mathematica* `Log[x]` denotes $\log_e x$; other bases are handled by `Log[b, x]`. And numerical functions such as `Log` apply to all the numbers in any level in a list, so `Log[data]` takes the logarithm of both coordinates.

If we fit a line to the log–log data (this minimizes the least-squares error), we get a slope very close to 3.

```
Fit[Log[10, data], {1, logn}, logn]
```

  -4.49712 + 3.00061 logn

This indicates that the time data is close to $10^{-4.49712} \, n^{3.00061}$, or $0.0000318 n^{3.00061}$. In fact, the pure cubic $0.000032 n^3$ fits pretty well, as we can discover by fitting a $c\,n^3$ function.

```
Fit[data, n³, n]
```

  0.0000319037 n³

```
Plot[0.000032 n³, {n, 5, 60},
    Epilog → {PointSize[0.025], Map[Point, data]},
    AxesLabel → {"n", "Time"}];
```

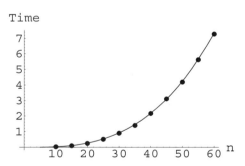

Figure 1.6. The timing data for squaring a matrix is very well approximated by the cubic function $0.000032 \, n^3$.

When you know the running time for a given algorithm on a given machine with a given input, big-O estimates will give you an idea of how much you can enlarge the input and still expect a result in reasonable time or within the hardware capacities of your machine. By itself, it is not terribly helpful in comparing two algorithms. Given an integer less than a million, it would be silly to try to use the number field sieve to try to factor it. Trial division works best.

Having said this, we shall now try to avoid big-O notation in this book. It is no substitute for a hands-on feel for how fast the algorithms run and, as pointed out above, it can be deceptive. But we have introduced it here because it does give a useful shorthand for describing the nature of the superiority of one algorithm over another, and anyone who pursues algorithmic number theory beyond the scope of this book will find that it assumes increasing importance. A slower order of growth means that there will be a crossover point someplace. It may not be attainable with current machines or implementations, but some day, in some situation, the algorithm with a lower order of growth will demonstrate its superiority.

Exercises for Section 1.4

1.39. Implement the right-to-left method of binary exponentiation (the computation of b^n (mod m)), based on the description at the beginning of Section 1.4. Here is a tabular description of the basic method, with no modular reduction. Successive squares are stored in, say, sq. The exponent, e starts out being n and is divided by 2 at each step, with its parity determining what happens to the answer being built up. If e is odd, the answer gets multiplied by the current value of square; if e is even, nothing is done.

PowerAlgorithm[b, 41, Method → BinaryRightToLeft]

Exponent	Parity	Repeated Squares	Answer
41	1	b	1
20	0	b^2	b
10	0	b^4	b
5	1	b^8	b
2	0	b^{16}	b^9
1	1	b^{32}	b^9
			b^{41}

AdditionChain[41, BinaryRightToLeft]

$\{1, 2, 4, 8, 9, 16, 32, 41\}$

Compare the performance of this method to powermodLR. Note that the addition chains from the two algorithms are different, but they have the same length. A straight time comparison is a little unfair because powermodLR uses a built-in *Mathematica* function to get the base-2 digits. So here is code that implements the left-to-right algorithm manually, thus making for a fairer comparison. Try these on triples (b, m, n) where b is small and m and n are very large, and you should see a clear victory for the left-to-right method. Note that one often has to repeat a calculation 5 or 10 or 100 times to get the times up into a meaningful range (1–10 seconds, say).

```
powermodLRFairer[a_, n_, m_] :=
  Module[{ans = a, digits = {}, nn = n},
    While[nn ≠ 0,
```

```
    digits = {Mod[nn, 2], digits};
    nn = Quotient[nn, 2]];
  digits = Flatten[digits];
  Do[ans = Mod[ans², m];
    If[digits[[i]] == 1, ans = Mod[ans * a, m]],
    {i, 2, Length[digits]}];
  ans]
```

A note on the code: the base-2 digits are built-up in `digits` in the form of a nested list via the `{Mod[nn, 2], digits}` phrase; later, `digits` is flattened out. This is much faster than, say, using `AppendTo`.

1.40. There is one integer between 2 and 20 that has an addition chain that is shorter then the one provided by either of the binary methods. Find it.

1.41. Here is another method for finding addition chains; it is a recursive procedure called the ***factor method***. Explain why the following rules give a valid algorithm for an addition chain. Work out the factor chain for 78 by hand. Then write a program that implements the factor method, and compare the results to the binary method.

1. The chain for 1 is {1}; the chain for 2 is {1, 2}; the chain for 3 is {1, 2, 3}.
2. The chain of a prime p is obtained by tacking p onto the end of the chain for $p - 1$.
3. The chain for n, a composite integer of the form $a \cdot b$, is obtained by starting with the chain for a and following it with the chain for b with each term multiplied by a. The duplicated entry — the a at the end of the first part and the beginning of the second part — is replaced by a single a.

Here are examples of the factor method in action on exponent 39. You can use these package functions to check your program.

AdditionChain[39, FactorMethod]

{1, 2, 3, 6, 12, 24, 36, 39}

PowerAlgorithm[b, 39, Method → FactorMethod]

The Recursive Calls	Answer
{3, 13}	b
2	b^2
13	b^3
{2, 6}	b^6
{2, 3}	b^{12}
3	b^{24}
2	b^{36}
	b^{39}

1.42. There is an ancient method for multiplying two large numbers without the use of multiliplication tables. It is sometimes refered to as Russian peasant multiplication. To multiply, for example, 361×425, we form two columns with the smaller number at the top of the left column and the larger number atop the right column. We find the numbers in the left column by taking half of the top entry, discarding any remainder, and continuing until we get to 1. We find the numbers in the right column by doubling the top entry and continuing until this column is as long as the left column.

361	425
180	850
90	1700
45	3400
22	6800
11	13600
5	27200
2	54400
1	108800

We now cross out the entries on the right for which the corresponding entry on the left is even, and then add the remaining entries in the right column:

361	425
180	
90	
45	3400
22	
11	13600
5	27200
2	
1	108800
Answer	153425

The last result, 153425, is the product of the original integers. Explain why this works.

1.43. Carry out timing experiments to obtain a function that describes the time complexity of the `PowerMod` function, using three random n-digit integers as input.

1.44. A *perfect shuffle* involves cutting a deck of cards exactly in half and then shuffling so that the cards alternate from each hand. An *out-shuffle* is a perfect shuffle in which the top card stays on top (and the bottom card stays on the bottom). An *in-shuffle* is a perfect shuffle in which the top card moves to the second position. Prove that it is always possible to use a sequence of in-shuffles and out-shuffles to move the top card to any position in the deck. Show how to use the binary expansion of j to determine the minimal sequence of in-shuffles and out-shuffles that will move the top card to position $j + 1$.

CHAPTER 2

Congruences, Equations, and Powers

2.0 Introduction

The Euclidean algorithm is the foundation of many important computational ideas. In this chapter we show how it can be used to solve certain Diophantine equations (meaning: equations where solutions in integers only are desired). Two other critical foundational results are the Chinese Remainder Theorem for solving certain systems of congruences and Fermat's theorem on powers modulo a prime. His theorem is the basis of some simple tests for primality, and that is pursued in the section on pseudoprimes.

2.1 Solving Linear Congruences

◼ Linear Diophantine Equations in Two Variables

A Diophantine equation, named after Diophantus of Alexandria (c. 250 CE) refers to any equation for which solutions in integers are sought. The most famous example is Fermat's Last Theorem, recently proved by Andrew Wiles, which says that, if $n > 2$, there is no integer solution to $x^n + y^n = z^n$ except for the trivial solution where x, y, or z is 0. Another very famous Diophantine equations is Catalan's equation: $x^a - y^b = 1$; the only solution known is $3^2 - 2^3 = 1$, and it is suspected that there are no others. Diophantine equations have a long history in the area of recreational mathematics, where puzzles such as the following are common: Suppose a piggy bank has 37 coins, all of which are nickels, dimes, and quarters, with at least one coin of each type. If the total value of the coins is \$2.25, then how many coins of each type are in the bank? In this section we discuss a method for getting a complete solution to a linear Diophantine equation. The algorithm will make short work of puzzles such as the one just given.

Although the general problem of solving Diophantine equations is extremely difficult, a comprehensive theory has been developed to handle the case of one or more linear equations, and our work with the Euclidean algorithm in Chapter 1 gives us a method for dealing with the simplest case, where the equation is linear and involves only two variables. Such equations have the form $ax + by = c$.

Proposition 2.1. Consider the equation $ax + by = c$ and let $d = \gcd(a, b)$. If d does not divide c, the equation has no solution in integers. If d divides c, then there are infinitely many solutions, which are obtained as follows. A particular solution is given by $(x_0, y_0) = (cs/d, ct/d)$, where s and t are the result of the extended Euclidean algorithm on the pair a, b. The general solution then has the form $x = x_0 + (b/d)n$, $y = y_0 - (a/d)n$, where n is any integer.

Proof. For a solution to exist, d must divide c, since d divides a and b. If d does divide c, then the extended Euclidean algorithm (Section 1.2) yields integers s and t such that $sa + tb = d$. Because c/d is an integer, multiplication by c/d yields $a(c/d)s + b(c/d)t = (c/d)d = c$, and thus we have our first solution, $x_0 = cs/d$ and $y_0 = ct/d$.

For the general case, we use the common algebraic technique of first studying the equation (called a *homogeneous equation*) with 0 as the right side. One solution is $(bn, -an)$, but a better solution is $((b/d)n, -(a/d)n)$, because any solution is a multiple of this last one. For suppose $ax_1 + by_1 = 0$. Then $(a/d)x_1 + (b/d)y_1 = 0$, so $(a/d)x_1 = -(b/d)y_1$. But a/d and b/d are relatively prime (Exercise 1.9), so a/d divides $-y_1$ (Proposition 1.9), say $-y_1 = n(a/d)$. Substituting into the equation relating x_1 and y_1 yields $x_1 = bn/d$, completing the analysis of the homogeneous equation.

Now suppose (x_2, y_2) is a solution to the general equation. Then it is easy to check that $(x_2 - x_0, y_2 - y_0)$ satisfies the homogeneous equation, and so there is an n such that $x_2 - x_0 = bn/d$ and $y_2 - y_0 = -na/d$. But this is the desired representation of (x_2, y_2). □

The preceding proof yields an algorithm to solve linear Diophantine equations in two variables.

Algorithm 2.1. The Diophantine Equation $ax + by = c$

The code that follows uses the built-in `ExtendedGCD` function to implement the algorithm of the preceding proof. The setup requires one parameter, which is used to represent the infinite family of solutions. An alternative approach would code the output in the form of a 2×2 matrix.

```
DiophantineSolve[{a_, b_}, c_, n_] := Module[{d, e},
   {d, e} = ExtendedGCD[a, b];
   If[Mod[c, d] == 0,
    Transpose[{c e, {b, -a}} / d].{1, n}, {}]]
```

```
gen = DiophantineSolve[{13, 51}, 500, n]
```

$\{2000 + 51\, n,\ -500 - 13\, n\}$

Note that the numbers are quite large. A change of variable of the form $n \rightarrow n - 39$ will reduce the size of the coefficients, where 39 is obtained as the nearest integer to $\frac{2000}{51}$.

```
gen = Expand[gen /. n → n - 39]
```

$\{11 + 51\, n,\ 7 - 13\, n\}$

And here are some explicit solutions.

```
Table[gen, {n, 1, 5}]
```

$\{\{62, -6\}, \{113, -19\}, \{164, -32\}, \{215, -45\}, \{266, -58\}\}$

◾ Linear Equations in Several Variables

There is a well-developed theory for solving linear Diophantine equations, and even systems of such equations can be handled. Exercise 2.4 shows how one linear equation in many variables can be handled by a method that reduces the number of variables until two are left. The CNT package contains `LinearDiophantineSolve`, which implements this idea. Here is how it can be used to solve $2x + 3y + 4z + 5w = 20$.

```
LinearDiophantineSolve[{2, 3, 4, 5}, 20] // MatrixForm
```

$$\begin{pmatrix} 0 & 1 & 0 & 0 \\ 0 & 0 & 1 & 0 \\ -20 & 2 & 3 & 5 \\ 20 & -2 & -3 & -4 \end{pmatrix}$$

The matrix represents the general solution in that the first column is a particular solution while the other three columns form the general solution to the homogeneous equation $2x + 3y + 4z + 5w = 0$. Thus multiplying the matrix by $\{1, n, m, r\}$ yields the complete representation of the general solution.

```
LinearDiophantineSolve[{2, 3, 4, 5}, 20].{1, n, m, r}
```

$\{n,\ m,\ -20 + 3\, m + 2\, n + 5\, r,\ 20 - 3\, m - 2\, n - 4\, r\}$

`LinearDiophantineSolve` contains an appeal to *Mathematica*'s standard `Algebra`InequalitySolve`` package to solve the inequalities that arise when the solutions are restricted to be positive or nonnegative.

Here is the complete list of all solutions to the sample equation subject to the constraint that no negative numbers appear.

```
LinearDiophantineSolve[{2, 3, 4, 5}, 20,
  SolutionType → NonNegative]
```

```
{{0, 0, 0, 4}, {0, 0, 5, 0}, {0, 1, 3, 1}, {0, 2, 1, 2},
 {0, 4, 2, 0}, {0, 5, 0, 1}, {1, 0, 2, 2}, {1, 1, 0, 3},
 {1, 2, 3, 0}, {1, 3, 1, 1}, {1, 6, 0, 0}, {2, 0, 4, 0},
 {2, 1, 2, 1}, {2, 2, 0, 2}, {2, 4, 1, 0}, {3, 0, 1, 2},
 {3, 2, 2, 0}, {3, 3, 0, 1}, {4, 0, 3, 0}, {4, 1, 1, 1},
 {4, 4, 0, 0}, {5, 0, 0, 2}, {5, 2, 1, 0}, {6, 0, 2, 0},
 {6, 1, 0, 1}, {7, 2, 0, 0}, {8, 0, 1, 0}, {10, 0, 0, 0}}
```

The function also allows the specification of additional assumptions, so that certain common problems can be easily solved. For example, recall the problem mentioned at the beginning of this section: A piggy bank has 37 coins, all of which are nickels, dimes, and quarters, with at least one coin of each type. If the total value of the coins is $2.25, then how many coins of each type are in the bank? We could look at all positive solutions and `Select` the ones for which $x + y + z = 37$, but the function can do that for us.

```
good[x_] := Apply[Plus, x] == 37
solutions = LinearDiophantineSolve[{5, 10, 25}, 225,
  SolutionType → Positive, Assumptions → good]
```

```
{{32, 4, 1}}
```

■ Linear Congruences

The notion of modular inverses makes short work of solving congruences such as $7x \equiv 2 \pmod{17}$. Just multiply both sides by the mod-17 inverse of 7.

```
PowerMod[7, -1, 17]
```

```
5
```

So $2 \cdot 5$, or 10, is a solution, and indeed $7 \cdot 10 = 2 + 68$. But this method requires the inverse. In fact, the congruence can be solved whether or not the inverse exists.

Proposition 2.2. Consider the modular equation $ax \equiv b \pmod{m}$, where a, b, and m are integers with $m > 0$.

(a) If $\gcd(a, m) = 1$, then the equation has the solution $a^{-1}b$, where the inverse is modulo m; the solution is unique modulo m.

(b) If $\gcd(a, m)$ fails to divide b, then the equation has no solutions.

(c) If $d = \gcd(a, m)$ and d divides b, then there are exactly d incongruent solutions modulo m.

Proof.   **(a)** It is clear that $a^{-1}b$ is a solution. And if x_0 satisfies $ax_0 \equiv b \pmod{m}$, then $a^{-1}ax_0 \equiv a^{-1}b \pmod{m}$, and $x_0 \equiv a^{-1}b \pmod{m}$ as claimed.

(b) If x_0 is a solution, then $ax_0 = b + km$ for some k, and so $\gcd(a, m)$ must divide b.

(c) The congruence $ax \equiv b \pmod{m}$ is equivalent to the Diophantine equation $ax + km = b$, and so we may apply Proposition 2.1. Therefore solutions for x are given by $x_0 + (m/d)n$. Two such solutions are congruent modulo m if and only if $(m/d)n_1 \equiv (m/d)n_2 \pmod{m}$. But, by Proposition 1.15, this is equivalent to: $n_1 \equiv n_2 \pmod{m/\gcd[m, m/d]}$, which reduces to $n_1 \equiv n_2 \pmod{m/(m/d)}$, or $n_1 \equiv n_2 \pmod{d}$. This means there are d incongruent solutions, obtained by letting n run through the residue classes modulo d. $\square$

Solving a single linear congruence in one variable is essentially the same problem as solving a single linear Diophantine equation in two variables, and is easy to implement. A function in the CNT package does the job. Here is how to solve $6123123x \equiv 6123123123 \pmod{9123123123123}$.

LinearCongruenceSolve[{6123123, 6123123123}, 9123123123123]

{2282503531938, 5323544572979, 8364585614020}

Mathematica's LinearSolve function can be coaxed to work modulo m, and can handle linear systems where the number of unknowns equals the number of equations. Ordinarily LinearSolve[A, c] solves the linear system $Ax = c$. If the Modulus option is used it works modulo m.

v = LinearSolve[
  A = {{5, 10, 3}, {1, 1, 1}, {1, 2, 5}}, {1, 2, 3}, Modulus → 53]

{27, 37, 44}

Mod[A.v, 53]

{1, 2, 3}

One can also appeal directly to Solve; this approach fails for large integers, but has the advantage that quadratic and higher equations can be handled.

{x, y, z} /. Solve[{
        5 x + 10 y + 3 z == 1,
        x + y + z == 2,
        x + 2 y + 5 z == 3, Modulus == 53}, x]

{{27, -16, -9}}

x /. Solve[{5 x^3 + x^2 == 17, Modulus == 7919}, x]

{1101, 5621, 7532}

```
Mod[5 x³ + x² /. x → {1101, 5621, 7532}, 7919]
{17, 17, 17}
```

▣ The Conductor

A special type of Diophantine problem concerns the collections of numbers that can be represented by a certain formula. For example, if one asks for the set of integers represented by $6x + 9y + 20z$, one has the Chicken McNugget® problem mentioned in Exercise 2.8. There are many fascinating problems of this form; for example, it is unsolved whether every fraction $4/n$ has the form $1/x + 1/y + 1/z$. For this sort of problem, even the general linear case is quite difficult. However, the 2-variable case is not hard. Given a and b with $\gcd(a, b) = 1$, there is always a point beyond which every integer is representable as $ax + by$. The least such point is called the **conductor** of a and b. For example, it is not hard to see that 2 and all greater integers are representable as $2x + 3y$ with nonnegative x and y, while 1 is not so representable. So conductor[2, 3] = 2.

Before studying the following proof, the reader should review the postage stamp problem at the beginning of Section 1.3. Indeed, the conductor problem also has its history in a problem of postage stamps, because it answers the question: Given denominations a and b, which postage amounts are feasible? The following table from Section 1.3 shows which denominations are representable (in bold) using 5¢ and 8¢ stamps.

0	1	2	3	4
5	**6**	7	**8**	9
10	**11**	12	**13**	14
15	**16**	**17**	**18**	19
20	**21**	**22**	**23**	**24**
25	**26**	27	**28**	**29**
30	**31**	**32**	**33**	**34**
35	**36**	**37**	**38**	**39**
⋮	⋮	⋮	⋮	⋮

Theorem 2.3. If a and b are relatively prime and positive, then the conductor of a and b is $(a-1)(b-1)$.

Proof. We need several facts:

- If a and b are relatively prime, then in each residue class modulo a, the smallest positive integer that can be represented by $ax + by$, x and y nonnegative integers, must be a multiple of b (otherwise subtract a; in the table these multiples are 0, 16, 32, 8, and 24).

- When we can find an integer in a given residue class modulo a that is representable by $ax + by$, then any larger integer in that residue class is also representable (just add multiples of a).

- If a and b are relatively prime, then $0b$, $1b$, ..., $(a-1)b$ lie in different residue classes modulo a (Exercise 2.24).

These facts tell us that the residue class that contains $(a-1)b$ is the residue class with the largest integer that cannot be represented. This nonrepresentable integer is $(a-1)b-a$, and so the conductor of a and b is $(a-1)b-a+1$, or $(a-1)(b-1)$. In the table, this is $27+1$, or 28. □

To completely understand the set of representable positive integers, we must also figure out what happens below the conductor. Exactly half of the integers between 0 and the conductor are representable (Exercise 2.7).

The conductor of any finite set of pairwise coprime positive integers always exists (Schur's theorem; see [Wil]). The CNT package contains an algorithm for the conductor of three integers. The algorithm cannot handle all cases, but it can do some. For example, 197 is the largest integer not of the form $12x+19y+31z$.

Conductor$[\{12, 19, 31\}]$

198

◼ An Important Quadratic Congruence

For any m, 1 and $m-1$ are solutions to $a^2 \equiv 1 \pmod{m}$; this is just a restatement of $(\pm 1)^2 = 1$. A very important fact is that for prime moduli there are no other solutions.

Proposition 2.4. If p is prime then the only solutions to $x^2 \equiv 1 \pmod{p}$ are 1 and $p-1$.

Proof. Suppose a is its own inverse modulo p. Then $a^2 - 1 \equiv 0 \pmod{p}$, whence p divides $a^2 - 1$. But then p divides $(a-1)(a+1)$ so, by Corollary 1.10, either p divides $a-1$ or p divides $a+1$. But this means that $a \equiv \pm 1 \pmod{p}$. □

Exercises for Section 2.1

2.1. Modify the code of Algorithm 2.1 so that it performs a change of variable in an attempt to make the integers in the output small.

2.2. A van that can hold no more than 20 people ferried a group of travelers to a train station, making five trips and taking an equal number of people in each group. Then they were joined by 7 people who walked over, and all were evenly divided among 14 railroad cars. How many people were traveling? Use the algorithm of the text, working out the details by hand.

2.3. Find a two-digit number such that the digit in the ones place times 8 is thirteen less than the digit in the tens place times 3.

2.4. (a) Implement an algorithm that uses the ideas of Algorithm 2.1 to get a parametric representation of all solutions of $ax + by + cz = B$. Use an inductive approach by letting w represent $(b/d)y + (c/d)z$ where $d = \gcd(b, c)$ and solving $ax + dw = B$. Then use the representation of w that this solution yields — it will involve a parameter, say n — to solve the original problem by solving $(b/d)y + (c/d)z = w$. Because b/d and c/d are relatively prime, the fact that w involves n is not an obstacle.

(b) Use part (a) to solve $5x + 10y + 25z = 100$. Then use the solution to find out how many ways one can make change for $1.00 using nickels, dimes, and quarters.

(c) Use recursion to extend the ideas of part (a) to apply to a general linear Diophantine equation $a_1 x_1 + a_2 x_2 + \cdots + a_n x_n = B$. There are well-developed methods based on linear algebra to solve Diophantine systems of n linear equations in m unknowns; see [NZM, §5.2].

2.5. A good argument can be made that the following puzzle, whose first appearance was in an article by Ben Ames Williams in the *Saturday Evening Post* in 1926, is one of the most studied Diophantine problems in history! The magazine received 20,000 solutions.

Five sailors on an island gather a pile of coconuts and decide to divide them among themselves the next day. During the night, one of the sailors awoke, counted the coconuts, gave one to the monkey that was hanging around camp, and took exactly $^1/_5$ of the rest for himself. Later that night, a second sailor awoke and did exactly the same thing. And so on, in turn, with none of the sailors being aware of what the others were doing. In the morning, all awoke and, after tossing one coconut to the waiting monkey, divided the remaining coconuts into five equal parts. How many coconuts had the sailors gathered?

Show how the conditions can be combined to get an equation of the form $ax + by = c$, where x is the total number of coconuts. Then use the methods of this section to solve the equation and find the smallest possibility for x.

2.6. Write a routine that accepts a, b, and m as input and returns the complete mod-m solution set to $ax \equiv b \pmod{m}$.

2.7. Theorem 2.3 states that if $\gcd(a, b) = 1$, then any number greater than or equal to $(a - 1)(b - 1)$ is representable as $ax + by$, with $x, y \geq 0$. Show that, among the numbers between 0 and $(a - 1)(b - 1) - 1$, inclusive, exactly half are representable. *Hint:* Use the ideas in the proof of Theorem 2.3.

2.8. What is the largest number of Chicken McNuggets that you cannot buy? Chicken McNuggets are sold in baskets of size 6, 9, and 20. (Do this without using the `Conductor` function in the `CNT` package.)

2.9. Show that Proposition 2.4 can fail if p is not prime.

2.10. Show that Proposition 2.4 can be generalized to the following: If p is an odd prime and $x^2 \equiv a \pmod{p}$ has one solution in $\mathbb{Z}_p$, then it has precisely two solutions: b and $p - b$.

2.11. Show that Proposition 2.4 can be generalized to: If p is an odd prime then $x^2 \equiv 1 \pmod{p^e}$ has two solutions modulo p^e: 1 and $p^e - 1$. What happens if the modulus is 2^e?

2.12. (Wilson's Theorem) Use Proposition 2.4 to prove that, for any prime p, $(p - 1)! \equiv -1 \pmod{p}$. Then prove the converse: If n satisfies $(n - 1)! \equiv -1 \pmod{n}$ then n is prime. *Hint*: Consider the two cases (1) n is divisible by two primes, and (2) n is a prime power. Use the formula (Exercise 1.24) for the highest power of a prime that divides a factorial.

2.2 The Chinese Remainder Theorem

It is easy to find an integer of the form $3k + 1$. And anyone can find an integer of the form $5m + 3$. But is there a single integer that has both forms? This problem is a special case of a method that has come to be known as the Chinese Remainder Theorem. Of course, a brute force approach would let k be 0, 1, 2, 3, ... in succession until we obtain a value of $3k + 1$ that is congruent to 3 (mod 5). Will this always work? Yes, and again that is part of the conclusion of the Chinese Remainder Theorem.

```
Mod[Table[3 k + 1, {k, 0, 19}], 5]

{1, 4, 2, 0, 3, 1, 4, 2, 0, 3, 1, 4, 2, 0, 3, 1, 4, 2, 0, 3}
```

The fifth entry ($k = 4$) yields the desired 3, so $3 \cdot 4 + 1$, or 13, is a solution to the sample problem. But in fact any number of congruences can be solved, so long as each pair of moduli are relatively prime.

Theorem 2.5. Chinese Remainder Theorem. Suppose $m_1, m_2, \ldots, m_r$ are positive integers that are relatively prime in pairs, and $M = m_1 m_2 \cdots m_r$. Consider the following system of congruences:

$$x \equiv a_1 \pmod{m_1}$$
$$x \equiv a_2 \pmod{m_2}$$
$$\vdots$$
$$x \equiv a_r \pmod{m_r}.$$

Then there is exactly one integer in $\{0, 1, 2, \ldots, M - 1\}$ that solves the system.

Before proving the result, we give a very important interpretation. The least nonnegative residues modulo n are usually denoted $\mathbb{Z}_n$; that is, $\mathbb{Z}_n = \{0, 1, \ldots, n - 1\}$. Moreover, under the operation of mod-n addition, $\mathbb{Z}_n$

forms a **group** (the operation is associative, there is an identity, 0, and every element x of the group has an inverse, namely the element congruent to $-x$). Now, given $x \in \mathbb{Z}_{mn}$, one can get integers in $\mathbb{Z}_m$ and $\mathbb{Z}_n$ by just looking at Mod$[x, m]$ and Mod$[x, n]$. The point of the CRT is that, provided gcd$(m, n) = 1$, any pair from $\mathbb{Z}_m$ and $\mathbb{Z}_n$, respectively, comes from a unique member of $\mathbb{Z}_{mn}$ under congruence mod m and mod n.

More precisely, the set of pairs with the first entry from one group, G, and the second from another, H, itself forms a group, with the operation defined componentwise; this is called the **direct product**, $G \times H$. So the pairs just mentioned are objects in $\mathbb{Z}_m \times \mathbb{Z}_n$. Modular reduction gives a function from $\mathbb{Z}_{mn}$ to $\mathbb{Z}_m \times \mathbb{Z}_n$; when gcd$(m, n) = 1$ the CRT says that the function is one-to-one and onto (and therefore is invertible). In short, when m and n are relatively prime, the two groups $\mathbb{Z}_m \times \mathbb{Z}_n$ and $\mathbb{Z}_{mn}$ are **isomorphic**, meaning that there is a one-to-one and onto function from one to the other that preserves the group operation. And of course the Chinese Remainder Theorem deals with any number of congruences, not just two. So it tells us that $\mathbb{Z}_{m_1} \times \mathbb{Z}_{m_2} \times \cdots \times \mathbb{Z}_{m_r}$ is isomorphic to $\mathbb{Z}_{m_1 m_2 \cdots m_r}$.

Proof. Consider $M / m_1 = m_2 m_3 \cdots m_n$; this integer is relatively prime to m_1, so we can let b_1 be the inverse of M / m_1 modulo m_1. Then look at $y_1 = a_1 b_1 (M / m_1)$; $y_1 \equiv 0 \pmod{m_i}$ for each $i > 1$, and $y_1 \equiv a_1 \pmod{m_1}$. In other words, y_1 is a solution to the first congruence, and contributes nothing as far as the other moduli are concerned. So, for a general solution, just form $y_2, y_3, \ldots, y_n$ in the same way, and sum all the y-values. The sum, which we can reduce modulo M without affecting any of the congruences, is a solution to the system.

It remains to show that the solution is unique modulo M. Suppose x and z are two solutions. Then $x \equiv z \pmod{m_i}$ for each i, so, by Proposition 1.7, $x \equiv z \pmod{M}$. $\square$

Here is a visual interpretation of the CRT. The columns are indexed by the integers mod 9 and the rows by the integers mod 4. For example, the pair $\{2, 7\}$ in $\mathbb{Z}_4 \times \mathbb{Z}_9$ corresponds to 34 in $\mathbb{Z}_{36}$. Note how the group operations are preserved: in $\mathbb{Z}_{36}$, $21 + 34 = 55 \equiv 19$. But adding up the components, $\{1, 3\} + \{2, 7\}$, yields $\{3, 10\}$, or $\{3, 1\}$, which is also 19.

CRTGrid[4, 9];

$\mathbb{Z}_4$ \ $\mathbb{Z}_9$	0	1	2	3	4	5	6	7	8
0	0	28	20	12	4	32	24	16	8
1	9	1	29	21	13	5	33	25	17
2	18	10	2	30	22	14	6	34	26
3	27	19	11	3	31	23	15	7	35

Algorithm 2.2 Chinese Remainder Theorem

The implementation that follows uses the formula inherent in the proof of Theorem 2.5 to solve a system of congruences of the form $x \equiv a \pmod{m}$, where a and m represent lists. The output is simply the least nonnegative solution; the user can add multiples of M to get the general solution. We first define a query function that decides if a list of integers is pairwise relatively prime. That function, `PairwiseCoprimeQ`, is used as a restrictor in the `CRT` definition so that if it fails the program will not apply. Note that `PairwiseCoprimeQ` avoids the obvious, but very slow approach of checking all pairs (see Exercise 1.11). See Exercises 2.15 and 2.20 for different ways of programming CRT that are faster, and more powerful in that they deal with the case of non-coprime moduli too.

```
PairwiseCoprimeQ[m_List] := Apply[LCM, m] == Apply[Times, m]


CRT[a_, m_?PairwiseCoprimeQ] :=
 Module[{M = Apply[Times, m]},
  Mod[Apply[Plus, a * PowerMod[M / m, -1, m] M / m], M]]
```

The following alternative approach is both shorter and faster. In general, replacing multiplication and addition with a dot product speeds things up. And it uses a pure function to act on the product of the moduli, thus avoiding the need to store that product in M.

```
CRT[a_, m_?PairwiseCoprimeQ] :=
 Mod[a.(PowerMod[# / m, -1, m] # / m), #] & [Times @@ m]
```

Here are some examples. The usage is such that CRT[{1,3,5},{3,5,7}] finds x congruent to 1 (mod 3), 3 (mod 5), and 5 (mod 7).

```
CRT[{1, 3, 5}, {3, 5, 7}]
```
```
103
```

The way the code is written, nothing happens if the moduli are not pairwise relatively prime.

```
CRT[{1, 3, 4}, {3, 5, 25}]
```
```
CRT[{1, 3, 4}, {3, 5, 25}]
```

A classic application of the CRT algorithm is to the following problem: Find five consecutive integers each of which is divisible by a square. We can approach this by seeking x such that $x \equiv 0 \pmod 4$, $x + 1 \equiv 0 \pmod 9$, $x + 2 \equiv 0 \pmod{25}$, and so on, where we are using squares of small primes for the moduli because they will be relatively prime. Isolating the x's leads to a system that is easily solved by CRT. The code that follows forms the squares of the first five primes by squaring the list, and then using CRT just defined. In fact, a `ChineseRemainder` function (in versions earlier than

4.0, this was called `ChineseRemainderTheorem`) is included in the standard *Mathematica* package `NumberTheory`NumberTheoryFunctions``.

```
x = CRT[Range[0, -4, -1], Prime[Range[5]]²]
```

1308248

```
consecutives = x + Range[0, 4]
```

{1308248, 1308249, 1308250, 1308251, 1308252}

```
Mod[consecutives, {2, 3, 5, 7, 11}²]
```

{0, 0, 0, 0, 0}

The sequence just found is by no means the smallest sequence of five integers, each of which is divisible by a square. Exercise 2.16 asks you to find the smallest such sequence.

A classic application of the CRT algorithm is its use in high-precision arithmetic. One could take a very large computation and work in a coordinate system based on a certain fixed set of primes; that is, one would be working with vectors consisting of the residues modulo these primes, and do arithmetic on those residues only. Then, at the very end, the CRT would be invoked to get the real answer. Here is a simple example showing how 100! can be computed using low-precision operations. First we select sufficiently many small primes. For the problem at hand it is sufficient to use the first 50 ten-digit primes, for their product is beyond 100!. We use $\pi(10^6)$ to get the number of primes under 10^6.

```
primes = Prime[π[10⁶] + Range[50]]
```

{1000003, 1000033, 1000037, 1000039, 1000081, 1000099, 1000117,
 1000121, 1000133, 1000151, 1000159, 1000171, 1000183, 1000187,
 1000193, 1000199, 1000211, 1000213, 1000231, 1000249,
 1000253, 1000273, 1000289, 1000291, 1000303, 1000313,
 1000333, 1000357, 1000367, 1000381, 1000393, 1000397,
 1000403, 1000409, 1000423, 1000427, 1000429, 1000453,
 1000457, 1000507, 1000537, 1000541, 1000547, 1000577,
 1000579, 1000589, 1000609, 1000619, 1000621, 1000639}

Now we start with a vector of fifty 1s and form 100! in each coordinate, reducing modulo the primes at each step.

```
vector = Table[1, {50}];
Do[vector = Mod[vector * i, primes], {i, 2, 100}];
vector
```

{371073, 427462, 989068, 576756, 619053, 955393, 80338, 523401,
 480115, 974383, 910868, 416971, 575000, 911639, 660763,
 46221, 655210, 379024, 558302, 177614, 641952, 983349,
 379088, 294345, 276429, 924165, 656771, 616678, 494696,
 129955, 950403, 403948, 305621, 875111, 987884, 982602,
 48647, 92881, 537812, 286612, 326210, 77332, 930653,
 646107, 209175, 58863, 641930, 42908, 198746, 534767}

The final step uses the CRT to get the real value of 100!.

ChineseRemainder[vector, primes]

933262154439441526816992388562667004907159682643816214685929 ⋮
638952175999932299156089414639761565182862536979208272237582 ⋮
25118521091686400000000000000000000000000000

Ideas such as these are occasionally used in massive high-precision computations. Indeed, CRT methods were used by David Bailey in one of his record-breaking computations of the digits of π.

Exercises for Section 2.2

2.13. Use the Chinese Remainder Theorem algorithm to find, by hand-calculation (no computer or calculator), all solutions to the system: $x \equiv 2 \pmod 3$, $x \equiv 3 \pmod 5$, $x \equiv 4 \pmod 7$.

2.14. A very short, though nonalgorithmic, proof of the Chinese Remainder Theorem exists. Show that if the moduli are pairwise relatively prime, then the function from $\mathbb{Z}_{m_1 m_2 \cdots m_r}$ to $\mathbb{Z}_{m_1} \times \mathbb{Z}_{m_2} \times \cdots \times \mathbb{Z}_{m_r}$ given by sending x to $(\mathrm{Mod}[x, m_1], \mathrm{Mod}[x, m_2], \ldots, \mathrm{Mod}[x, m_r])$, is one-to-one. Show why this implies that the system of equations has a unique solution modulo the product of the m_i.

2.15. When the moduli are at all large there is a faster approach to the Chinese Remainder Theorem than Algorithm 2.2. Implement the algorithm described below and compare its performance to Algorithm 2.2 in the case that the a-values are given by $\{1, 2, 3, \ldots, r\}$ and the moduli are the first r primes past the 100000th (the moduli can be described as `Prime[ Range[1, r] + 100000]`). Try both small and large values of r, doing multiple runs in the case of small values to get reasonable timing values.

(a) Show that the case of two equations with coprime moduli

$$x \equiv a_1 \pmod{m_1}$$
$$x \equiv a_2 \pmod{m_2}$$

has a solution given by $a_1 + (a_2 - a_1) m_1 (m_1^{-1} \bmod m_2)$ (*Hint*: The algebra underlying this verification is simple, but to see where the formula comes from look at the grid for $\mathbb{Z}_4 \times \mathbb{Z}_9$ on page 50 and observe that once one knows the entry in $\{0, 1\}$ one can easily get any entry in the grid.)

(b) Given r equations $x \equiv a_i \pmod{m_i}$ with pairwise relatively prime moduli, generate a solution by replacing the first two equations by the single equation $x \equiv a \pmod{m_1 m_2}$ where a is given by part (a). Continue such reductions until the final answer is obtained. Try to avoid using recursion in your program, as an iterative approach will be faster.

2.16. Find the smallest sequence of 5 consecutive integers each of which is divisible by a square.

2.17. Suppose $\gcd(m, n) = 1$ and let $f : \mathbb{Z}_m \times \mathbb{Z}_n \to \mathbb{Z}_{mn}$ be the one-to-one and onto function given by the Chinese Remainder Theorem. Prove that the product $f(a, b)f(s, t)$ in $\mathbb{Z}_{mn}$ is the same as $f(as, bt)$. This shows that f preserves the group operation.

2.18. Let $m = 2^a p_1{}^{b_1} \cdots p_r{}^{b_r}$, where $a \geq 0$, $r \geq 0$, $b_i \geq 1$. How many incongruent solutions are there to $x^2 \equiv 1 \pmod{m}$? *Hint*: Use Exercise 2.11 and the Chinese Remainder Theorem. Verify your result by an exhaustive search in several small cases.

2.19. Write a program that handles the general Chinese Remainder Theorem situation, including the case that the moduli are not relatively prime. Your program should return the empty set if the equations have no solution (such as $x \equiv 1 \pmod 5$, $x \equiv 3 \pmod{25}$), and should return the least nonnegative solution otherwise. Here are some key ideas:

(a) Prove that the system

$$x \equiv a_1 \pmod{m_1}$$
$$x \equiv a_2 \pmod{m_2}$$

has a solution if and only if $a_1 - a_2$ is divisible by $\gcd(m_1, m_2)$, and that when there is a solution it is unique modulo $\mathrm{LCM}[m_1, m_2]$.

(b) Find an explicit formula for a solution to the system in (a) when such exists.

(c) Deal with r equations by reducing the first two to a single equation, thus reducing the system to a smaller one.

2.20. (Dan Lichtblau, Wolfram Research, Inc.)   An even faster approach to a CRT algorithm is by a divide-and-conquer method. Use the method of Exercise 2.15 when there are seventeen or fewer equations. But when there are more, break the system into two nearly equal halves and use recursion. With proper care this can be made to work in the noncoprime case as well, using the ideas of Exercise 2.19 for the base case. Implement this method and make timing comparisons to the other methods discussed in this chapter.

2.21. We will see in Chapter 6 that 2 is a square modulo those primes, and only those primes, congruent to $\pm 1 \pmod 8$. Here are the first six such primes.

```
pr = Select[Prime[Range[18]], Mod[#, 8] == 7 || Mod[#, 8] == 1 &]
```

```
{7, 17, 23, 31, 41, 47}
```

We will learn later a formula for getting $\sqrt{2}$ modulo a prime, but for now we can do a brute force search.

```
sqrts = Table[Select[Range[pr[[i]]], Mod[#², pr[[i]]] == 2 &], {i, 6}]
```

```
{{3, 4}, {6, 11}, {5, 18}, {8, 23}, {17, 24}, {7, 40}}
```

As a check, observe that $5^2 \equiv 18^2 \equiv 2 \pmod{23}$. Now let M be the product of the primes in question.

```
M = Apply[Times, pr]
```

```
163500169
```

Show that x is a square root of 2 (mod M) if and only if $x \equiv a_p \pmod{p}$ for each p in `pr`, where each a_p is a mod-p square root of 2. Use this result to show that there are 64 square roots of 2 (mod M), and then use the CRT to find them all.

2.22. Let p_k denote the kth prime. Implement the following algorithm and explain why it returns $2p_k + 1$ consecutive composite integers. Use the Chinese Remainder Theorem to find b such that b is divisible by $\prod_{i=1}^{k} p_i$, $b \equiv -1 \pmod{p_{k+1}}$, and $b \equiv 1 \pmod{p_{k+2}}$. Then $\{b - p_k, \ldots, b - 1, b, b + 1, \ldots, b + p_k\}$ are all composite.

2.23. Find five consecutive integers each of which is divisible by the 100th power of an integer.

2.3 PowerMod Patterns

▰ Fermat's Little Theorem

Perhaps the most important function in elementary number theory is the modular power: $a^i \pmod{m}$. When one looks at a few of these powers many extremely interesting patterns jump out. Indeed, you should experiment with the tabular and visual functions described shortly to see how many patterns you can find. There are lots of them. In this section we will restrict ourselves to prime moduli; there are many important patterns in the composite case too, as we will discover in Section 3.4.

Here are two typical tables, with m equal to the primes 7 and 11. Each row gives the modular powers of a single a; so the second row of the first table shows $2, 2^2, 2^3, 2^4, 2^5, 2^6$, all modulo 7. Whenever $p - 1$ occurs as a power, -1 is shown instead; the most important entries are the 1s, shown in bold. How many patterns can you find in this array? Examine other arrays with prime modulus and see if the patterns you see persist.

PowerGrid[7]
PowerGrid[11]

i =	1	2	3	4	5	6	7	8	9	10
a = 1	1	1	1	1	1	1	1	1	1	1
a = 2	2	4	8	5	-1	9	7	3	6	1
3	3	9	5	4	1	3	9	5	4	1
4	4	5	9	3	1	4	5	9	3	1
5	5	3	4	9	1	5	3	4	9	1
6	6	3	7	9	-1	5	8	4	2	1
7	7	5	2	3	-1	4	6	9	8	1
8	8	9	6	4	-1	3	2	5	7	1
9	9	4	3	5	1	9	4	3	5	1
10	-1	1	-1	1	-1	1	-1	1	-1	1

i =	1	2	3	4	5	6
a = 1	1	1	1	1	1	1
a = 2	2	4	1	2	4	1
3	3	2	-1	4	5	1
4	4	2	1	4	2	1
5	5	4	-1	2	3	1
6	-1	1	-1	1	-1	1

n = 7

n = 11

Of course, the pattern that leaps out of the data is the fact that the rightmost column is all 1s. Indeed, this property tells us that there is no need to examine any columns beyond the $(p-1)$st, because entries will just repeat. And it is clear that there is no need to examine rows beyond the $(p-1)$st because $a \equiv b \pmod{m}$ implies $a^k \equiv b^k \pmod{m}$. The fact that $a^{p-1} \equiv 1 \pmod{p}$ is known as Fermat's Little Theorem. This identity is subtle and important; the proof that follows is short, but that doesn't take away from the inobvious nature of this result.

Theorem 2.6. Fermat's Little Theorem. If a is not divisible by p, then $a^{p-1} \equiv 1 \pmod{p}$. For any a, $a^p \equiv a \pmod{p}$.

Proof. A set of m numbers that are distinct modulo m is called a ***complete residue system*** **(mod m)**. For example, $\{0, 1, 2, 3, \ldots, m-1\}$ is such a system. And so is $\{0, 3, 4, 5, 6, \ldots, m-1, m+1, m+2\}$. Any such system, when reduced modulo m, is just a permutation of $\{0, 1, 2, 3, \ldots, m-1\}$. We need the following fact: If a is relatively prime to m and X is a complete residue system mod m, then so is the set obtained by multiplying each member of X by a. The proof is a straightforward application of the modular cancellation law and is left as an exercise.

Now, if p is prime, then two complete residue systems (mod p) are $X = \{0, 1, 2, 3, \ldots, p-1\}$ and $Y = \{a, 2a, 3a, \ldots, (p-1)a\}$. Because, viewed modulo m, X and Y are just permutations of each other, the products of their entries are congruent modulo m. This means that

$$a^{p-1}(p-1)! \equiv (p-1)! \pmod{p}$$

Because p is prime, it cannot divide $(p-1)!$. This means that $(p-1)!$ and p are relatively prime, so $(p-1)!$ can be cancelled from the preceding congruence. [Think of multiplying both sides by the mod-p inverse of $(p-1)!$; in fact, $(p-1)!$ is the same as $-1 \pmod{p}$ by Wilson's theorem (see Exercise 2.12), but that is not needed for this proof.] The cancellation leaves us with $a^{p-1} \equiv 1 \pmod{p}$. The additional result of the second assertion of the theorem requires only checking the case that a is divisible by p, and then both sides are 0 (mod p).   □

▣ More Patterns in Powers

Many other patterns are visible in the data. Indeed, we can visualize the power grid by using colors and the `VisualPowerGrid` function. The default scheme uses red for +1, yellow for −1, and different shades of blue for the other residues. If the powers of a run through all possibilities — which is the same as saying that the powers are all distinct — then a is called a **_primitive root_** for p. The tabular data presented earlier can be examined for primitive roots, while the varying shades that we use make their recognition more difficult. However, the `ShowPrimitiveRoots` option causes the a-values that are primitive roots to be shown in red (or with an arrow, as in Figure 2.1). Some color output of `VisualPowerGrid` appears in color plate 1; black and white images appear in Figure 2.1.

```
VisualPowerGrid[13, ShowPrimitiveRoots → True];
VisualPowerGrid[17, ShowPrimitiveRoots → True];
```

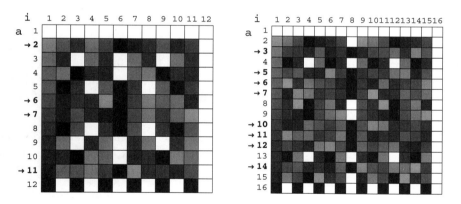

Figure 2.1. Two graphic displays of the power grid, for the prime moduli 13 and 17. White denotes +1, black denotes −1, and other values are represented by shades of gray. The arrows indicate the rows consisting of distinct entries (the primitive roots). Many patterns are visible. Some of them always hold; others will be disproved by looking at more examples.

The reader is encouraged to experiment with this function and discover what patterns he or she can. There are very many patterns here, both when the modulus is prime and when it is composite. Questions to think about: Which columns (rows) consist only of ±1s? Which rows (columns) consist of distinct values? Are there any patterns to the +1s and −1s? Some patterns that you will find will be true and not hard to prove. Others will be false, though a counterexample might be tricky to find. Others may involve unsolved problems.

Of course, number theory is notorious for generating simple-sounding statements that are difficult to prove. However, the fact that the central column consists only of ± 1s is easy to understand, for it follows immediately from Proposition 2.4.

Finally, we observe that Fermat's Little Theorem gives an alternate way to look at modular inverses. If $\gcd(a, p) = 1$ then, because $a^{p-1} \equiv 1 \pmod{p}$, we see that the mod-$p$ inverse of a is given by a^{p-2}. This is interesting, and a conceptually very simple algorithm, but in fact the method for getting inverses based on the Euclidean algorithm is faster.

Exercises for Section 2.3

2.24. Prove that if $\gcd(a, m) = 1$ and X is a complete residue system modulo m, then so is the set obtained by multiplying each member of X by a.

2.25. Write a program that accepts primes as inputs and counts the number of primitive roots.

2.26. Identify as many patterns as you can in the power grids. For each one, try to gain some evidence for its truth by either finding a proof or doing lots of numerical computations without using grids. To get started, consider: Which columns (rows) consist only of ± 1s? Which rows (columns) consist of distinct values? Are there any patterns to the $+1$s and -1s?

2.27. Compare the speed of the following two methods of getting modular inverses: (1) via the ideas of the extended Euclidean algorithm; (2) via the formula $a^{-1} \equiv a^{p-2} \pmod{p}$. For a fair comparison you should use large numbers, and make the programs as similar as possible in terms of their use of built-in functions.

2.28. Fermat's Little Theorem says that if p is prime then p divides $2^{p-1} - 1$. If a prime p is such that p^2 divides $2^{p-1} - 1$, then p is called a **Weiferich prime**. There are two Weiferich primes under 10^{12}. Find them.

2.29. Prove Fermat's Little Theorem by fixing a prime p and using induction on a, starting with $a = 1$. *Hint*: Use the binomial theorem to expand $(a + 1)^p$. You will need the auxiliary fact (Exercise 1.36): $\binom{p}{i} \equiv 0 \pmod{p}$ when $1 < i < p$.

2.30. Prove that if $a^b \equiv 1 \pmod{p}$, then $a^{\gcd(b, p-1)} \equiv 1 \pmod{p}$. *Hint*: Consider integers s and t such that $s\,b + t(p - 1) = \gcd(b, p - 1)$.

2.31. Let p be prime. Show that there is no prime less than p that divides $2^p - 1$. Use this to prove that if $p < m < p^2$ and m divides $2^p - 1$, then m is prime.

2.32. Find a prime p between $\sqrt{267497}$ and 267497 such that 267497 divides $2^p - 1$.

2.4 Pseudoprimes

Fermat's Little Theorem tells us that $2^{n-1} \equiv 1 \pmod{n}$ whenever n is prime. Because the modular power can be computed very quickly even when n is huge, it would be wonderful if $2^{n-1} \not\equiv 1 \pmod{n}$ whenever n is not prime; this would be an extremely fast and elegant test for primality. Unfortunately, this simple criterion fails to characterize the primes. Here are some functions to test how well this works; `TwoPseudoprimeQ` returns `True` if $2^{n-1} \equiv 1 \pmod{n}$ but n is not prime. We use `TwoPseudoprimeQ` for this function because such numbers are called **2-pseudoprimes**. Because we will want to use this as a prime test too, we define a separate function to carry out the power computation.

```
TwoPseudoprimeTest[n_] := PowerMod[2, n - 1, n] == 1
TwoPseudoprimeQ[n_] := TwoPseudoprimeTest[n] && !PrimeQ[n]
```

Now we look at the first 1000 integers, and discover that there are three that fool our criterion; the first one is 341. More generally, an integer n is called a **b-pseudoprime** if $b^{n-1} \equiv 1 \pmod{n}$ and n is not prime.

```
Select[Range[1000], TwoPseudoprimeQ]
```

{341, 561, 645}

The CNT package has a function that checks whether n is a b-pseudoprime. Because `PseudoprimeQ` has two arguments, we must use the pure function construction: `f[#, 2] &` refers to the function that, on input n returns `f[n, 2]`, in other words, # acts as a universal variable.

```
Select[Range[1000], PseudoprimeQ[2, #] &]
```

{341, 561, 645}

If there were only finitely many 2-pseudoprimes, we could incorporate a list into a program and so would have a foolproof and fast test for primality. But, even though they do become rarer as numbers get larger, there are infinitely many of them.

Proposition 2.7. There are infinitely many 2-pseudoprimes.

Proof. First a general fact: If d divides n then $2^d - 1$ divides $2^n - 1$. This can be proved by algebraic factorization of $x^{dk} - 1$ (Exercise 2.34), but it is smoother to use congruence arithmetic:

$$2^n - 1 = 2^{dk} - 1 = (2^d)^k - 1 = (2^d - 1 + 1)^k - 1 \equiv 1^k - 1 = 0 \pmod{2^d - 1}$$

Now we can build our infinite list of 2-pseudoprimes by showing that if n is one, then so is $m = 2^n - 1$. We know that n is composite, so the fact just proved tells us that m is composite too. So it remains to show that

$2^{m-1} \equiv 1 \pmod{m}$. Because n is a 2-pseudoprime, $2^n = 2 \cdot 2^{n-1} \equiv 2 \pmod{n}$, which means that n divides $2^n - 2$. Then, by the fact proved at the start of the proof, $2^n - 1$ divides $2^{2^n-2} - 1$, which equals $2^{m-1} - 1$. This construction gives the infinite sequence of 2-pseudoprimes: $341, 2^{341} - 1, 2^{2^{341}-1}, \ldots$. □

It never hurts to check.

```
m = 2^341 - 1;
{Mod[m, 2^11 - 1], PowerMod[2, m - 1, m]}

{0, 1}
```

It is useful to know that arithmetic can be suppressed in *Mathematica* by the use of `HoldForm`. Here is how one can generate a compact list of seven pseudoprimes. Can you estimate the number of base-10 digits in each of these numbers?

```
f[x_] := HoldForm[2^x - 1]
NestList[f, 341, 6]
```

$$\left\{ 341, \; 2^{341} - 1, \; 2^{2^{341}-1} - 1, \; 2^{2^{2^{341}-1}-1} - 1, \right.$$
$$\left. 2^{2^{2^{2^{341}-1}-1}-1} - 1, \; 2^{2^{2^{2^{2^{341}-1}-1}-1}-1} - 1, \; 2^{2^{2^{2^{2^{2^{341}-1}-1}-1}-1}-1} - 1 \right\}$$

Naturally one can also look at b-pseudoprimes for any b, the definition being simply: n is not prime and $b^{n-1} \equiv 1 \pmod{n}$. Here is a table of all the b-pseudoprimes under 1000, with $b = 2, 3, 5, \ldots, 19$. It is common to consider prime values of b only (see Exercise 2.40), though one can investigate b-pseudoprimes when b is not prime.

`PseudoprimeTable[8, 1000]`

b												
2	341	561	645									
3	91	121	286	671	703	949						
5	4	124	217	561	781							
7	6	25	325	561	703	817						
11	10	15	70	133	190	259	305	481	645	703	793	
13	4	6	12	21	85	105	231	244	276	357	427	561
17	4	8	9	16	45	91	145	261	781			
19	6	9	15	18	45	49	153	169	343	561	637	889	905	906

It looks like our focus on $b = 2$ was wise, because that seems to have fewer counterexamples than any other b. Of course, we would be especially pleased if there was a b for which there were no pseudoprimes whatsoever, or perhaps only finitely many. But that is false (Exercise 2.43). Worse, there are certain numbers that can never have their primality verified by a pseudoprime test. Notice how 561, which factors as $3 \cdot 11 \cdot 17$, appears in each of the lists (except $b = 3$, 11, or 17). This means that 561 is a b-pseudoprime for every possible b, except those that have a factor in common with 561. So the only way that 561's compositeness could be discovered would be by using $b = 3$, 11, or 17, in which case a simple divisibility test would show that 561 is not prime. Numbers such as 561 are called Carmichael numbers. Precisely: A ***Carmichael number*** n is an integer that is a b-pseudoprime for every b that is relatively prime to n. It is now known (though this is a recent result; it was conjectured by Carmichael in 1912 and not proved until 1992, by Alford, Granville, and Pomerance [AGP]) that there are infinitely many Carmichael numbers.

It is easy to see why 561 is a Carmichael number. Let $\gcd(b, 561) = 1$. To show that $b^{560} \equiv 1 \pmod{561}$ it is sufficient (by Proposition 1.17) to show that this congruence holds modulo 3, 11, and 17. But $b^{560} \equiv b^{10 \cdot 56} \equiv 1^{56} = 1 \pmod{11}$ by Fermat's Little Theorem. Similarly $b^{560} \equiv b^{16 \cdot 35} \equiv 1 \pmod{17}$ and $b^{560} \equiv b^{2 \cdot 280} \equiv 1 \pmod{3}$. In short, $n = 561$ has the form $\prod p_i$ where the p_i are distinct odd primes and each $p_i - 1$ divides $n - 1$. In fact every Carmichael number has this form, with three or more prime factors (see Section 3.7).

Here are the first seven Carmichael numbers.

```
Select[Range[2, 10000], CarmichaelQ]
```

```
{561, 1105, 1729, 2465, 2821, 6601, 8911}
```

Even though the 2-psp test ("psp" is used to abbreviate "pseudoprime") fails to fully characterize primality, it is immensely important as a first tool. For if the simple modular power fails to be 1 (mod n), we know *with certainty* that n is composite. If n is large we may have no idea how to find any factors of n, but we can be absolutely certain that n fails to be prime. For example, let n be the 3571st Fibonacci number, which has 746 digits; then $2^{n-1} \not\equiv 1 \pmod{n}$, as we now show.

```
n = Fibonacci[3571];
N[n]
```

$8.818167227258249 \times 10^{745}$

```
PowerMod[2, n - 1, n] // N
```

$4.155047893611567 \times 10^{745}$

This is not 1, and so we know with certainty that F_{3571} is not prime, even though we have no idea what any of its factors are. We can try to find some small factors by taking the gcd of n with the product of all primes less than 30000, but this turns up nothing. We use $\pi(30000)$ (which is 3245) to figure out how many primes to use.

```
GCD[n, Apply[Times, Prime[Range[π[30000]]]]]
```

```
1
```

■ Using the Pseudoprime Test

Because pseudoprimes are relatively rare, we can use a pseudoprime test to find large numbers that are, almost surely, prime. Note that, when looking for prime Fibonacci numbers, one looks only at F_p where p is prime, for reasons mentioned in Exercise 1.6.

```
Select[Prime[Range[2, 100]],
 TwoPseudoprimeTest[Fibonacci[#]] &]
```

```
{5, 7, 11, 13, 17, 23, 29, 43, 47,
 83, 131, 137, 359, 431, 433, 449, 509}
```

We can look for the first few primes beyond a googol. The package function `ExponentForm` splits a power of 10 away from an integer, which is best for displaying numbers near a googol.

```
Select[Range[10^100 + 1, 10^100 + 1000, 2], TwoPseudoprimeTest] //
 ExponentForm
```

```
{10^100 + 267, 10^100 + 949}
```

We again emphasize that this does not *prove* that $10^{100} + 267$ is prime. But it is reasonable to believe it is. On the other hand, this computation tells us with absolute certainty that, with at most two exceptions, all the numbers in the first 1000 past 10^{100} are composite.

The Mersenne numbers (numbers of the form $2^n - 1$) occupy a special place in the history of primes because the world record largest primes have almost always been Mersenne primes. The current largest prime is $2^{6972593} - 1$. There is a special-purpose test that finds Mersenne primes (see Section 9.2), but we can try a pseudoprime test.

```
Mersenne[n_] := 2^n - 1
Select[Range[100], TwoPseudoprimeTest[Mersenne[#]] &]
```

```
{2, 3, 5, 7, 11, 13, 17, 19, 23, 29, 31, 37,
 41, 43, 47, 53, 59, 61, 67, 71, 73, 79, 83, 89, 97}
```

Ouch! There are too many here! For example, $2^{11} - 1 = 2047 = 23 \cdot 89$. Recall from Proposition 2.7 (and Exercise 2.36) that numbers of the form $2^p - 1$ have a tendency to be 2-pseudoprimes. Indeed, of the 25 hits that the 2-psp test found, the imposters outnumber the actual primes 15 to 10. It is easy to unmask the composites by using a 3-psp test.

```
Select[Range[100],
 PowerMod[3, Mersenne[#] - 1, Mersenne[#]] == 1 &]
```

```
{3, 5, 7, 13, 17, 19, 31, 61, 89}
```

These are in fact the true Mersenne primes among the first 100 Mersenne number, with one exception: the index 2 is missing from this list. That is because $2^2 - 1$ is 3 which is prime, but hides from the 3-psp test because 3^2 is not 1 (mod 3). Using a 5-psp test would get them all in this range. We repeat that combining pseudoprime tests for different bases, while it does cut down the number of failures, can never lead to a perfect test because of Carmichael numbers.

Of course, when investigating large numbers for probable primality we could just use the built-in `PrimeQ` function. It uses more sophisticated, but still very fast, pseudoprime tests, which will be explained in Chapters 4 and 8.

Exercises for Section 2.4

2.33. Consider the following primality test: Fix an initial segment of the primes, such as 2, 3, 5, 7, and combine the b-psp test for each b in this list. For several such initial segments find the first n for which this test gives an incorrect answer.

2.34. Prove that if d divides n then $2^d - 1$ divides $2^n - 1$, via a direct argument based on factoring $x^{dk} - 1$.

2.35. Consider the sequence of 2-pseudoprimes

$$341, \ 2^{341} - 1, \ 2^{(2^{341}-1)} - 1, \ 2^{2^{(2^{341}-1)}-1} - 1,$$

$$2^{2^{2^{(2^{341}-1)}-1}-1} - 1, \ 2^{2^{2^{2^{(2^{341}-1)}-1}-1}-1} - 1, \ 2^{2^{2^{2^{2^{(2^{341}-1)}-1}-1}-1}-1} - 1$$

For each entry, estimate the number of base-10 digits it has. You cannot be exact here because the numbers are so large. Just come up with a concise description.

2.36. Show that $2^p - 1$ is a 2-pseudoprime whenever p is prime and $2^p - 1$ is composite.

2.37. Prove the converse of the divisibility fact used in Proposition 2.7: If $2^d - 1$ divides $2^n - 1$, then d divides n.

2.38. (a) (D. H. Lehmer) If n is a 2-pseudoprime, then n must be odd and $2^n \equiv 2 \pmod{n}$. Find an even integer n (other than 2) such that $2^n \equiv 2 \pmod{n}$.

(b) Use the result of part (a) to resolve the following. We saw in the proof of Proposition 2.7 that if $2^n - 1$ is prime, then n is prime. True or False: If $2^n - 1$ is a 2-pseudoprime then n is either prime or a 2-pseudoprime.

2.39. True or False: If n is a 3-psp, then so is $3^n - 1$.

2.40. Prove that if n is composite then the least b such that n is not a b-psp is prime.

2.41. Prove that if $2^n + 1$ is prime, then n must be a power of 2. *Hint*: When n has an odd divisor $2^n + 1$ has an algebraic factorization.

2.42. The ***Fermat numbers***, F_n, are numbers of the form $2^{2^n} + 1$. The first few are prime and it was once suspected that all might be prime, but now it is suspected that, except for the first few, they are all composite. Show that whenever a Fermat number is composite, it is a 2-pseudoprime. Prove that F_5 is composite.

2.43. (M. Cipolla, 1904) Show that for any b there are infinitely many b-pseudoprimes by showing that for any odd prime p that does not divide $b(b^2 - 1)$,

$$\frac{b^p - 1}{b - 1} \frac{b^p + 1}{b + 1}$$

is a b-pseudoprime. Note that this expression has the neat form: $1 + b^2 + b^4 + \cdots + b^{2p-2}$.

2.44. A ***pseudoprime witness*** for a composite n is an integer b such that $\gcd(b, n) = 1$, $1 \le b \le n - 1$, and $b^{n-1} \not\equiv 1 \pmod{n}$. Of course, the existence of such a witness proves that n is composite. Do some computations to support the following formula due to Monier, Baillie, and Wagstaff [BW]: If n is odd, the number of psp witnesses for n is

$$\phi(n) - \prod_{p \text{ a prime divisor of } n} (\gcd(n - 1, p - 1) - 1).$$

See Exercise 3.39 for more information on this formula.

2.45. Use the 2-psp test to get the likely primes in the first 10000 integers beyond a googol. How does this compare with the heuristic value for the expected number of primes in this interval?

2.46. A Mersenne number in base 2 has the form $111\ldots111_2$. One can look at similar numbers in base 10 and see how many are prime. These numbers are sometimes called ***repunits***, with ***repunit***$_n$ denoting the concatenation of n 1s; they are easy to define arithmetically.

```
Repunit[n_] := (10ⁿ - 1) / 9
Repunit[7]

1111111
```

(a) Prove that if repunit$_n$ is prime, then n is prime.

(b) Find some (probably) prime repunits. For a long time only 5 were known, but in late 1999 H. Dubner discovered a sixth, since he found that repunit$_{49081}$ is probably prime (see [Rib]).

Euler's ϕ Function

3.0 Introduction

The goal of this chapter is to explore the structure of the power grid for composite moduli. In particular, we will be looking for rows that contain all possible residues: Which moduli have such rows? How many such rows are there? What can be said about the positions of these rows? The answers will depend on arithmetic properties of the modulus.

Getting these answers will lead us to investigate two important arithmetic functions: Euler's function $\phi(n)$, which counts the number of integers in $\{1, ..., n\}$ that are relatively prime to n, and the sum of divisors function, $\sigma(n)$, which is the sum of the positive integers that divide n. The ϕ-function is particularly significant. It provides information about the order of the residues modulo n — the least power that is congruent to 1 — and will play a role in many algorithms for factorization and prime testing, as well as the RSA public-key cryptosystem. The σ function will be used to explore perfect numbers, but it too has deeper applications, some of which will be discussed in Chapter 9.

3.1 Euler's ϕ Function

The **_Euler phi function_**, denoted $\boldsymbol{\phi(n)}$ and available in *Mathematica* as EulerPhi, or just ϕ if the CNT package is loaded, is central to much of number theory. It counts the number of integers in $\{1, 2, ..., n\}$ that are relatively prime to n. For example, $\phi(12) = 4$ because of the 4-element set: $\{1, 5, 7, 11\}$. Here is a table showing ϕ-values of the first 100 integers. Not too many patterns are apparent from this data — it is always instructive for the reader to try to recognize what patterns there are — but there are in fact many important rules (and fascinating unsolved problems) involving ϕ.

FunctionTable[ϕ, 100, Columns $\rightarrow$ 5]

n	$\phi(n)$	n	$\phi(n)$	n	$\phi(n)$	n	$\phi(n)$	n	$\phi(n)$
1	1	21	12	41	40	61	60	81	54
2	1	22	10	42	12	62	30	82	40
3	2	23	22	43	42	63	36	83	82
4	2	24	8	44	20	64	32	84	24
5	4	25	20	45	24	65	48	85	64
6	2	26	12	46	22	66	20	86	42
7	6	27	18	47	46	67	66	87	56
8	4	28	12	48	16	68	32	88	40
9	6	29	28	49	42	69	44	89	88
10	4	30	8	50	20	70	24	90	24
11	10	31	30	51	32	71	70	91	72
12	4	32	16	52	24	72	24	92	44
13	12	33	20	53	52	73	72	93	60
14	6	34	16	54	18	74	36	94	46
15	8	35	24	55	40	75	40	95	72
16	8	36	12	56	24	76	36	96	32
17	16	37	36	57	36	77	60	97	96
18	6	38	18	58	28	78	24	98	42
19	18	39	24	59	58	79	78	99	60
20	8	40	16	60	16	80	32	100	40

For convenience we list several facts about ϕ in the following theorem. We use $d \mid n$ to abbreviate "d divides n".

Theorem 3.1. (a) If $n \geq 2$, then $1 \leq \phi(n) \leq n - 1$.

(b) If $n > 2$, $\phi(n)$ is even.

(c) If n is prime then $\phi(n) = n - 1$.

(d) If $\phi(n) = n - 1$, then n is prime.

(e) If p is prime, $\phi(p^e) = p^e - p^{e-1} = p^{e-1}(p - 1)$.

(f) If $\{p_i\}$ are the prime divisors of n, then $\phi(n) = n \prod_i \left(1 - \frac{1}{p_i}\right)$.

(g) $\phi(n) \geq n \prod_{i=1}^{r} \left(1 - \frac{1}{p_i}\right) \geq \frac{n}{\log_2(2n)}$ where p_1, p_2, ..., p_r are the first r primes and n has r distinct prime factors.

(h) $\sum_{d \mid n} \phi(d) = n$.

The proofs of (f)–(h) will be given shortly. Facts (a)–(e) are straightforward. The definition yields (a) immediately. For (b), if $\gcd(a, n) = 1$ then $\gcd(n - a, n) = 1$ too, and $n - a \neq a$ if $n > 2$. Assertion (c) holds because all candidates (except n itself) are coprime to a prime; for (d) observe that if n

has a divisor, then that divisor is excluded from the ϕ-count, making it submaximal. And (e) is proved by counting the integers in $\{1, 2, ..., p^e\}$ that are excluded from the ϕ-count; there are p^{e-1} of them: $p, 2p, 3p, ..., p^{e-1}p$.

In 1932, D. H. Lehmer asked if condition (c) could be weakened and still characterize the primes. Precisely: Is there a composite n such that $\phi(n)$ divides $n-1$. It is known that any example must be larger than 10^{20} (see [Guy, problem B37] and [Rib, p. 36]).

Understanding ϕ and many other functions of number theory requires the notion of a multiplicative function. A function f from positive integers to positive integers is called ***multiplicative*** if $f(mn) = f(m)f(n)$ whenever $\gcd(m, n) = 1$. The main point to note about such functions is that a value $f(n)$ can be computed by just multiplying together the values $f(p^e)$ for all prime powers p^e in the prime factorization of n. Of course, the difficulty of the general factoring problem means that this is useless for very large numbers. But it does provide an efficient way of computing $f(n)$ when the factorization of n can be quickly obtained (which is the case for $n < 10^{20}$).

Proposition 3.2. ϕ is a multiplicative function.

Proof. Recall the mapping $F : \mathbb{Z}_{mn} \to \mathbb{Z}_m \times \mathbb{Z}_n$ given by $F(a) = (\text{Mod}[a, m], \text{Mod}[a, n])$. The Chinese Remainder Theorem, which uses the fact that $\gcd(m, n) = 1$, states that this is one-to-one and onto. Now define $\mathbb{Z}_n^*$ to consist of those integers in $\{0, 1, 2, 3, ..., n-1\}$ that are relatively prime to n. The size of $\mathbb{Z}_n^*$ is $\phi(n)$. Therefore, we need only prove that, given relatively prime m and n, the sets $\mathbb{Z}_{mn}^*$ and $\mathbb{Z}_m^* \times \mathbb{Z}_n^*$ have the same size. Suppose $k \in \mathbb{Z}_{mn}^*$; then $\gcd(k, mn) = 1$ so $\gcd(k, m) = \gcd(k, n) = 1$, and $F(k) \in \mathbb{Z}_m^* \times \mathbb{Z}_n^*$. This means that F may be viewed as a function from $\mathbb{Z}_{mn}^*$ to $\mathbb{Z}_m^* \times \mathbb{Z}_n^*$. Because the original F is one-to-one, so is the restriction. So all that remains is to show that the restriction of F is onto.

Suppose that $(a, b) \in \mathbb{Z}_m^* \times \mathbb{Z}_n^*$. The Chinese Remainder Theorem yields a unique x in $\mathbb{Z}_{mn}$ such that $F(x) = (a, b)$. We must show that x lies in $\mathbb{Z}_{mn}^*$. But if some prime divided both x and mn, then it would divide one of m and n, a contradiction. So F is onto and proves that $\mathbb{Z}_{mn}^*$ and $\mathbb{Z}_m^* \times \mathbb{Z}_n^*$ have the same size. □

Knowing that ϕ is multiplicative suffices to prove Theorem 3.1(f), because it follows from the explicit formula in Theorem 3.1(e). The formula in Theorem 3.1(f) is easily implemented, as follows. We will not use this function explicitly since we can just use ϕ or `EulerPhi`.

```
phi[n_] := n Apply[Times, 1 - 1/Map[First, FactorInteger[n]]]
```

```
{phi[100], ϕ[100]}
```

```
{40, 40}
```

In general we cannot compute $\phi(n)$ when n is very large, because the algorithm requires factoring n and, moreover, it is known that there is no way to avoid this in general (Exercise 3.3). However, if n can be factored then there is no problem.

```
{φ[100!], φ[10^100]} // ExponentForm
```

```
{112287575737249544081242426905513539497182967348719447547068`
   244023512514148985543894293521677050772263812830066323523 9`
   912361741516 8 10^28 , 4 10^99 }
```

To complete the proof of Theorem 3.1 we explain the formulas in parts (g) and (h). We start with (h), which is easier.

Proof of Theorem 3.1(h). Let d be a divisor of n. The integers m in $\{1, 2, ..., n\}$ for which $\gcd(m, n) = d$ are precisely those integers of the form dx where $1 \leq x \leq n/d$ and x is relatively prime to n/d. From the definition of ϕ, there are $\phi(n/d)$ such values of x. As d runs through the divisors of n, each of the integers $1, 2, ..., n$ is counted exactly once and therefore $\sum_{d|n} \phi(n/d) = n$. As d runs through the divisors of n, so does n/d, which means that the preceding sum is identical to $\sum_{d|n} \phi(d) = n$. □

Now on to part (g) which gives a lower bound on $\phi(n)$.

Proof of Theorem 3.1(g). The second inequality is left as an exercise. Let $q_1, ..., q_r$ be the distinct primes dividing n, in increasing order. Then

$$\phi(n) = n \prod_{i=1}^{r} \left(1 - \frac{1}{q_i}\right) \geq n \prod_{i=1}^{r} \left(1 - \frac{1}{p_i}\right),$$

where the last inequality follows from $q_i \geq p_i$. But this is exactly what is wanted. □

The lower bound just proved helps us compute the inverse of ϕ. For example, suppose that we want to find all integers x such that $\phi(x) = 100$. Let x have the distinct prime factors $q_1, ..., q_r$. Then

$$\phi(x) = x \prod_{i=1}^{r} \left(1 - \frac{1}{q_i}\right)$$

$$\geq x \prod_{i=1}^{r} \left(1 - \frac{1}{p_i}\right) \geq \prod_{i=1}^{r} p_i \prod_{i=1}^{r} \left(1 - \frac{1}{p_i}\right) = \prod_{i=1}^{r} (p_i - 1).$$

When $r = 5$ this last product is $1 \cdot 2 \cdot 4 \cdot 6 \cdot 10$, or 480, which is greater than 100. So $r \leq 4$. This means that $\phi(x) \geq x \prod_{i=1}^{4} (1 - 1/p_i)$ so, if $\phi(x) = 100$, $x \leq 100 / \prod_{i=1}^{4} (1 - 1/p_i) \approx 437.5$. Thus, we can simply check the interval $[101, 438]$ to find appropriate values of x.

```
Select[Range[101, 437], φ[#] == 100 &]
```

```
{101, 125, 202, 250}
```

It is a little faster to compute all the ϕ-values at once and then see where the 100s are in that list, as follows.

100 + Flatten[Position[ϕ[Range[101, 437]], 100]]

{101, 125, 202, 250}

Turning this into a general program is a good exercise. But there is a much faster way, though it requires a greater programming effort based on using the prime factorization of n to deduce which primes can appear in x. This fancier method is implemented in the CNT package as PhiInverse, and the package enhances Solve so that it can handle the inverse of ϕ.

<u>PhiInverse</u>[1000]

{1111, 1255, 1375, 1875, 2008,
 2222, 2500, 2510, 2750, 3012, 3750}

x /. Solve[ϕ[x] == 720, x]

{779, 793, 803, 905, 925, 1001, 1045, 1085, 1107, 1209, 1221,
 1281, 1287, 1395, 1425, 1448, 1485, 1558, 1575, 1586, 1606,
 1612, 1628, 1672, 1708, 1736, 1810, 1850, 1900, 2002, 2090,
 2170, 2172, 2196, 2214, 2232, 2376, 2418, 2442, 2508,
 2562, 2574, 2604, 2700, 2772, 2790, 2850, 2970, 3150}

The values that turn up as $\phi(x)$ lead to a famous unsolved problem called Carmichael's conjecture. From the table of ϕ-values on page 66 you can see that whenever a number arises as a ϕ-value, it shows up more than once. For example, $\phi(13) = 12$, which shows up again as $\phi(21)$. Some numbers do not ever appear; Theorem 3.1 tells us that odd numbers, except 1, do not appear, but it is not hard to see that certain even numbers are also missing, such as 14 (see Exercise 3.1). Carmichael conjectured that no integer shows up exactly once as a ϕ-value. We can make a stronger conjecture by defining the ***ϕ-multiplicity*** of n to be the number of x-values for which $\phi(x) = n$.

Carmichael's Conjecture. The ϕ-multiplicity is never 1; in other words, if $\phi(x) = n$ then there is some y different from x such that $\phi(y) = n$.

Sierpiński had conjectured that every number other than 1 does occur as a multiplicity. Computations show that every multiplicity under 311 (except 1) does occur (Exercise 3.9), and in 1998 Kevin Ford [For] proved the complete Sierpiński conjecture.

The PhiMultiplicity function from the CNT package just uses PhiInverse. A good programming exercise is to speed up this process by precomputing a giant table of ϕ-values, perhaps 100,000 of them, and using it to generate multiplicities; that would require a machine with adequate memory. The table that follows shows that there are eleven integers whose ϕ-value is 48, and 48 is the smallest number with this property. The table also shows that every multiplicity between 2 and 11 occurs. The notation V_ϕ is used because this function is also known as the **valence** of ϕ.

```
FunctionTable[PhiMultiplicity, 50,
    FunctionHeading → "V_φ (n) ", FontSize → 9]
```

n	$V_\phi(n)$	n	$V_\phi(n)$	n	$V_\phi(n)$	n	$V_\phi(n)$	n	$V_\phi(n)$
1	2	11	0	21	0	31	0	41	0
2	3	12	6	22	2	32	7	42	4
3	0	13	0	23	0	33	0	43	0
4	4	14	0	24	10	34	0	44	3
5	0	15	0	25	0	35	0	45	0
6	4	16	6	26	0	36	8	46	2
7	0	17	0	27	0	37	0	47	0
8	5	18	4	28	2	38	0	48	11
9	0	19	0	29	0	39	0	49	0
10	2	20	5	30	2	40	9	50	0

Regarding Carmichael's conjecture, it is not too hard to show that a counterexample — a value of n such that $V_\phi(n) = 1$ — must be incredibly large. Schlafly and Wagon [SW] showed that such an n must be greater than $10^{10,920,000}$ (recently extended by Kevin Ford [For] to $10^{10^{10}}$); moreover, there is every reason to believe that the bound could be extended indefinitely, which would mean, if only it could be proved, that the conjecture is true.

Finally, we mention another elementary thread of ideas surrounding ϕ. D. J. Newman proved recently that, given any positive integers a and b, there is an n such that $\phi(an + b) < \phi(an)$. However such integers are not always easy to find. In particular Newman observed that no n under 2,000,000 satisfies $\phi(30n + 1) < \phi(30n)$. Greg Martin discovered the first number that satisfies Newman's equation. It has 1116 digits! To see Martin's number, let p_i denote the ith prime. Then let $z = p_4 p_5 p_6 \cdots p_{382} p_{383} p_{385} p_{388}$. Then $z \equiv 1 \pmod{30}$. Let $n = (z - 1)/30$. Then, as proved by Martin [Mar], $\phi(30n + 1) < \phi(30n)$.

Exercises for Section 3.1

3.1. Prove that $\phi(n)$ is never equal to 50.

3.2. Which integers n have the property that n is divisible by $\phi(n)$?

3.3. Suppose that $n = 1696\,10714\,26634\,37950\,88165\,66546\,93070\,58932\cdot.98221$ and it is known that n is a product of two primes. Suppose further that $\phi(n)$ is known to be $1696\,10714\,26634\,37950\,86324\,91471\,72767\,86428\cdot.51100$. Find the prime factors of n. *Hint*: Perhaps consider the analogous problem with some smaller numbers. Or consider the more general problem where one is given $\phi(n)$ and the information that n is a product of two primes. This result says that computing $\phi(n)$ in general is at least as difficult as factoring products of two primes. Because the latter is considered to be computationally infeasible, there is likely no efficient algorithm for computing $\phi(n)$ in general.

3.4. Prove that $\prod_{i=1}^{r}\left(1-\frac{1}{p_i}\right) \geq \frac{1}{\log_2(2n)}$, where p_i is the ith prime and n has r distinct prime factors. This completes the proof of Theorem 3.1(g).

3.5. (a) Use the method outlined in the text, based on identifying an interval that captures all candidates, to write a program that, given n, finds the set of x for which $\phi(x) = n$.

(b) Now program the inverse-ϕ problem more efficiently by first analyzing the prime factorization of n (the power of 2 in this factorization plays a critical role) to determine the primes that can show up in x. Then generate a list of all candidates, and finish by checking this list to find the solutions. The set of candidates gets large, but you should be able to handle n-values up to 10000 or more. It seems reasonable to use recursion to built up the list of candidates. Use the package function `PhiInverse` or your work in part (a) to check your results.

3.6. Let $A(x)$ be the average value of $\phi(n)$ up to x: $A(x) = (1/x)\sum_{n=1}^{x}\phi(n)$. Then $A(x)$ is well approximated by a linear function, cx. Perform computations in an attempt to find the best value of c in such a model. Create some graphics to show how well the model fits the data. *Hint:* c involves π^2. *Mathematica Tip:* Using `Sum` or `Plus` to add up numbers causes *Mathematica* to form the entire list of summands, which can be deadly when that list is large. Using a `Do`-loop as follows is much better: (`sum = 0.; Do[sum = sum + EulerPhi[i], {i, x}]; sum / x`). While it is instructive for you to program the average function yourself, the `CNT` package does include two functions relevant to this sort of work: `Average` and `AverageList`. Use `?Average` and `?AverageList` to learn about them.

3.7. Do some computations in an attempt to estimate the long-term average of $\phi(n)/n$. It turns out to have the form $6/\pi^2$. Using this one can give a heuristic argument that if n is restricted to even integers, the long-term average of $\phi(n)/n$ is $4/\pi^2$. Do some experiments to confirm this. Finally, try to estimate the long-term average of $\phi(p-1)/(p-1)$, where p is prime. Such results show that $p-1$ is not a typical even number.

3.8. If f maps integers to integers, define $\textbf{\textit{log}}_f(n)$ to be the number of times f must be iterated, starting with n, to reach a fixed point. If $f(x) = \lfloor x/10 \rfloor$, then $\log_f(n)$ is the number of digits in n. Now, the only fixed point of ϕ is 1, so $\log_\phi(n)$ is the number of times ϕ must be applied to get n down to 1. For example, $\log_\phi(100) = 6$.

```
FixedPointList[ϕ, 100]
{100, 40, 16, 8, 4, 2, 1, 1}
```

Investigate the $\log_\phi$ function. Prove that $\log_\phi(n) \leq \lceil \log_2(n) \rceil$. Let $\boldsymbol{exp_\phi(n)}$ be the first x such that $\log_\phi(x) = n$. It was thought for many years that $\exp_\phi(n)$ would have to be prime. Find a counterexample.

3.9. Investigate the multiplicity function $V_\phi(n)$, which is the number of times n occurs as $\phi(x)$. Try to show that all multiplicities up to 50 (or 100, or 200 or more), occur (except for 1). You could use the package function `PhiMultiplicity`, but it is faster, if you have enough memory, to precompute a large number of ϕ-values, say 50,000 or 100,000, and use that data set together with facts about bounds on ϕ proved in this section, to deduce multiplicity information. For a really large computation, speed is less of a priority than memory consumption. Our own computations show that every multiplicity between 2 and 311 does occur (but Kevin Ford [For] recently proved the long-standing conjecture that all multiples (except 1) definitely occur). What is the first n for which $\phi^{-1}(n)$ has size 312?

3.2 Perfect Numbers and Their Relatives

■ The Sum of Divisors Function

The special properties of certain numbers have fascinated both professionals and amateurs for millenia. And no property has exerted a greater fascination than perfection: A number n is **perfect** if it is equal to the sum of its divisors, excluding n itself. The first perfect number is 6, which equals $1 + 2 + 3$. The next one is 28. In fact, there is a simple formula for the even perfect numbers, but to understand it we must first investigate the sum-of-divisors function. For $k \geq 0$, $\sigma_k(n)$ is defined to be the sum of the kth powers of all the positive integers that divide n. When the subscript is omitted it is assumed to be 1; $\sigma(n)$ is just the sum of n's divisors. So $\sigma(6) = 12$. When $k = 0$, σ_0 counts a 1 for every divisor, and so gives the number of divisors. That function often goes by the special name $\tau(n)$.

These functions are built into *Mathematica* as `DivisorSigma[k,n]`, but the CNT package allows us to use σ and σ_k. Let us also introduce σ^- for the reduced σ function, where the number itself is excluded; then n is perfect if $\sigma^-(n) = n$.

> `{Divisors[28], `$\underline{\sigma}$`[28], `$\underline{\sigma}^-$`[28]}`
>
> `{{1, 2, 4, 7, 14, 28}, 56, 28}`

A formula for τ is easy to derive, and the reader might experiment with some small numbers in an attempt to discover it. There is a relatively simple formula for σ, but understanding it requires that we first show that σ is a multiplicative function. That follows from the following more general result. But note that σ^- is not multiplicative ($\sigma^-(2) = 1$ and $\sigma^-(3) = 1$, but $\sigma^-(6) = 3$), which explains why it is a second-string function.

Proposition 3.3. If f is multiplicative then so is F, defined by

$$F(n) = \sum_{d|n} f(d).$$

Proof. Observe first that any divisor r of mn, where m and n are relatively prime, splits into cd where c divides m, d divides n, and $\gcd(c, d) = 1$. For a proof just look at the prime factorization of mn and let c and d be the gcd of r with m and n, respectively. Moreover, given c and d dividing m and n, respectively, their product is a divisor of mn.

Now, if $\gcd(m, n) = 1$, we must prove that $F(mn) = F(m)F(n)$, or $\sum_{d|mn} f(d) = (\sum_{d|m} f(d))(\sum_{d|n} f(d))$. But this follows immediately from algebraic expansion of the right side: if c_i are the divisors of m and d_j are the divisors of n, then the right side is

$$[f(c_1) + f(c_2) + f(c_3) + \cdots][f(d_1) + f(d_2) + f(d_3) + \cdots] =$$
$$\sum_{i,j} f(c_i)f(d_j) = \sum_{i,j} f(c_i d_j) = \sum_{d|mn} f(d). \quad \square$$

The sequence of equalities just given, indeed, the entire proof, can be given concrete form using symbolic calculation. As an example, let m and n be 4 and 15, respectively. The phrase f[m_] f[n_] :> f[m n] (the :> symbol is the same as :>) causes the substitution to apply to all products, which is all right since the pairs that arise are all relatively prime; f[1]^2 requires a special case.

```
F[n_] := Apply[Plus, Map[f, Divisors[n]]]
left = F[60]

f[1] + f[2] + f[3] + f[4] + f[5] + f[6] +
 f[10] + f[12] + f[15] + f[20] + f[30] + f[60]

right = F[4] F[15]

(f[1] + f[2] + f[4]) (f[1] + f[3] + f[5] + f[15])

right = Expand[right]

f[1]^2 + f[1] f[2] + f[1] f[3] + f[2] f[3] +
 f[1] f[4] + f[3] f[4] + f[1] f[5] + f[2] f[5] +
 f[4] f[5] + f[1] f[15] + f[2] f[15] + f[4] f[15]

right = right /. {f[m_] f[n_] :> f[m n], f[1]^2 -> f[1]}

f[1] + f[2] + f[3] + f[4] + f[5] + f[6] +
 f[10] + f[12] + f[15] + f[20] + f[30] + f[60]

left == right

True
```

Proposition 3.4. The function $\sigma_k(n)$ is a multiplicative function of n. In particular, σ and τ are multiplicative.

Proof. This follows from Proposition 3.3 because $\sigma_k(n) = \sum_{d|n} d^k$ and $f(d) = d^k$ defines a function of d that is obviously multiplicative. $\square$

Corollary 3.5. If n's prime factorization is $\prod p_i^{e_i}$, then $\tau(n) = \prod_{i=1}^{n} (e_i + 1)$ and $\sigma(n) = \prod_{i=1}^{n} \frac{p_i^{e_i+1}-1}{p_i-1}$.

Proof. Because σ and τ are multiplicative, we need only establish the formulas in the case of prime powers. This is easy for τ because we can list the $e+1$ divisors of p^e: $\{1, p, p^2, ..., p^e\}$. For σ we must consider $1 + p + p^2 + \cdots + p^e$; but this is a geometric progression and sums to $(p^e - 1)/(p - 1)$, as desired. $\square$

Note that modern software can handle symbolic sums such as the one that just arose.

$$\sum_{i=0}^{e} p^i$$

$$\frac{p^{e+1} - 1}{p - 1}$$

■ Perfect Numbers

Now that we see how to compute σ we can blindly search for some perfect numbers. A quick computation gets us four of them. Can you see the pattern they follow?

```
perfect = Select[Range[10000], σ[#] == # &]
{6, 28, 496, 8128}
```

Factoring the numbers is a first investigative step.

```
Map[FactorForm, FactorInteger[perfect]]
{2 3 , 2² 7 , 2⁴ 31 , 2⁶ 127 }
```

The connection between even perfect numbers and primes of the form $2^s - 1$ (Mersenne primes) such as 7, 31, and 127 was known to the ancient Greeks. Here it is.

Proposition 3.6 (Euler–Euclid Formula). If $2^s - 1$ is prime, then $2^{s-1}(2^s - 1)$ is perfect. Moreover, every even perfect number arises in this way.

Proof. Showing that Mersenne primes lead to perfect number is a straight-forward application of the formula for $\sigma(n)$ and we leave the details to an exercise. For the harder direction, suppose that n is even and perfect. Write $n = 2^{s-1}t$ where $s > 1$ and t is odd. Because $\sigma(n) = 2n$, we know that $2^s t = \sigma(2^{s-1}t) = (2^s - 1)\sigma(t)$. So

$$\sigma(t) = \frac{2^s t}{2^s - 1} = t + \frac{t}{2^s - 1}\,.$$

It follows that $t/(2^s - 1)$ is an integer and therefore t and $t/(2^s - 1)$ each divide t. The expression for $\sigma(t)$ then implies that there can be no other divisors of $t/(2^s - 1)$. This means that t is prime and $t/(2^s - 1) = 1$, so $n = 2^{s-1}(2^s - 1)$ and $2^s - 1$ is prime. □

The Euclid–Euler formula tells us that even perfect numbers are in one-to-one correspondence with Mersenne primes. So if there are infinitely many Mersenne primes, a famous unsolved problem, then there will be infinitely many even perfect numbers, and vice versa. But this brings us to one of the oldest unsolved mysteries of mathematics, as it goes back to the ancient Greeks.

The Odd Perfect Number Problem.   Is there an odd perfect number?

This problem has received a lot of attention and it is known that if there is an odd perfect number it must have 29 prime factors (not necessarily distinct), one of them is greater than 10^{20}, and the number must be greater than 10^{300}. The famous nineteenth-century mathematician J. J. Sylvester thought hard about it and, summarizing the many conditions that an odd perfect number would have to satisfy, came to the conclusion that "The existence of [an odd perfect number] — its escape, so to say, from the complex web of conditions which hem it in on all sides — would be little short of a miracle." Yet, the problem is quite sensitive and perhaps some combination of 100 odd primes is perfect; see Exercise 3.12 for a near miss discovered by Descartes. For more information on this conjecture see [KW].

■ Amicable, Abundant, and Deficient Numbers

If m and n are different integers such that the divisors of n (excluding n itself) sum to m and vice versa, then the pair (m, n) is called an ***amicable pair***. The first amicable pair is 220 and 284.

σ^- **[220]**

284

σ^- **[284]**

220

For   an   amicable   pair, $\sigma(m) + \sigma(n) = 2(m + n)$;   alternatively $\sigma^-(\sigma^-(n)) = n$.

```
AmicableQ[n_] := ((s = σ⁻[n]) ≠ n && σ⁻[s] == n)
Select[Range[2000], AmicableQ]
```

```
{220, 284, 1184, 1210}
```

Amicable numbers are not particularly important in number theory, but they do have a long history. Perhaps the most amusing application comes from a fifteenth-century Spanish book that recommended their use in formulating a love potion: Serve the object of your affection 220 grains of rice while you eat 284, for example. The experimental support is inconclusive.

The following problem has been much studied. Start with n, form $\sigma^-(n)$, $\sigma^-(\sigma^-(n))$, and so on. The sequence of numbers obtained in this way is called an ***aliquot sequence***. What will happen? Well, starting with 25 gets to $1 + 5$, a perfect number. Perhaps iterating σ^- will yield perfect numbers, maybe even an odd perfect number. Indeed, any fixed point of this process is a perfect number; a cycle of length 2 is an amicable pair. This iteration is interesting for several reasons. First, it is tied to the distinction between deficient and abundant numbers. A ***deficient number*** is one for which $\sigma^-(n) < n$; an ***abundant number*** has so many divisors that they add up to more than n: $\sigma^-(n) > n$.

```
DeficientQ[n_] := σ⁻[n] < n
AbundantQ[n_] := σ⁻[n] > n
```

```
Select[Range[3, 100], AbundantQ]
```

```
{12, 18, 20, 24, 30, 36, 40, 42, 48, 54,
 56, 60, 66, 70, 72, 78, 80, 84, 88, 90, 96, 100}
```

Computing problems related to σ illustrate diverse ideas of the theory of algorithms. Naturally we are most pleased to have algorithms, such as the Euclidean algorithm, that work quickly even on very large inputs. The algorithms for computing $\phi(n)$, factoring n, or determining the nth prime are fine when n is small, but are useless when n is large. But at least algorithms for these problems do exist, even if they take too long. If you want to know the nth prime then, in theory, there is a simple way to do it. Start counting with 2, 3, 5, and stop when the count reaches n. But if you want the 1000th Mersenne prime then there might be no algorithm. For if there turn out to be only finitely many Mersenne primes we may never know when we have come to the last one, and so we would not know when to tell our computer to stop looking and report "there is no 1000th Mersenne prime."

The unifying idea for these sorts of thoughts is the unsolvability of the halting problem, a fundamental result of computer science. This means that there is no computer program that, on input another computer program P and an integer n, will determine whether program P halts on input *n*. In essence this is because there is very little the halt-checking program can do other than run *P* on *n*. But it won't know when to stop and declare "*P*(*n*) does not halt" because there is always the possibility that just one more step will yield a halt. Of course, particular programs might be amenable to analysis, but general programs are too complex to be understood in a universal way.

It is interesting to come up with specific algorithms whose halting status is unresolved, and even more interesting to come up with particular instances. The aliquot sequence problem is, perhaps, one such. It is not known whether there can be an unbounded aliquot sequence. If the sequence does terminate, then it ends in 1, a perfect number, or a cycle, of which {220, 284} is the simplest one.

Aliquot Sequence Problem. Is there an integer *n* so that the aliquot sequence $n, \sigma^-(n), \sigma^-(\sigma^-(n)), \ldots$ is unbounded?

There was a time when the fate of 138 was unknown. Thanks to a decent factoring algorithm, 138 does not present a great challenge to *Mathematica*. The largest entry is 179931895322, and then the deficient numbers take over and the sequence dies out.

```
ali = AliquotSequence[138];
```

```
{138, 150, 222, 234, 312, 528, 960, 2088, 3762, 5598,
  6570, 10746, 13254, 13830, 19434, ≪99 terms omitted≫,
  40381357656, 60572036544, 100039354704,
  179931895322, 94685963278, 51399021218, 28358080762,
  18046051430, 17396081338, 8698040672, 8426226964,
  ≪37 terms omitted≫, 6154, 3674, 2374, 1190, 1402,
  704, 820, 944, 916, 694, 350, 394, 200, 265, 59, 1}
```

Figure 3.1(a) shows the entire sequence. Logarithms are used (via the Log-ListPlot function from a standard package) so that the image is visually appropriate.

But when we go beyond 138, things get much worse very quickly. No one knows what happens to the aliquot sequence of 276. This number is seriously abundant in the sense that iterations of σ^- lead, most of the time, to larger numbers. And when the numbers get large, factoring them is difficult. A 10-minute computation, using an upper bound of 10^{35} to shut it off, gets us 433 members of the sequence. Current knowledge is that this sequence has at least 628 terms. We repeat that this means that we have a conceptually simple algorithm (iterate σ until it hits 1 or enters a cycle) and a single instance (start with 276), but we do not know, and may never know, whether the algorithm halts on this input.

(a)

(b)

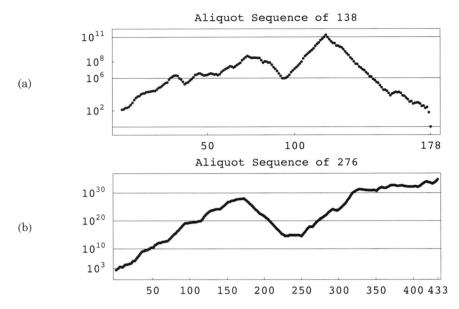

Figure 3.1. These diagrams show the sequences of values obtained by iterating the sum-of-divisors function on 138 and 276. The sequence for 138 terminates, but the ultimate fate of the sequence for 276 is not known.

```
ali = AliquotSequence[276, 10^35];
Length[ali]
```

433

```
Short[ali]
```

{276, 396, ≪429≫, 3086958132561459605202020358760354220,
 663486884280564997864994408769009948}

There are many facets of perfect numbers and the σ function that lead to interesting questions. For example, are there any triply perfect numbers: $\sigma(n) = 3n$?

```
Flatten[Position[σ[Range[1000]] / Range[1000], 3]]
```

{120, 672}

This leads to yet another computationally interesting question: Which rational numbers arise as values of $\sigma(n)/n$? Of course, 2 arises and we just saw that 3 shows up. But given a rational, there is no obvious way of telling whether it shows up. It has been conjectured by C. W. Anderson that the set of rationals that arise in this way is noncomputable. A computation up to 30,000 shows that, for example, $^5/_3$ and $^7/_5$ are missing. Perhaps they show up later.

```
sigmaRatios = Table[σ[n] / n, {n, 30000}];
```

```
Union[Select[sigmaRatios, Denominator[#] == 3 &]]
```
$$\left\{ \frac{4}{3}, \frac{7}{3}, \frac{8}{3}, \frac{10}{3} \right\}$$

```
Union[Select[sigmaRatios, Denominator[#] == 5 &]]
```
$$\left\{ \frac{6}{5}, \frac{8}{5}, \frac{9}{5}, \frac{12}{5}, \frac{13}{5}, \frac{14}{5}, \frac{16}{5}, \frac{18}{5} \right\}$$

There are many more questions regarding the behavior of perfect numbers and their relatives. Chapter B in [Guy] is an excellent source of additional problems of both a theoretical and computational flavor.

Exercises for Section 3.2

3.10. Prove that σ_k is a multiplicative function and derive an explicit formula for it in terms of the prime factorization of n.

3.11. (Euclid) Show that whenever $2^n - 1$ is prime, $2^{n-1}(2^n - 1)$ is a perfect number.

3.12. (René Descartes) Find the flaw in the following proof that an odd perfect number exists. Let $n = 3^2 \cdot 7^2 \cdot 11^2 \cdot 13^2 \cdot 22021$. Then

$$\sigma(n) = (1 + 3 + 3^2)(1 + 7 + 7^2)(1 + 11 + 11^2)(1 + 13 + 13^2)(1 + 22021) =$$
$$397171152378 = 2 \cdot 3^2 \cdot 7^2 \cdot 11^2 \cdot 13^2 \cdot 22021 = 2n.$$

3.13. Find all integer solutions to $1/x + 1/y = 1/100$. Express the number of solutions in terms of one of the functions introduced in this chapter.

3.14. Is there an odd abundant number?

3.15. A *unit fraction* is a fraction of the form $1/n$. A *proper unit fraction* excludes $1/1$. An *odd unit fraction* has an odd denominator. It is easy to write 1 as a sum of distinct unit fractions: $1 = 1/2 + 1/3 + 1/6$. Write the number 1 as the sum of finitely many distinct odd unit fractions (not using 1 itself).

3.16. An integer n, necessarily abundant, is called *pseudoperfect* if it is a sum of some of its divisors (excluding n itself). Find an abundant number that is not pseudoperfect. The following problem is unsolved: Is every odd abundant number pseudoperfect?

3.17. True or False: If n is an even perfect number and $k > 1$, then nk is abundant?

3.18 In late 1997, a new record was set by Mariano Garcia, who found an amicable pair that, at that time, was the largest one known (he has since found a larger pair). Garcia's pair is $\{m, n\}$ given below.

$$r = 2131 \cdot 51971 \cdot 168605317 \cdot 15378049151$$
$$p = 5744511433402789623743 13859$$
$$c = 2^{11} \, p^{89}$$
$$q = 136272576607912041393307632916794623$$
$$m = c \, r((p + q) \, p^{89} - 1)$$
$$n = c \, q((p - r) \, p^{89} - 1)$$

Each number in the pair has 4829 digits. Show, by computing σ of m and n, that the pair is amicable. Do not ask a computer for $\sigma(m)$ or $\sigma(n)$ directly, since that will get hung up on factoring. Use the formula for σ directly, after confirming that all prime-looking numbers in the formulas are probably prime via a 2-pseudoprime test. *Warning*: Do not try the love potion based on m grains of rice!

3.19. An ***aliquot cycle*** is a sequence $n, \sigma^-(n), \sigma^-(\sigma^-(n))$, ..., n. An amicable pair is an example of a 2-cycle. Write a progam that searches for aliquot cycles.

3.20. (H. Wilf) Develop and implement a formula that computes $\sigma(n, 1)$, the sum of the divisors of n that are congruent to 1 (mod 4). Here is an

1. Write n as $2^g N M$ where the primes in N are 1 (mod 4) and the primes in M are 3 (mod 4).
2. Show that $\sigma(n, 1) = \sigma(N)\sigma(M, 1)$.
3. Use recursion to solve the problem for M: Write M as $q^e M_1$ and look at divisors of M that involve q^0, q^1, q^2, and so on to get

$$\sigma(M, 1) = (1 + q^2 + q^4 + \ldots + q^{2c})\sigma(M_1, 1) +$$
$$(q + q^3 + q^5 + \ldots + q^{2d+1}) \, [\sigma(M_1) - \sigma(M_1, 1)]$$

   where $2c$ is the largest even number less than or equal to e and $2d + 1$ is the largest odd number less than or equal to e.
4. Use the algebraic rule $1 + w + w^2 + \cdots + w^a = (1 - w^{a+1})/(1 - w)$ with w equal to $-q$ and also q^2 to turn the preceding formula into the sum-free recursive formula that follows.

$$\sigma(M, 1) = \frac{(1 - (-q)^{e+1})}{1 + q} \sigma(M_1, 1) + \frac{q(1 - q^{2d+2})}{1 - q^2} \sigma(M_1)$$

5. Put the previous steps together to get a program that computes $\sigma(n, 1)$.
6. Check your implementation by comparing your results to the result of a straightforward sum using

```
Select[Divisors[n], Mod[#, 4] == 1 &].
```

7. Show how the sum of divisors congruent to 3 (mod 4) can be obtained from $\sigma(n, 1)$.
8. Show how the preceding work can be extended to get the sum of the kth powers of divisors of n that are congruent to 1 (mod 4); it simply requires adding k to various places in the treatment above.

3.21. Find and implement a formula for $\sigma_k^{alt}(n)$, which adds up $(-1)^d\, d^k$ over all divisors d of n. *Hints*: σ_k^{alt} is not multiplicative, but it is close; make use of σ_k, because that is already well understood and implemented in *Mathematica* as `DivisorSigma`.

3.3 Euler's Theorem

The Euler ϕ-function allows us to generalize Fermat's Little Theorem (Theorem 2.7) to composite numbers. Fermat's Little Theorem can be viewed as saying that modular powers for a prime modulus p are periodic with period $p - 1$. The two power grids in Figure 3.2 (see also color plate 2) show that there is periodicity for composite moduli too, so long as bases coprime to the modulus are used. That gets done with the `ReducedResiduesOnly` option.

```
VisualPowerGrid[10, ReducedResiduesOnly → True];
VisualPowerGrid[18, ReducedResiduesOnly → True];
```

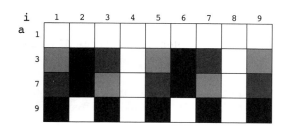

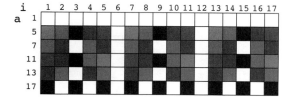

Figure 3.2. Two power grids for moduli 10 (top) and 18. Only values of a that are relatively prime to the modulus are shown, and the periodicity at $\phi(10) = 4$ and $\phi(18) = 6$ is evident.

The periodicity occurs at $\phi(n)$. In the case that n is prime, this is just $n-1$, and that is Fermat's Little Theorem. This result, called Euler's Theorem, has a proof that is essentially identical to that for the prime case.

Theorem 3.7. Euler's Theorem. If $m > 0$ and $(a, m) = 1$ then

$$a^{\phi(m)} \equiv 1 \pmod{m}.$$

Proof. Consider $\mathbb{Z}_m^*$, the set of integers in $\mathbb{Z}_m$ that are relatively prime to m; suppose that $\mathbb{Z}_m^* = \{b_1, \ldots, b_{\phi(m)}\}$. This is a **complete reduced residue system**, meaning that the entries are distinct modulo m and they represent all the residue classes of integers coprime to p. It follows exactly as for complete residue systems that $\{a b_1, \ldots, a b_{\phi(m)}\}$ is a complete reduced residue system also. Here is a quick example. The CNT package has a Reduced-Residues[n] function that can be accessed by the standard notation $\mathbb{Z}_{15}^*$. The simplest way to enter this is by the following keystrokes (CTRL 5 goes to the opposite position in a subsuperscript, dsZ stands for double-struck Z): ESC dsZ ESC CTRL – 15 CTRL 5 *.

> ```
> {Z*₁₅, multiplied = Mod[7 Z*₁₅, 15], Sort[multiplied]}
> ```
>
> ```
> {{1, 2, 4, 7, 8, 11, 13, 14},
>  {7, 14, 13, 4, 11, 2, 1, 8}, {1, 2, 4, 7, 8, 11, 13, 14}}
> ```

Therefore the products of the two systems are congruent:

$$a^{\phi(m)} \prod_{j=1}^{\phi[m]} b_j \equiv \prod_{j=1}^{\phi[m]} b_j \pmod{m}.$$

The product is relatively prime to a, and so can be cancelled, leaving $a^{\phi(m)} \equiv 1 \pmod{m}$. □

One standard application of Euler's Theorem is to hand calculations of large powers. If you want to know $10^{1000} \pmod{21}$, use the fact that $\phi(21)$ is 12 and $10^{12} \equiv 1 \pmod{21}$. Then $10^{1000} = 10^{83 \cdot 12 + 4} = (10^{12})^{83} 10^4 \equiv 10^4 = 100^2 \equiv (-4)^2 = 16 \pmod{21}$. The following corollary shows the general rule underlying this sort of computation.

Corollary 3.8. Suppose that s is relatively prime to each of a and b and $a \equiv b \pmod{\phi(m)}$. Then $s^a \equiv s^b \pmod{m}$.

Proof. Write a as $k\phi(m) + b$. Then $s^a = s^{k\phi(m)+b} = s^{\phi(m)k} s^b \equiv s^b \pmod{m}$. □

Euler's Theorem can be restated as: If $\gcd(a, n) = 1$, then $a^{\{0,1,2,3,\ldots\}}$ (mod n) is periodic with period $\phi(n)$. But an investigation of the power grids, or some numerical computation, will show that there is some interesting periodicity at work even in the cases where $\gcd(a, n) \neq 1$.

Proposition 3.9. The power sequence $a^{\{\phi(n), \phi(n)+1, \phi(n)+2, \ldots\}}$ (mod n) is periodic with period $\phi(n)$.

Proof. Factor n as $p^{\alpha} q^{\beta} \cdots$. It suffices to show that $a^{\phi(n)} \equiv a^{2\phi(n)}$ (mod n), because all others follow by multiplication by a. And it suffices to work modulo p^{α}, for the other powers follow similarly, and then the desired congruence will be true modulo n.

Case 1: p divides a. In this case $a^{\phi(n)} \equiv 0$ (mod p^{α}). To see this, it suffices to know that $\phi(n) \geq \alpha$. But $\phi(n) \geq \phi(p^{\alpha}) = p^{\alpha} - p^{\alpha-1} \geq 2^{\alpha-1} \geq \alpha$.

Case 2: p does not divide a. Then $a^{\phi(n)} \equiv a^{\phi(p^{\alpha})\phi(n/p^{\alpha})} \equiv 1^{\phi(n/p^{\alpha})} \equiv 1 \pmod{p^{\alpha}}$, using Euler's Theorem. $\square$

A consequence of this proposition is that when looking at power grids, and including all bases a, one need never look at power grids beyond the 2nth column.

In fact, there are some values of n for which the statement of Euler's Theorem holds in as strong a form as is possible for *any* base a. Can you discover which n have this property? For which n is it true that $a^{\{1,2,3,\ldots\}}$ (mod n) is periodic with period $\phi(n)$? The details are left to Exercise 3.23.

Exercises for Section 3.3

3.22. Find an example to show that Proposition 3.9 cannot be improved by starting earlier, with the power $a^{\phi(n)-1}$.

3.23. Find as many integers n as you can for which, for any integer a, $a^{\{1,2,3,\ldots\}}$ (mod n) is periodic with period $\phi(n)$. Try to find a characterization of the set of n that work, and then prove that if n is not in this set, the power sequence is not periodic.

3.24. Show how Euler's Theorem can be used to solve a congruence of the form $ax \equiv b$ (mod n) where n is composite. Start by thinking about the example $34x \equiv 14$ (mod 90). *Hint*: Isolate x by multiplying both sides by an appropriate power of 34.

3.4 Primitive Roots for Primes

■ The Order of an Integer

Euler's Theorem leads naturally to the mod-m order of an integer. If a and m are relatively prime, then the **mod-m order** of a is the least positive integer e such that $a^e \equiv 1 (\bmod\, m)$; we denote this by $\mathbf{ord_m(a)}$. If a and m have a common divisor greater than 1, then no power of a will ever be $1 \,(\bmod\, m)$, and the order is undefined. Euler's Theorem says that $\mathrm{ord}_m(a) \le \phi(m)$. A natural question is whether the order equals $\phi(m)$. If it does, then a is called a **primitive root** for m. For example, 2 is a primitive root for 13.

```
PowerMod[2, Range[12], 13]

{2, 4, 8, 3, 6, 12, 11, 9, 5, 10, 7, 1}
```

But 3 is not.

```
PowerMod[3, Range[12], 13]

{3, 9, 1, 3, 9, 1, 3, 9, 1, 3, 9, 1}
```

The next lemma emphasizes the central property of primitive roots: the eth powers, $e = 1, \ldots, \phi(m)$ are all distinct and take on all values in $\mathbb{Z}_m^*$.

Lemma 3.10. If g is a primitive root for m then every integer that is relatively prime to m is congruent to g^i for some $i = 1, 2, \ldots, \phi(m)$. Equivalently, $\{g^i : i = 1, \ldots, \phi(m)\}$ is a complete reduced resude system modulo m.

Proof. The $\phi(m)$ integers $g, g^2, g^3, \ldots, g^{\phi(m)}$ are distinct modulo m, for if $g^i \equiv g^j \,(\bmod\, m)$, with $i < j$, then $g^{j-i} \equiv 1 \,(\bmod\, m)$, contradicting the hypothesis on g. The powers are all relatively prime to m and because there are $\phi(m)$ of them, they must exhaust all the possiblities in $\mathbb{Z}_m^*$. □

Primitive roots and the structure of the order function play an important role in prime testing — finding a primitive root can be a crucial step in proving that an integer is prime — and we will study them in detail in this section and the next. Our first lemma tells us that when we know $\mathrm{ord}_m(a)$, we know all about the powers of a that are $1 \,(\bmod\, m)$: they are precisely the eth powers where e is a multiple of $\mathrm{ord}_m(a)$.

Lemma 3.11. Suppose that $\gcd(a, m) = 1$. Then $a^x \equiv 1 \,(\bmod\, m)$ if and only if $\mathrm{ord}_m(a)$ divides x.

Proof. All orders and congruences are modulo m. The reverse direction is clear: $a^x = a^{\mathrm{ord}(a)\cdot k} = \left(a^{\mathrm{ord}(a)}\right)^k \equiv 1$. For the other direction, suppose $a^x \equiv 1$.

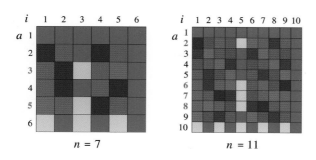

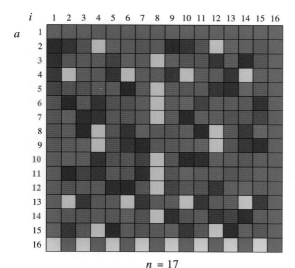

Color Plate 1. Prime Power Grids. Three views of the power function a^i modulo some primes (7, 11, and 17). Red denotes $+1$, yellow denotes -1, and the other shades denote the other values. Several patterns are visible, such as Fermat's Little Theorem (the red columns at the right) and the criterion for quadratic residues (the red–yellow column in the middle). If a is a primitive root (all powers are distinct), then it is shown in red.

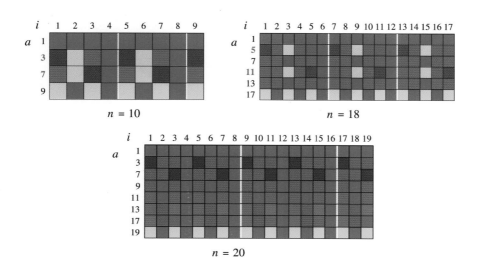

Color Plate 2. Composite Power Grids. Views of the power function modulo some composites (10, 18, and 20). Only the resides coprime to n are shown. Euler's Theorem is visible in the periodicity that occurs at $\phi(n)$ (which is 4, 6, and 8, respectively). In the case of 20, the periodicity actually occurs at the 4th power, illustrating the phenomenon that the least universal exponent $\lambda(n)$ can be less than $\phi(n)$.

Color Plate 3. Three Gaussian Components. Three sets of Gaussian primes reachable from $1 + i$. The yellow squares are the ones reachable using steps of size $\sqrt{2}$ or less; the blue network consists of those reachable in steps bounded by 2; and the the red network corresponds to a step-bound of $\sqrt{8}$.

Write x as $q \cdot \text{ord}(a) + r$, where $0 \le r < \text{ord}(a)$. Then $1 \equiv a^x = a^{q \cdot \text{ord}(a) + r} = a^{q \cdot \text{ord}(a)} a^r \equiv a^r$, a contradiction to minimality of $\text{ord}(a)$ unless $r = 0$. $\square$

Note that this result can be interpreted as: $a^x \equiv a^0 \pmod{m}$ if and only if $x \equiv 0 \pmod{\text{ord}_m(a)}$. More generally (proof left as exercise): $a^i \equiv a^j \pmod{m}$ if and only if $i \equiv j \pmod{\text{ord}_m(a)}$.

Corollary 3.12. Suppose that $\gcd(a, m) = 1$. Then $\text{ord}_m(a)$ divides $\phi(m)$.

The result of this corollary can be used to get an algorithm for computing orders. Just check all the divisors of $\phi(m)$ and take the first one that works.

As an example, the order of 5 modulo 17 must be one of the divisors of $\phi(17)$, which is 16. So the only choices are 1, 2, 4, 8, or 16. But $5^2 = 25 \equiv 8$, $5^4 \equiv 8^2 = 64 \equiv -4$, and $5^8 \equiv (-4)^2 = 16 \pmod{17}$, so the order of 5 must be 16. And this means that 5 is a primitive root for 17.

Theorem 3.13. Order-of-Powers Theorem. If $\text{ord}_n(a) = t$, then for any u, $\text{ord}_n(a^u) = t / \gcd(t, u)$.

Proof. All orders and congruences are modulo n. Let $s = \text{ord}(a^u)$, $d = \gcd(t, u)$, $t_1 = t/d$, $u_1 = u/d$. Then t_1 and u_1 are relatively prime. Now, $(a^u)^{t_1} = a^{ut/d} = (a^t)^{u_1} \equiv 1 \pmod{n}$, so s divides t_1. Our goal is $s = t_1$, so we must show that t_1 divides s. We have $a^{us} = (a^u)^s \equiv 1$, so t divides us. This means that $t_1 d$ divides $u_1 ds$, so t_1 divides $u_1 s$. But t_1 and u_1 are relatively prime, so t_1 divides s. $\square$

Corollary 3.14. If m has a primitive root, then m has exactly $\phi(\phi(m))$ primitive roots.

Proof. If r is a primitive root, the powers r^u, $u = 1, 2, ..., \phi(m)$, form a complete reduced residue system. So we need only check which of these are primitive roots. By the Order-of-Powers Theorem, the order of r^u is $\phi(m)/\gcd(\phi(m), u)$. This is $\phi(m)$ whenever $\gcd(\phi(m), u) = 1$, and there are $\phi(\phi(m))$ such powers u. $\square$

We saw earlier that 2 is a primitive root for 13. Here is the list of all such primitive roots. There are four of them because $\phi(13) = 12$ and $\phi(12) = 4$.

PrimitiveRoots[13]

{2, 6, 7, 11}

The Order-of-Powers Theorem has a consequence for our knowledge of pseudoprimes. Recall that there are numbers, such as 561, that are b-pseudoprimes for every b coprime to 561. This aberrant behavior can never happen for a prime power. The proof that follows uses the result, proved in the next subsection, that every prime has a primitive root.

Proposition 3.15. Let $n = p^j$ where p is an odd prime, and let b be a primitive root for n. Then n is not a b-pseudoprime.

Proof. Assume $j \geq 2$ because primes are not pseudoprimes. Because b is a primitive root, $\text{ord}_n b = \phi(p^j) = p^{j-1}(p-1)$, which is divisible by p. But $n - 1 = p^j - 1$ is not divisible by p and so $\phi(p^j)$ cannot divide $n - 1$, whence $b^{n-1} \not\equiv 1 \pmod{n}$. □

▣ Primes Have Primitive Roots

We now prove the first important theorem about primitive roots (discovered by Gauss, who also found the characterization of numbers admitting primitive roots that is proved in the next section).

Theorem 3.16. *Every prime has a primitive root.*

The proof of Theorem 3.16 requires some preliminary work.

Theorem 3.17. Lagrange's Theorem. *If p is prime, then a polynomial of degree n ($n \geq 1$) with integer coefficients and leading coefficient not divisible by p has at most n solutions in $\mathbb{Z}_p$.*

Proof. Because $\mathbb{Z}_p$ is a field, this theorem is a special case of the more general result of algebra that a polynomial of degree n with coefficients in a field has at most n roots in the field. We prove the case at hand by induction. If $n = 1$, then the polynomial is $a_1 x + a_0$ where p does not divide a_1. A root in $\mathbb{Z}_p$ corresponds to a solution of $a_1 x \equiv -a_0 \pmod{p}$, and we know that this has a unique solution $-a_0 a_1^{-1}$.

Now assume that the result is true for degree $n - 1$ and assume, to get a contradiction, that $f(x) = a_n x^n + a_{n-1} x^{n-1} + \cdots + a_1 x + a_0$ has $n + 1$ incongruent roots modulo p. Let the roots be $\beta_0, \ldots, \beta_n$. Then

$$
\begin{aligned}
f(x) - f(\beta_0) &= a_n(x^n - \beta_0{}^n) + a_{n-1}(x^{n-1} - \beta_0{}^{n-1}) + \cdots + a_1(x - \beta_0) \\
&= a_n(x - \beta_0)(x^{n-1} + x^{n-2}\beta_0 + \cdots + x\beta_0{}^{n-2} + \beta_0{}^{n-1}) + a_{n-1}(x - \beta_0) \\
&\quad (x^{n-2} + x^{n-3}\beta_0 + \cdots + x\,\beta_0{}^{n-3} + \beta_0{}^{n-2}) + \cdots + a_1(x - \beta_0) \\
&= (x - \beta_0)g(x), \quad \text{where } g \text{ has degree } n - 1.
\end{aligned}
$$

Now, look at the n numbers β_k, $k > 0$. We know that p divides $f(\beta_k) - f(\beta_0)$, so p divides $(\beta_k - \beta_0)g(\beta_k)$. Because β_k and β_0 are distinct modulo p, p divides $g(\beta_k)$. This contradicts the inductive hypothesis. □

Theorem 3.18. *If p is prime and d divides $p - 1$, then the dth degree polynomial $x^d - 1 = 0$ has exactly d solutions mod-p.*

Proof. Write $p - 1$ as dk. Then

$$
x^{p-1} - 1 = (x^d - 1)(x^{d(k-1)} + x^{d(k-2)} + \cdots + x^d + 1) = (x^d - 1)g(x).
$$

By Fermat's Little Theorem, we know that each of the $p-1$ numbers $1, 2, \ldots, p-1$ is a solution of $x^{p-1} - 1 \equiv 0 \pmod{p}$. Thus each of these numbers is a mod-p root of either $x^d - 1$ or $g(x)$. But $g(x)$ has degree $d(k-1)$, and so, by Lagrange's Theorem, has at most $d(k-1)$ roots. This means that $x^d - 1$ has at least $(p-1) - (d(k-1))$ roots. But this quantity is $p - 1 - (p-1) + d$, or d, as desired. $\square$

Solve can handle modular equations when the modulus is not too large. Here is an example having 37 solutions, as predicted by the preceding theorem, because $7918 = 2 \cdot 37 \cdot 107$

```
solns = x /. Solve[{x^37 == 1, Modulus == 7919}, x]
Length[solns]
```

```
{1, 6, 36, 216, 220, 755, 788, 886, 1015, 1296,
 1320, 1489, 1568, 2771, 2901, 3262, 3423, 3734, 3909,
 4443, 4530, 4611, 4700, 4728, 4864, 4930, 5316, 5427,
 5823, 6090, 6101, 6566, 6725, 7061, 7616, 7720, 7776}
```

```
37
```

Theorem 3.19. If p is prime, then the number of incongruent integers whose order mod-p is d is $\phi(d)$ if d divides $p-1$, 0 otherwise.

Proof. For each d dividing $p-1$, let $F(d)$ be the number of integers in $\{1, \ldots, p-1\}$ of order d. If $F(d) > 0$, then d divides $p-1$ by Corollary 3.12. Therefore $p-1 = \sum_{d|p-1} F(d)$. But Theorem 3.1(h) says that $p-1 = \sum_{d|p-1} \phi(d)$. So if we can show $F(d) \leq \phi(d)$ when d divides $p-1$ we will be done, for that will mean that $F(d)$ must equal $\phi(d)$ in each case. Fix d, a divisor of $p-1$. If $F(d) = 0$, we have the inequality. If not, there is at least one b such that $\text{ord}_p(b) = d$. This means that each b^i, $i = 1, \ldots, d$ satisfies $x^d - 1 \equiv 0$ and hence, by Theorem 3.18, there are no other solutions. But among these d solutions, there are exactly $\phi(d)$ of them that have order d: this follows from the Order-of-Powers Theorem, because they are the ones whose exponent is coprime to d. $\square$

Now the proof that every prime has a primitive root is complete.

Proof of Theorem 3.16. If p is prime, then Theorem 3.19 states that the number of integers of order $p-1$ is $\phi(p-1)$. This is positive, so there is at least one b having mod-p order equal to $p-1$. $\square$

■ Repeating Decimals

A rational number is a number of the form r/s where r and s are integers. The rationals all have decimal expansions that are eventually periodic, and any eventually periodic decimal represents a rational number. We prove this now, and then discuss a connection between decimal expansions and primitive roots.

Proposition 3.20. Every rational number has a decimal expansion that consists of an initial part (the **preperiod**) followed by a periodic part. Conversely, every eventually periodic decimal expansion represents a rational number.

Proof. We will only sketch the main ideas. Filling in all the details is a good exercise. If r/s is rational, imagine performing the long division as you learned in grade school. At each step there will be a remainder less than s. Therefore, after finitely many steps, the remainders must repeat; this means that the resulting decimal expression — the quotient — will start repeating at that point.

For the other direction, suppose that $x = m.abc...defg...hefg...hefg...h...$ where m is an integer, the **preperiod** $abc...d$ has length r, and the periodic part $efg...h$ has length s. Then, using the formula for the sum of a geometric series,

$$x = m + \frac{abc...d}{10^r} + \frac{\frac{efg...h}{10^{r+s}}}{1 - \frac{1}{10^s}} = m + \frac{1}{10^r}\left(abc...d + \frac{efg...h}{10^s - 1}\right).$$

This last is a sum of rational numbers, and so is rational.   □

Mathematica has some functions for dealing with decimal expansions of rationals. `RealDigits` returns the preperiod and period and `FromDigits` does the opposite.

```
RealDigits[4111 / 33000]
```

```
{{{1, 2, 4}}, {5, 7}}
```

```
FromDigits[{{{1, 2, 4}}, {5, 7}}]
```

$$\frac{4111}{33000}$$

Now, some rationals have periods that are as long as possible: for example, the period of $1/7$ is 6 which, by the proof of Proposition 3.20, is the largest possible.

```
RealDigits[1 / 7]
N[1 / 7, 30]
```

```
{{{}}, {1, 4, 2, 8, 5, 7}}
```

```
0.142857142857142857142857142857
```

This leads to the question: Which rationals r/s have the property that the periodic part of the decimal expansion has length $s - 1$? We will not prove it here, but it turns out that the answer is: r/s has decimal period $s - 1$ if and only if s is prime and 10 is a primitive root for s. Here is a computation that finds all these fractions of the form $1/s$, where $s \le 200$.

```
Select[Range[2, 200], Length[RealDigits[1/#][[2]]] == # - 1 &]
Select[Range[2, 200], OrderMod[10, #] == # - 1 &]
```

```
{7, 17, 19, 23, 29, 47, 59, 61, 97,
 109, 113, 131, 149, 167, 179, 181, 193}
{7, 17, 19, 23, 29, 47, 59, 61, 97,
 109, 113, 131, 149, 167, 179, 181, 193}
```

And here is the result that describes the whole situation in base 10; similar results hold for base b (see [Ros, §10.1] for proofs).

Theorem 3.21. Suppose that $x = r/s$ where r and s are positive integers and $r < s$. Write s as $2^a 5^b y$ where $\gcd(y, 10) = 1$. Then the length of the periodic part of the decimal representation of x is $\text{ord}_y(10)$ and the length of the preperiodic part is $\max(a, b)$.

Exercises for Section 3.4

3.25. If b has order 360 (mod m), what is the order of b^{150}?

3.26. Suppose $\gcd(a, m) = 1$. Prove that $a^i \equiv a^j$ (mod m) if and only if $i \equiv j$ (mod $\text{ord}_m(a)$).

3.27. If g is a primitive root modulo 67, what powers of g represent the other primitive roots modulo 67? Do this by hand.

3.28. Show that if p is prime and d is relatively prime to $p - 1$, then every positive integer less than p is congruent to the dth power of some other integer.

3.29. (D. Brizolis [Guy, p. 248]) Investigate by computations the following two assertions, which are unproved:

(a) If p is prime, then p has a primitive root g such that $\gcd(g, p - 1) = 1$.

(b) If p is prime, then p has a primitive root g such that, for some i, $g^i \equiv i$ (mod p).

3.30. For each d in $A = \{1, 2, 4, 13, 26, 52\}$ find those values of a for which $x^d \equiv a$ (mod 53) has exactly d solutions modulo 53. For each d in A, what are the orders of the corresponding values of a? What are necessary and sufficient conditions on the order of a so that $x^d \equiv a$ (mod p) has exactly d incongruent solutions when d divides $p - 1$?

3.31. Implement the formula in the proof of Proposition 3.20 via a program that returns the exact rational represented by a decimal expression with a given preperiod and period.

3.32. Implement a routine that, given a rational, returns the lengths of the preperiodic and periodic parts in its decimal expansion.

3.5 Primitive Roots for Composites

Which integers have primitive roots? There is a characterization, but it is a little tricky and hard to guess from the data. Here is how to find all the integers under 100 that have primitive roots (where we use the `Primitive-Roots` function from the `CNT` package). Of course, all the primes show up. Can you detect any patterns to the nonprimes in the list?

```
Select[Range[100], PrimitiveRoots[#] ≠ {} &]
```

```
{2, 3, 4, 5, 6, 7, 9, 10, 11, 13, 14, 17, 18, 19, 22, 23, 25, 26,
 27, 29, 31, 34, 37, 38, 41, 43, 46, 47, 49, 50, 53, 54, 58, 59,
 61, 62, 67, 71, 73, 74, 79, 81, 82, 83, 86, 89, 94, 97, 98}
```

We will get the complete characterization (Theorem 3.25) by working in steps.

Lemma 3.22. The only powers of 2 that have primitive roots are 2 and 4.

Proof. The small cases are easy: 1 is a primitive root for 2 and 3 is a primitive root for 4. For powers of 2 beyond 4 we prove the stronger result that for any odd s, if $k \geq 3$ then $s^{2^{k-2}} \equiv 1 \pmod{2^k}$. This shows that s cannot be a primitive root, because $\phi(2^k) = 2^{k-1}$. Use induction on k. We already know that $s^2 \equiv 1 \pmod 8$, so assume $k \geq 3$ and $s^{2^{k-2}} \equiv 1 \pmod{2^k}$ for any odd integer s. Then 2^k divides $s^{2^{k-2}} - 1$. Squaring a number doubles the exponent of 2, so 2^{2k} divides $s^{2^{k-1}} - 2s^{2^{k-2}} + 1$. We will use only the weaker fact that 2^{k+1} divides $s^{2^{k-1}} - 2s^{2^{k-2}} + 1$. Working modulo 2^{k+1}, the inductive hypothesis reduces this last expression as follows.

$$s^{2^{k-1}} - 2s^{2^{k-2}} + 1 = s^{2^{k-1}} - 2s^{2^{k-2}} + 2 - 1 = s^{2^{k-1}} - 2\left(s^{2^{k-2}} - 1\right) - 1 \equiv s^{2^{k-1}} - 1$$

as required. □

Lemma 3.23. If p is an odd prime and g is a primitive root for p, but not for p^k, then $g + p$ is a primitive root for p^k.

Proof. Let $e = \operatorname{ord}_{p^k}(g)$. Then p^k divides $g^e - 1$, so p divides this as well, which means that e is a multiple of $\operatorname{ord}_p(g)$, which is $p - 1$ by hypothesis. Because g is not a primitive root for p^k, e divides but is not equal to $(p-1)p^{k-1}$. This implies that

$$g^{(p-1)p^{k-2}} \equiv 1 \pmod{p^k}.$$

Because $g + p$ is also a primitive root for p, its order modulo p^k must be divisible by $p - 1$ and must divide $(p-1)p^{k-1}$. It suffices to show that $\operatorname{ord}_{p^k}(g + p)$ is not a divisor of $(p-1)p^{k-2}$. We need a fact about binomial

3.5. PRIMITIVE ROOTS FOR COMPOSITES **91**

coefficients: $\binom{a\,p^b}{j}p^j$ is divisible by p^{b+2} for $j \geq 2$. Because p is odd, this is clear when $j = 2$. For $j \geq 3$, the expression has the form $m\,p^b\,p^j\,/\,j!$, and it is easy to see that there are at least two more factors of p in p^j than there are in $j!$. We use the binomial theorem as follows, where the preceding fact is used in the first step to eradicate all but the first two terms of the expansion modulo p^k.

$$(g+p)^{(p-1)p^{k-2}} \equiv g^{(p-1)p^{k-2}} + (p-1)p^{k-2}\,g^{(p-1)p^{k-2}-1}\,p$$
$$\equiv 1 - g^{(p-1)p^{k-2}}\,p^{k-1} \pmod{p^k}$$

Because p does not divide any power of g, the right side of this last congruence is not 1, and so $\mathrm{ord}_{p^k}(g+p)$ does not divide $(p-1)p^{k-2}$. This means that the order of $g+p$ must be $(p-1)p^{k-1}$, which is $\phi(p^k)$. □

Lemma 3.24. If p is an odd prime and g is a primitive root for p^k, then either g (if g is odd) or $g + p^k$ (if g is even) is a primitive root for $2\,p^k$.

Proof. Assume that g is odd. As in the proof of Lemma 3.23, $\mathrm{ord}_{2p^k}(g)$ must be divisible by $\mathrm{ord}_{p^k}(g)$. But the latter is $\phi(p^k)$, and because $\phi(p^k) = \phi(2\,p^k)$, g is a primitive root for $2p^k$. If g is even, then $g + p^k$ is odd and the previous case applies. □

The preceding three lemmas complete the proof of one direction of Gauss's characterization, which we now state.

Theorem 3.25. An integer m admits a primitive root if and only if m is 2, 4, a power of an odd prime, or twice a power of an odd prime.

All that remains is to show that all other integers do not have primitive roots. The numbers not yet covered have the form mn where m and n are relatively prime and larger than 2. One more lemma will finish the job.

Lemma 3.26. Suppose that b, m, and n are pairwise relatively prime. Then $\mathrm{ord}_{mn}(b) = \mathrm{lcm}[\mathrm{ord}_m(b), \mathrm{ord}_n(b)]$.

Proof. Let $e = \mathrm{ord}_{mn}(b)$ and $f = \mathrm{lcm}[\mathrm{ord}_m(b), \mathrm{ord}_n(b)]$. Then $b^e \equiv 1 \pmod{mn}$ and so $b^e \equiv 1$ modulo each of m and n. This means that each of $\mathrm{ord}_m(b)$ and $\mathrm{ord}_n(b)$ divides e, so their lcm does too. So f divides e. And $b^f \equiv 1$ modulo each of m and n, and therefore $b^f \equiv 1 \pmod{mn}$ so e divides f. Therefore $e = f$. □

This lemma tells us that the order of any element modulo mn must divide $\mathrm{lcm}[\phi(m), \phi(n)]$. If m and n are larger than 2, then both ϕ-values are even, and so $\mathrm{lcm}[\phi(m), \phi(n)] < \phi(m)\phi(n) = \phi(mn)$, and this means the order of any b is less than $\phi(mn)$. □

Recall that when n admits a primitive root, there are precisely $\phi(\phi(n))$ of them. Here's an example.

```
n = 2 13²; prs = PrimitiveRoots[n]
{Length[prs], ϕ[ϕ[n]]}
```

{7, 11, 15, 33, 37, 41, 45, 59, 63, 67, 71, 85, 93, 97, 111, 115, 119, 123, 137, 141, 145, 149, 163, 167, 171, 175, 189, 193, 197, 201, 215, 219, 223, 227, 241, 245, 253, 267, 271, 275, 279, 293, 297, 301, 305, 323, 327, 331}

{48, 48}

Exercises for Section 3.5

3.33. What are the possible orders for an integer modulo 1423, and how many positive integers less than 1423 are there of each order?

3.34. Show that if there is no primitive root for n, then the order of each element divides $\phi(n)/2$.

3.35. Let p be an odd prime. Show that Lemma 3.23 implies that there are at least $\phi(p-1)(p^{k-1}+1)/2$ primitive roots modulo p^k.

3.36. Calculate the number of primitive roots modulo p^k for various values of an odd prime p and integer $k \geq 2$. How do these numbers compare with the minimum established in Exercise 3.35?

3.37. Prove a stronger form of Lemma 3.23: If g is a primitive root for p but not for p^k and if a is not divisible by p, then $g + ap$ is a primitive root for p^k.

3.38. Use Exercise 3.37 to prove that there are at least $\phi(p-1)p^{k-2}(p-1)$ primitive roots for p^k when $k \geq 2$. How does this new minimum compare with the actual number of primitive roots for various values of odd primes p and integers $k \geq 2$?

3.39. In Exercise 2.44, you were asked to do computations to support the Monier–Baillie–Wagstaff formula that the number of witnesses for n is $\phi(n) - \prod_{p|n} \gcd(n-1, p-1)$. You should now be able to prove this formula. *Hint*: Show that if p is a prime divisor of the odd integer n and p^e is the highest power of p that divides n, then the number of residues a, modulo p^e, for which $a^{n-1} \equiv 1 \pmod{p^e}$ is the gcd of $n-1$ and $p-1$. Then use the Chinese Remainder Theorem to show that the number of residues a, modulo n, for which $a^{n-1} \equiv 1 \pmod{n}$ is $\prod_{p|n} \gcd(n-1, p-1)$.

3.6 The Universal Exponent

▥ Universal Exponents

A ***universal exponent*** for m is an integer e such that $a^e \equiv 1 \pmod{m}$ for every a relatively prime to m. Euler's Theorem tells us that $\phi(m)$ is a universal exponent. Sometimes $\phi(m)$ is the smallest one, but most of the time there is a smaller universal exponent Consider $m = 8$. Then $\phi(8) = 4$, but 2 serves as a universal exponent because $1^2, 3^2, 5^2, 7^2$ are all congruent to 1 (mod 8). Or look at $m = 15$: a universal exponent is 4 even though $\phi(15)$ is 8.

```
PowerMod[Z*₁₅, 4, 15]
φ[15]

{1, 1, 1, 1, 1, 1, 1, 1}

8
```

Define $\lambda(m)$ to be the ***least universal exponent*** for m, also called ***Carmichael's lambda***. Some traditional applications of ϕ really depend only on λ, so it is worth knowing how to compute it. There is a formula that is very similar to the one for ϕ.

Theorem 3.27. $\lambda(2) = 1$; $\lambda(4) = 2$; $\lambda(2^s) = 2^{s-2}$ if $r \geq 3$; and $\lambda(p^k) = \phi(p^k)$ if p is an odd prime. In general, write n as $2^s \prod_{i=1}^{r} p_i^{a_i}$, where each p_i is an odd prime; then $\lambda(n) = \mathrm{lcm}[\lambda(2^r), \phi(p_1^{a_1}), \ldots, \phi(p_r^{a_r})]$.

Proof. Simple inspection takes care of 2 and 4. For 2^s we already know that $s - 2$ is a universal exponent by the proof of Lemma 3.22. We need a little bit more, namely, that $s - 2$ is the *least* universal exponent. So we need a t such that $\mathrm{ord}_{2^s}(t) = r - 2$. It turns out that $t = 5$ works; verification is left as an exercise (Exercise 3.40). The case of powers of odd primes is easy because such numbers have primitive roots, which proves that $\phi(p^k)$ is the least universal exponent.

The general formula for $\lambda(n)$ follows immediately from the following auxiliary fact, which allows us to paste together the results for prime powers.

If $n = ab$ where a and b are relatively prime and less than n, then

- there is an integer s such that $\mathrm{ord}_n(s) = \mathrm{lcm}[\lambda(a), \lambda(b)]$
- $\lambda(n) = \mathrm{lcm}[\lambda(a), \lambda(b)]$

Proof of fact: We know that the desired special value s exists for prime powers, because a primitive root does the job. So we can use induction, where the inductive hypothesis gets us s and t such that $\mathrm{ord}_a(s) = \lambda(a)$ and

$\operatorname{ord}_b(t) = \lambda(b)$. Use the Chinese Remainder Theorem to find u congruent to s (mod a) and t (mod b). Then by Lemma 3.26,

$$\operatorname{ord}_{ab}(u) = \operatorname{lcm}[\operatorname{ord}_a(u), \operatorname{ord}_b(u)] = \operatorname{lcm}[\operatorname{ord}_a(s), \operatorname{ord}_b(t)] = \operatorname{lcm}[\lambda(a), \lambda(b)].$$

Now, it is clear that $\operatorname{lcm}[\lambda(a), \lambda(b)]$ is a universal exponent for n because it is a multiple of universal exponents for a and b. But the existence of s shows that it is the least universal exponent.   □

Carmichael's λ function is built into *Mathematica* (version 4) as `CarmichaelLambda`. Our `CNT` package allows us to use just λ.

    λ[102]

    16

Note that $\lambda(n)$ is even unless $n = 2$, if m divides n then $\lambda(m)$ divides $\lambda(n)$, and $\lambda(n)$ always divides $\phi(n)$. In fact, $\lambda(n)$ can be quite a bit less than $\phi(n)$; here are some values for which λ is less than one tenth of ϕ.

    Select[Range[1000], λ[#] < φ[#] / 10 &]

    {240, 252, 273, 315, 364, 399, 455, 468, 480, 481, 504, 513, 520,
     532, 544, 546, 560, 585, 624, 630, 651, 665, 680, 684, 693, 703,
     720, 728, 741, 756, 777, 780, 793, 798, 816, 819, 825, 840,
     855, 868, 880, 903, 910, 936, 945, 949, 960, 962, 988, 999}

■ Power Towers

Here is a pretty application of λ to a computational problem: What is $2^{2^{2^{2^{2^{2^{2^{2}}}}}}}$ (mod 719)? (We use the typographical convention that a^{b^c} always means $a^{(b^c)}$; in other words, the bracketing is top-down.) This is a simple-sounding problem, but the numbers that arise, especially if instead of eight 2s we had one hundred or more 2s, are so huge that some sophisticated ideas are needed to answer it. First we show that ϕ can be replaced by λ in Corollary 3.21.

Lemma 3.28. *If s is relatively prime to m and $a \equiv b$ (mod $\lambda(m)$), then $s^a \equiv s^b$ (mod m).*

Proof. Write a as $k\lambda(m) + b$. Then $s^a = s^{k\lambda(m)+b} = (s^{\lambda(m)})^k s^b \equiv s^b$ (mod m).   □

Here is how the lemma can be used to answer the tower-of-2s question. Let's introduce the notation $\boldsymbol{a \uparrow n}$ for the height-n tower $a^{a^{a^{\cdot^{\cdot}}}}$; then the sample problem asks for $2 \uparrow 7 \pmod{719}$. First we need the result of iterating λ six times on 719.

```
NestList[λ, 719, 6]
```

{719, 718, 358, 178, 88, 10, 4}

This means that $\lambda(10) = 4$, $\lambda(88) = 10$, and so on. Now we work down from the top, using the lemma (and `PowerMod`) at each stage.

$$2 \equiv 2 \ (\text{mod } 4)$$
$$2^2 \equiv 2^2 = 4 \ (\text{mod } 10)$$
$$2^{2^2} \equiv 2^4 = 16 \ (\text{mod } 88)$$
$$2 \uparrow 4 \equiv 2^{16} = 65536 \equiv 32 \ (\text{mod } 178)$$
$$2 \uparrow 5 \equiv 2^{32} = 126 \ (\text{mod } 358)$$
$$2 \uparrow 6 \equiv 2^{126} = 66 \ (\text{mod } 718)$$
$$2 \uparrow 7 \equiv 2^{66} = 596 \ (\text{mod } 719)$$

The noncoprime case adds some minor complications, but we can program a function `TowerMod[a, n, m]` that will give us, in an instant, the results of mind-boggling computations such as $123456789 \uparrow 100 \ (\text{mod } 10^{100})$.

Algorithm 3.1. Modular Towers

Here is how to turn the ideas of the preceding example into an algorithm for computing $a \uparrow n \ (\text{mod } m)$. First we need the $\log^*$ function, where $\textbf{\textit{log}}^*(\textbf{\textit{a}}, \textbf{\textit{s}})$ is the largest integer n such that $a \uparrow n \le s$. Here is a recursive implementation, followed by a line that allows us to use a *-superscript. We temporarily set `$MaxExtraPrecision` to a high value by using `Block`, because it can happen that a lot of extra precision is needed to resolve the inequality. We also define $a \uparrow n$ to be the pure tower $a^{a^{\cdot^{\cdot}}}$ by redefining the built-in `UpArrow`.

```
LogStar[a_, n_] := Block[{$MaxExtraPrecision = 10000},
   If[n < a, 0, LogStar[a, Log[a, n]] + 1]]
Log*[a_, n_] := LogStar[a, n]

Unprotect[UpArrow];
UpArrow[a_, n_] := If[n == 0, 1, a^a↑(n-1)]
Protect[UpArrow];
```

$\left\{2 \uparrow 4, \ 2^{2^{2^2}}\right\}$

{65536, 65536}

$\{\underline{\textbf{Log}}^*[3, 3 \uparrow 3], \ \underline{\textbf{Log}}^*[3, (3 \uparrow 3) - 1]\}$

{3, 2}

To get $a \uparrow n \pmod{m}$ in general, assume $n \geq 2$, for otherwise it is easy. Then we can split m into a product of prime powers and paste the answers together using the Chinese Remainder Theorem. Thus, it is sufficient to assume $m = p^e$, where p is prime. Let $g = \gcd(a, p^e)$; then $g = p^r$ for some r.

Case 1. $g = 1$. This is easy: p^e is coprime to a, so recursion and Lemma 3.28 get the result as $a^{a\uparrow(n-1) \pmod{\lambda[p^e]}} \pmod{m}$.

Case 2. $g > 1$. Subcase 1. $n - 1 > \log^*[a, \log_g(p^e)]$

The answer is simply 0. For we know that $a \uparrow (n-1) > \log_g(p^e)$, which means that $g^{a\uparrow(n-1)} > p^e$. But g is a power of p, so this means that $g^{a\uparrow(n-1)} \equiv 0 \pmod{p^e}$, and therefore $a^{a\uparrow(n-1)} \equiv 0 \pmod{p^e}$ because g divides a.

Subcase 2. $n - 1 \leq \log^*[a, \log_g(p^e)]$

In this case the answer is $g^{a\uparrow(n-1)} (a/g)^{a\uparrow(n-1) \bmod \lambda[p^e]} \pmod{p^e}$. To see this, write $a \uparrow n$ as $a^{a\uparrow(n-1)}$ and split it into $g^{a\uparrow(n-1)} (a/g)^{a\uparrow(n-1)}$. The first factor is easy to compute because the case definition means it is at most p^e; therefore we can get it nonmodularly by just iterating powers. For the second factor the following point is critical: $\gcd(a/g, p^e) = 1$. Given this, we can work modulo $\lambda(p^e)$ as in case 1. Why is this gcd assertion true? We must be sure that g eats up all the powers of p that lie in a. For this it suffices to know that $a \leq p^e$. Here is a proof:

$$a \leq a \uparrow (n-1) \leq \log_g(p^e) = \log_{p^r}(p^e) \leq e \leq p^e.$$

We now give complete `TowerMod` code. The first three lines handle some small cases. The fourth line factors m, gets the results for the prime powers, and uses the Chinese Remainder Theorem to combine these results (this code requires the `CNT` package for `ChineseRemainder`, λ, and `LogStar`). The last section of code implements the crux of the method in the case that the modulus is p^e, where p and e are treated as separate arguments, in part to avoid unnecessary refactoring.

```
TowerMod[a_, n_, 1] := 0
TowerMod[a_, 0, m_] := 1 /; m > 1
TowerMod[a_, 2, m_] := PowerMod[a, a, m]

TowerMod[b_, n_, m_] :=
 Module[{fi = FactorInteger[m]}, ChineseRemainder[Map[
    TowerModPrimePower[b, n, #] &, fi], Power @@ Transpose[fi]]]

TowerModPrimePower[_, 0, {p_, _}] := 1 /; p > 1
TowerModPrimePower[_, _, 1] := 0
TowerModPrimePower[a_, 2, {p_, e_}] := PowerMod[a, a, p^e]
```

```
TowerModPrimePower[a_, n_, {p_, e_}] :=
  Module[{g, pp = pᵉ}, g = GCD[a, pp]; If[g > 1 &&
      n - 1 > LogStar[a, Log[g, pᵉ]], 0, Mod[If[g == 1, 1, gᵃ↑⁽ⁿ⁻¹⁾]
        PowerMod[a / g, TowerMod[a, n - 1, λ[pᵉ]], pp], pp]]]
Attributes[TowerMod] = Listable;
```

Here is an application to the sample problem

```
TowerMod[2, Range[10], 719]
```

{2, 4, 16, 107, 624, 257, 596, 507, 507, 507}

In order to handle very large numbers we have to change the default recursion limit from 256 to a larger value.

```
$RecursionLimit = 1000;
TowerMod[123456789, 100, 10¹⁰⁰]
```

271107080101886985421805864950045993811195866850701956456551﹕
6360755938313499646999752691133563109509

If you examine sequences of the form $a\uparrow 1$, $a\uparrow 2$, $a\uparrow 3$, ..., all modulo m, you will see that they become constant, and in fewer than $\log_2 m$ steps (Exercise 3.46 asks you to prove this).

```
TowerMod[311, Range[10], 100000]
```

{311, 11911, 87911, 47911,
  47911, 47911, 47911, 47911, 47911, 47911}

▣ The Form of Carmichael Numbers

Recall that a Carmichael number, such as 561 ($= 3 \cdot 11 \cdot 17$), is a composite integer n for which $b^{n-1} \equiv 1 \pmod{n}$ for every b coprime to n. It turns out that such numbers are very restricted in the form they can have. An alternate route to the following proposition can be obtained by using the result of 3.39.

Proposition 3.29. If n is a Carmichael number then $n = p_1 p_2 \cdots p_r$ where each p_i is prime and $n - 1$ is divisible by each $p_i - 1$.

Proof. We know from Theorem 3.27 that there is some b such that $\mathrm{ord}_n(b) = \lambda(n)$. Because $b^{n-1} \equiv 1 \pmod{n}$, this means that $\lambda(n)$ divides $n - 1$. Because $n > 2$, $\lambda(n)$ is even and so n must be odd. Now suppose that some p^e divides n with p odd and $e \geq 2$. Then $\lambda(p^e)$ divides $\lambda(n)$, which means that $\phi(p^e) = p^{e-1}(p - 1)$ divides $n - 1$. But this is a contradiction because p divides n. This proves the first part. For the rest, use the fact that $\lambda(p_i)$ divides $\lambda(n)$, so $p_i - 1$ divides $n - 1$. $\square$

Proposition 3.30. If n is a Carmichael number then $n = p_1 p_2 \cdots p_r$ where each p_i is prime and $r \geq 3$.

Proof. By the preceding proposition, we need only consider the case that n is a product of two primes, say $n = pq$. Suppose that n is a Carmichael number and let g be a primitive root for p. Then the order of g is $p - 1$, which must divide $pq - 1$. Therefore, $1 \equiv pq \equiv q \pmod{p - 1}$. This means that $q \geq p$. But a completely symmetric argument shows that $p \geq q$. Therefore, $p = q$, a contradiction. $\square$

The preceding result gives us an efficient way to search for Carmichael numbers. The code that follows looks at products of three primes.

```
Select[Flatten[Table[Prime[i] Prime[j] Prime[k],
    {i, 2, 20}, {j, i + 1, 20}, {k, j + 1, 20}]], CarmichaelQ]
```

```
{561, 1105, 2465, 1729, 2821, 8911, 6601, 29341}
```

Exercises for Section 3.6

3.40. Show that if $k \geq 3$, then $5^{2^{k-3}} \not\equiv 1 \pmod{2^k}$. This shows that $\lambda(2^k)$ cannot be less than 2^{k-2}, and so, by the first part of the proof of Theorem 3.27, $\lambda(2^k) = 2^{k-2}$. *Hint*: Use induction on k to show that, if $k \geq 3$, $5^{2^{k-3}} \equiv 1 + 2^{k-1} \pmod{2^k}$.

3.41. Prove that $\lambda(m)$ divides $\lambda(n)$ whenever m divides n.

3.42. Prove that $\lambda(n)$ always divides $\phi(n)$.

3.43. For which n does $\lambda(n) = \phi(n)$?

3.44. Write a program that, on input n, computes $\lambda(n)$.

3.45. Recall the definition of $\log_f(n)$ from Exercise 3.8. Show that $\log_\lambda(n) \leq \log_\phi(n)$. It then follows from Exercise 3.8 that $\log_\lambda(n) \leq \lceil \log_2(n) \rceil$.

3.46. Prove that the sequence $a \uparrow 1$, $a \uparrow 2$, $a \uparrow 3$, ..., all modulo m, is eventually constant. Do this by showing that whenever $n \geq c = \lceil \log_2(m) \rceil$, $a \uparrow n \equiv a \uparrow c \pmod{m}$. *Hint*: It suffices to prove it for the case that m is a prime power, p^e. Then break it further into cases according as a and p are relatively prime. Use Exercise 3.45.

3.47. Implement a `TowerLimit[a, m]` function that returns the limiting value of $a \uparrow n \pmod{m}$.

3.48. Find a Carmichael number that is a product of 4 primes.

CHAPTER 4

Prime Numbers

4.0 Introduction

Prime numbers have fascinated counters for thousands of years. Indeed, the search for patterns in prime numbers is one of the oldest problems of mathematics. Moreover there are many extremely elementary problems about primes that are still unsolved (one example: Are there infinitely many primes of the form $n^2 + 1$?). While prime numbers have shown up in applications from time to time, interest in them increased greatly with the discovery in the 1970s that they could be used to generate methods of encrypting data that were almost surely unbreakable. In this chapter we will discuss several aspects of the theory of prime numbers focusing on the following questions:

- What patterns govern the appearance of the primes among the integers at large?
- Is it possible to quickly tell whether a very large number is prime?
- How can we generate large numbers that are certifiably prime?
- How can we quickly find, for modest values of n, the nth prime exactly?

Here are three examples of functions that do some of these things. First we check the primality of a 101-digit number. PrimeQ is known to be valid up to 10^{16}, and no example for which it fails is known; thus, one should interpret the following computationas saying that 10^{267} is almost certainly prime.

```
PrimeQ[10^100 + 267]
```
```
True
```

If the primes are listed, starting from 2, as p_1, p_2,..., then here is $p_{123456789}$.

```
Prime[123456789]
```
```
2543568463
```

And here is how one can generate a large prime whose primality can be certified. There is no doubt as to its status.

CertifiedPrime[100]

238185411887402483621303885887112731219071682708477299244592·
6064205414067244958160368224765565033219

4.1 The Number of Primes

▦ We'll Never Run Out of Primes

Primes are the multiplicative building blocks of arithmetic. Finding the factorization of an integer into primes, recognizing primes when you encounter them, understanding how they are distributed: These are fundamental abilities that are required throughout number theory. We begin their study with a result that comes down to us from Euclid; there is no largest prime.

Theorem 4.1. There are infinitely many primes.

Proof. Let us assume that there is a largest prime number, call it p_r, so that the primes are given by

$$p_1 = 2, \ p_2 = 3, \ p_3 = 5, \ ..., \ p_r \, .$$

Let P be their product, perhaps a huge number but surely finite. Consider $P + 1$, and factor it into primes. Because all primes divide P, none of them divides $P + 1$ (any prime that divides both P and $P + 1$ would have to divide their difference, which is 1). Therefore, $P + 1$ is not divisible by any prime, a contradiction.   □

If we know the first n primes, then taking their product and adding 1 to it will give us a number that must be divisible by a prime not in our list. There is no guarantee that this product plus one is prime. In fact, it usually is not. The first such example is

$$2 \cdot 3 \cdot 5 \cdot 7 \cdot 11 \cdot 13 + 1 = 30031 = 59 \cdot 509 \, .$$

The next question to arise naturally is: How thickly are the primes spread among the integers? Are they plentiful or relatively scarce? The answer to this question is important because if they are scarce, that makes factorization easy: We can just trial-divide by the few primes that are around. On the other hand, a plentiful supply of primes is good for the many applications of number theory that require selecting a set of large primes. We saw one of these in Section 2.2 where the Chinese Remainder

Theorem was used with prime moduli to do high-precision arithmetic. The answer is that primes are plentiful, very plentiful.

The prime-counting function $\pi(x)$ gives the number of primes less than or equal to x. It is included in *Mathematica* as `PrimePi[x]`, but the CNT package allows $\pi[x]$ to be used as well. Computing $\pi(x)$ exactly is difficult when x is large (the basic algorithms are described at the end of Section 4.3. *Mathematica*'s implementation goes up to $9 \cdot 10^{13}$, but the CNT package includes the special values $\pi(10^{14})$, $\pi(10^{15})$, ..., $\pi(10^{20})$). The following computations show that there are 78,498 primes under one million and 2,220,819,602,560,918,840 primes under one hundred quintillion.

```
{π[10⁶], π[10²⁰]}
```

```
{78498, 2220819602560918840}
```

It is one of the most amazing facts of mathematics that the primes, though they appear randomly placed in the integers, have a growth rate that is incredibly smooth. (See the graphs in Figure 4.1, which were generated by simply plotting $\pi(x)$.)

Naturally the appearance of such a smooth curve (it is not straight: note the slight curvature it exhibits) makes us desperately want to know a formula that approximates it. This question has played a central role in number theory ever since Gauss examined tables of primes and came up with a conjecture about the long-term behavior of $\pi(x)$. We can make some similar investigations by asking how many primes there are in intervals of size 1000 near various powers of 10. Here is some raw data followed by a tabular view of same.

```
data =
  Table[{i, Count[Range[10ⁱ, 10ⁱ + 999], _?PrimeQ]}, {i, 0, 20}]
```

```
{{0, 168}, {1, 165}, {2, 159}, {3, 135}, {4, 106},
 {5, 81}, {6, 75}, {7, 61}, {8, 54}, {9, 49}, {10, 44},
 {11, 47}, {12, 37}, {13, 34}, {14, 30}, {15, 24},
 {16, 20}, {17, 27}, {18, 23}, {19, 28}, {20, 24}}
```

n	Number of Primes	n	Number of Primes	n	Number of Primes
0	168	7	61	14	30
1	165	8	54	15	24
2	159	9	49	16	20
3	135	10	44	17	27
4	106	11	47	18	23
5	81	12	37	19	28
6	75	13	34	20	24

Note that we cannot use `PrimePi` to make this computation because it does not work well beyond 10^{13}. Instead we are actually counting primes and living with the small risk of an incorrect output from `PrimeQ` beyond 10^{16}. There is a pattern to this data: as the power of 10

Plot[π[x], {x, 1, 10⁸}]

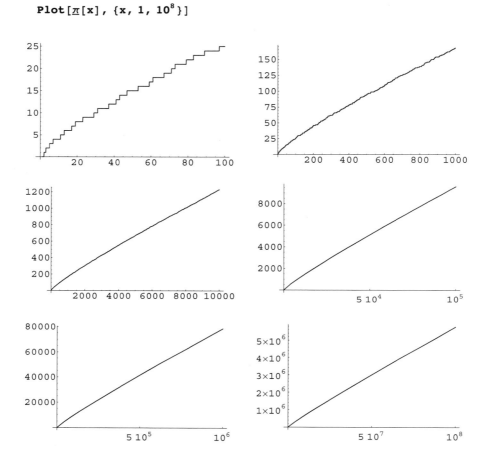

Figure 4.1. These graphs of $\pi(x)$ show that the growth rate of the primes is remarkably smooth.

doubles, the number of primes gets cut roughly in half. For example, near 10^4 there are 106 primes; near 10^8 there are 54. Because the first entry is the power of 10, this indicates that the relationship might be $k/(\log_{10} x)$, which is the same as $c/(\log x)$ [log x denotes natural logarithm, often written as ln x or $\log_e x$] . Now, the amazing thing is that the best value of c to make this function match the data is 1. In other words, the best logarithm to use is the logarithm to the base e — the chance of a number near x being prime is close to $1/\log x$. For example, $1/\log(10^8) = 0.0542868$, and because we are looking at intervals of size 1000, this would predict 54.3 primes in such an interval near 10^8; sure enough, there are 54 primes in $[10^8, 10^8 + 1000)$.

We can phrase this observation as: The probability that a number near x is prime is $1/(\log x)$. This would mean that $\pi(x)$ should be well approxi-

mated by $\int_2^x \frac{1}{\log t} dt$. This quantity is called the ***logarithmic integral***[1] of
x, denoted ***li(x)***, and available in *Mathematica* as `LogIntegral[x]`. The
CNT package defines the alias `li[x]` for `LogIntegral`. This function is not
too hard to compute but there is a rough approximation that is easier to
remember and to compute: $x/(\log x)$. This fraction is asymptotic to li(x),
where f is ***asymptotic*** to g means that $\lim_{x \to \infty} f(x)/g(x) = 1$ (Exercise 4.3).
Figure 4.2 shows graphs of the three functions $\pi(x)$, $x/(\log x)$, li(x).

```
Plot[{π[x], x / Log[x], li[x]}, {x, 2, 10000},
  PlotStyle → {{GrayLevel[0.8], AbsoluteThickness[3]},
    {AbsoluteThickness[2]}, {}}];
```

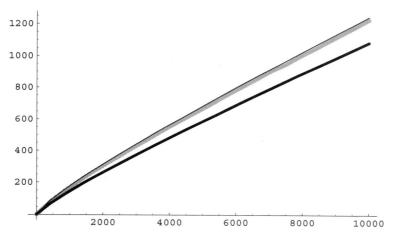

Figure 4.2. The graphs of $\pi(x)$ (gray), $x/(\log x)$ (thick), and li(x) (thin).

The fact that $\pi(x)$ is asymptotic to a simple function is a famous result.
It was conjectured by several mathematicians at the end of the eighteenth
century and not proved until the end of the nineteenth when Jacques
Hadamard (1865–1963) and Charles-Jean de la Vallée-Poussin (1866–1962)
independently completed the details of a proof that had been sketched by
Bernhard Riemann almost 40 years earlier. The result is now known as the
Prime Number Theorem.

Theorem 4.2. Prime Number Theorem. The number of primes less
than or equal to n is asymptotically equal to $n/(\log n)$.

---

[1]An alternative definition of the logarithmic integral (and the one used by
Mathematica's `LogIntegral` function) uses the interval from 0 to n, which can be
done even though $1/(\log t)$ has a vertical asymptote at $t = 1$. The difference between
starting at 0 or at 2 is only a little more than 1.045. For practical purposes, it does
not matter which definition is used.

The logarithmic integral is asymptotically equal to $n/(\log n)$, but is in fact a better approximation to $\pi(n)$. The table that follows compares the values of $\pi(n)$ with the nearest integers to $n/(\log n)$ and li(n); the error is given as a percentage. Note how well li(x) approximates $\pi(x)$.

n	$\pi(n)$	$n/\log n$	% error	li(n)	% error
10^2	25	22	12.	30	20.
10^3	168	145	14.	178	5.952
10^4	1229	1086	12.	1246	1.383
10^5	9592	8686	9.4	9630	0.3962
10^6	78498	72382	7.8	78628	0.1656
10^7	664579	620421	6.6	664918	0.05101
10^8	5761455	5428681	5.8	5762209	0.01309
10^9	50847534	48254942	5.1	50849235	0.003345
10^{10}	455052511	434294482	4.6	455055615	0.0006821
10^{11}	4118054813	3948131654	4.1	4118066401	0.0002814
10^{12}	37607912018	36191206825	3.8	37607950281	0.0001017
10^{13}	346065536839	334072678387	3.5	346065645810	0.00003149
10^{14}	3204941750802	3102103442166	3.2	3204942065692	9.825×10^{-6}
10^{15}	29844570422669	28952965460217	3.0	29844571475288	3.527×10^{-6}
10^{16}	279238341033925	271434051189532	2.8	279238344248557	1.151×10^{-6}
10^{17}	2623557157654233	2554673422960305	2.6	2623557165610822	3.033×10^{-7}
10^{18}	24739954287740860	24127471216847324	2.5	24739954309690415	8.872×10^{-8}
10^{19}	234057667276344607	228576043106974646	2.3	234057667376222382	4.267×10^{-8}
10^{20}	2220819602560918840	2171472409516259138	2.2	2220819602783663484	1.003×10^{-8}

While the proof of the Prime Number Theorem requires far more background than we can expect for this book, we will show with a slightly less accurate estimate that this formula makes sense. Its implications are profound. The number of 100-digit primes is closely approximated by li$(10^{100}) -$ li(10^{99}), which can be computed very quickly.

num = li[10.^{100}] - li[10^{99}]

3.92135×10^{97}

$$\frac{10^{100} - 10^{99}}{\textbf{num}}$$

229.513

So, one out of every 230 hundred-digit numbers is prime. When you consider that even numbers, multiples of 3, and multiples of 5 are all easy to recognize and none of them will be prime, the odds of choosing a prime just by randomly selecting one of the remaining hundred-digit integers goes up to one in 77. It is easy to find a hundred-digit prime. Generate hundred-digit numbers that are not divisible by 2, 3, or 5 and use the 2-pseudoprime test to check whether or not each is prime. Of course, actually *proving* that you have a prime is another matter. The CNT package has a RandomPrime function to get random d-digit primes, and these are certifiably prime, as discussed in Section 4.2.

<u>**RandomPrime**</u>[200]

615928921103364710401005258292446906431469251367927269346665\
177597056127483334924736672536984726193396670137468921634\
04704394348100782783846667487049206369338785579752512880290\
5734891784281324398871

More exact estimates of $\pi(x)$ quickly lead to deep mathematics, especially the notorious Riemann Hypothesis. When Riemann was investigating how one might prove the Prime Number Theorem, he used the function of s given by $\sum_{n=1}^{\infty} 1/n^s$, which today is known as the ***Riemann zeta function***. Written in this way, this function is defined only when the real part of s is strictly larger than 1, but Riemann showed that the function could be extended uniquely to all complex values of s. The Prime Number Theorem relies on the fact that this function is never 0 when the real part of s is 1. Riemann conjectured that this function is never 0 when the real part of s is larger than $^1/_2$, a statement that is equivalent to saying that for any $\varepsilon > 0$, there is a constant K_ε so that $|\pi(x) - \text{li}(x)| < K_\varepsilon x^{0.5+\varepsilon}$. This is the Riemann Hypothesis. It implies a regularity in the distribution of primes that has many important implications. For more information on these topics see [BS], [Rie], [Wag], and [Zag].

■ The Sieve of Eratosthenes

If you want to generate a table of all primes less than a given number, then the most efficient algorithm known today is the same as that known to Eratosthenes of Cyrene, now part of Libya. He was a Greek mathematician who lived from 276–194 BCE and taught in Alexandria in what is now Egypt. To find all primes less than or equal to n, we list all the integers from 2 to n. We then work our way down the list. The first integer (namely 2) must be prime. We cross off all multiples of 2 that are larger than 2. The first integer after 2 that has not been crossed off (namely 3) must be prime. We cross off all multiples of 3 that are larger than 3. We continue in this manner. When we have found a new prime, we cross off all multiples of that new prime that are larger than the prime itself and then move on to the next integer that has not been crossed off; again, it must be prime. The reason that this is efficient, as Eratosthenes realized, is that we do not have continue all the way up to n. When we have found a prime larger than $\sqrt{n}$, all of the remaining integers that have not been crossed off must be prime. Any composite integer must have a factor less than or equal to its square root. So if a positive integer does not have any divisors larger than 1 and less than or equal to its square root, then it must be prime. The following table (using blanks instead of 0s) shows how three sieving steps give us all the primes under 40; they are in the last column.

EratosthenesTable[40]

2	2	2	2
3	3	3	3
4			
5	5	5	5
6			
7	7	7	7
8			
9	9		
10			
11	11	11	11
12			
13	13	13	13
14			
15	15		
16			
17	17	17	17
18			
19	19	19	19
20			
21	21		
22			
23	23	23	23
24			
25	25	25	
26			
27	27		
28			
29	29	29	29
30			
31	31	31	31
32			
33	33		
34			
35	35	35	
36			
37	37	37	37
38			
39	39		
40			

Algorithm 4.1. The Sieve of Eratosthenes

This is the sieve of Eratosthenes, which finds all primes less than or equal to n. It begins by setting $L = \{2, 3, \ldots, n\}$. Starting with $i = 1$, if the ith entry of L is not 0, then it must be prime, call it p. Starting with p^2 in position $i^2 + 2i$, every pth entry of L is reset to 0. We increment i by 1 and repeat until $p > \sqrt{n}$; this requires $\pi(\sqrt{n})$ steps. The primes are all of the nonzero entries left in L.

```
Eratosthenes[n_] := Module[{L = Range[2, n], i = 1},
   While[L[[i]]² ≤ n,
     If[L[[i]] ≠ 0, Do[L[[k]] = 0, {k, i² + 2 i, n - 1, L[[i]]}]]; i++];
   Rest[Union[L]]]
```

Eratosthenes[500]

```
{2, 3, 5, 7, 11, 13, 17, 19, 23, 29, 31, 37, 41, 43, 47, 53,
 59, 61, 67, 71, 73, 79, 83, 89, 97, 101, 103, 107, 109,
 113, 127, 131, 137, 139, 149, 151, 157, 163, 167, 173,
 179, 181, 191, 193, 197, 199, 211, 223, 227, 229, 233,
 239, 241, 251, 257, 263, 269, 271, 277, 281, 283, 293,
 307, 311, 313, 317, 331, 337, 347, 349, 353, 359, 367,
 373, 379, 383, 389, 397, 401, 409, 419, 421, 431, 433,
 439, 443, 449, 457, 461, 463, 467, 479, 487, 491, 499}
```

This algorithm has some serious drawbacks. If n is very large, it requires a lot of memory. And if you want to use it to prove that n is prime, it would take approximately $\sqrt{n}$ steps. But it does possess a great strength that will come into play in the most powerful known factorization algorithms, the quadratic sieve and the number field sieve. All these sieves share the feature that they eliminate division and almost all multiplications from the step that sifts for the desired objects.

The sieve of Eratosthenes also plays a role in our understanding of the distribution of primes and the Prime Number Theorem. In number theory one often uses as a guide probabilistic statements about divisibility such as:

- About half the numbers between 2 and x are divisible by 2.
- About $\frac{1}{p}$ of the numbers between 2 and x are divisible by p.

These two examples are, when x is large, essentially true; but there is much more subtlety involved when one combines many such statements. For example, the sieve of Eratosthenes gets at the primes below x by removing the multiples of 2 (except 2 itself), then the multiples of 3 in what remains, then the multiples of 5 in what remains, and so on up to the last prime under $\sqrt{x}$. One might therefore think that the amount remaining at the end—the number of primes—would be approximately $x \prod_{p \leq \sqrt{x}} (1 - \frac{1}{p})$.

But this is false! It turns out that the true sieve is *more efficient* in its deletion of numbers than one would expect from the random model.

The reason for this surprising efficiency is that the deletions that occur in the true sieve are not really independent. For example, consider the integers to 10,000. The random view is that, say, $\frac{1}{11}\frac{1}{23}\frac{1}{41}$ of the numbers under x are divisible by 11, 23, and 41. But in reality there are *no such numbers*, because $11 \cdot 23 \cdot 41 > 10000$. This means that there is less overlap in the deletions, and so, at least most of the time, more numbers will be deleted than expected.

We can make a step-by-step comparison of the true sieve to the numbers predicted by the random model. Let us avoid issues related to small integers by considering a small interval around a large number; we'll use

$[10^8 - 5000, 10^8 + 5000]$. There are 1229 sieving primes — {2, 3, 5, ..., 9973} — and 548 primes in the interval. Here are the number of integers remaining after each sieving step. Figure 4.3 shows the difference between the predicted remainders using the random model and the actual remainders. The data show that the random model overestimates the number of integers not sieved out by 61, an error of 11%.

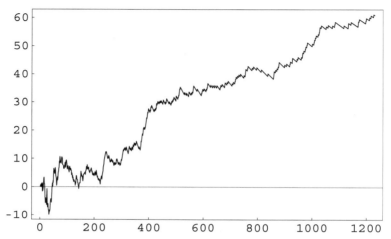

Figure 4.3. When sieving the integers in $[10^8 - 50000, 10^8 + 50000]$ for primes, the random model overestimates the number of true survivors at each step after the first few. Overall there is an overestimate of 61, a large overestimate because there are only 548 primes in the interval.

The following theorem shows exactly what has to be done to the random model to get a correct prediction. Recalling that the prime density near x is $1/(\log x)$, the theorem of Mertens (see [HW, thm. 4.29] for a proof) shows that by extending the product from $x^{0.5}$ to $x^{0.56}$ one gets the right answer.

Theorem 4.3. (F. Mertens, 1874) Let c be the constant $e^{-\gamma}$, where γ is Euler's constant, 0.577...; the numerical value of c is about 0.56.

(a) $\prod_{p \leq \sqrt{x}} \left(1 - \dfrac{1}{p}\right) \sim \dfrac{2c}{\log x}$

(b) $\prod_{p \leq x^c} \left(1 - \dfrac{1}{p}\right) \sim \dfrac{1}{\log x}$

Note that the assumption that the random model was a good predictor of the prime distribution might lead to a prediction, using part (a), that $\pi(x)$ would be asymptotic to $1.12x/(\log x)$. But the Prime Number Theorem tells us that the constant should not be there. One idea used to prove the Prime Number Theorem is to show directly (i.e., without the complex analysis used in the original 1896 proof) that $\pi(x)$ is asymptotic to the product $x \prod_{p \leq x^c} \left(1 - \frac{1}{p}\right)$. This is the essence of the celebrated Erdős–Selberg proof of the Prime Number Theorem in 1949.

◼ Chebyshev's Theorem and Bertrand's Postulate

An early approximate result is one due to Pafnuty Chebyshev (1821–1894). He found constants c_1 and c_2 such that

$$c_1 \frac{n}{\log n} < \pi(n) < c_2 \frac{n}{\log n}$$

for sufficiently large n. He was able to prove this result with $c_1 = 0.92$ and $c_2 = 1.106$. Since then improvements have been made on these constants. In 1989, N. Costa Pereira proved this result for $c_1 = 1 - 1/2976$, $c_2 = 1 + 1/2976$. To give the flavor of this approach, we present a proof for $c_2 = 1.7$. A proof for $c_1 = 2/3$ is outlined in Exercise 4.12.

Theorem 4.4. Chebyshev's Theorem. For any $x > 200$,

$$\frac{2}{3} \frac{x}{\log x} < \pi(x) < 1.7 \frac{x}{\log x} .$$

Proof. We will prove the upper bound, leaving the lower bound to Exercise 4.12. The condition on x is not necessary for the upper bound. The proof breaks into six short steps and is totally elementary.

(a) The result is true for $x < 1300$. This could be done by computing $\pi[\mathtt{x}]$, but it is more satisfying to do it from scratch, using an Eratosthenes sieve to get the primes below 1300. The following code looks at the primes and their positions in the list, and verifies that the result is true.

```
data = Eratosthenes[1300];
Apply[And, Table[x = data[[i]];
    Position[data, x][[1, 1]] < 1.7 x
                                 ------,  {i, Length[data]}]]
                                 Log[x]
```
```
True
```

(b) $\binom{2n}{n} < 4^n$. This follows from $4^n = (1 + 1)^{2n}$, which is the sum of *all* the binomial coefficients of the form $\binom{2n}{k}$. Note that this also proves $\binom{2n+1}{n} < 4^n$, because $\binom{2n+1}{n} = \binom{2n}{n} + \binom{2n}{n-1} < 4^n$.

(c) If $n \geq 1300$, then $3.09 \frac{n}{\log n} + 1 < 1.7 \frac{2n+1}{\log(2n+1)}$. The condition on n allows us to replace the left side of the inequality with $3.1 \frac{n}{\log n}$; we can also replace $2n + 1$ by $2n$ in the numerator of the right side, thus reducing the inequality to $\frac{3.1}{1.7 \cdot 2} < \frac{\log n}{\log(2n+1)}$. This is valid when $n = 1300$ and the derivative of the right side is positive, so it is valid beyond 1300.

(d) If $n \geq 1300$, then $3.09 \frac{n}{\log n} < 1.7 \frac{2n}{\log(2n)}$. To prove this, write $\log(2n)$ as $\log n + \log 2$ and simplify the difference to find the unique point where the sides are equal.

```
  2 * 1.7          3.09
------------- - ---------  // Together
Log[n] + Log[2]   Log[n]

 -2.14182 + 0.31 Log[n]
------------------------
Log[n] (Log[2] + Log[n])


E^2.14182/0.31

1001.34
```

(e) $\binom{2n}{n}$ is divisible by every prime p between n and $2n$. This is clear because any such prime appears in the numerator of the binomial coefficient and cannot be cancelled by the denominator.

Now we can complete the proof by induction, where we assume the result is true for n (where $2n \geq 1300$) and prove it for $2n$ and $2n + 1$. By (b) and (e):

$$4^n > \binom{2n}{n} \geq \prod_{n < p \leq 2n} p > \prod_{n < p \leq 2n} n = n^{\pi(2n) - \pi(n)} .$$

Take logarithms to get: $\pi(2n) - \pi(n) < 1.39 \frac{n}{\log n}$. But the inductive hypothesis (for n) gives $\pi(n) < 1.7 \frac{n}{\log n}$. So, using (d), $\pi(2n) < 3.09 \frac{n}{\log n} < 1.7 \frac{2n}{\log(2n)}$. For the case of $2n + 1$, combine the result for $2n$ with (c):

$$\pi(2n + 1) \leq \pi(2n) + 1 < 3.09 \frac{n}{\log n} + 1 < 3.10 \frac{n}{\log n} < 1.7 \frac{2n + 1}{\log(2n + 1)} . \quad \square$$

A related result is Bertrand's Postulate that for any integer $n \geq 2$, there always is a prime between n and $2n$. This result was conjectured by Joseph Louis François Bertrand (1822–1900) and proven by Chebyshev in 1850. A proof of this result due to Paul Erdős is outlined in Exercise 4.13. This result has a counterpoint in the observation that there are arbitrarily long strings of consecutive composite integers. To prove this just look at $k! + 2, k! + 3, \ldots, k! + k$, which is a sequence of $k - 1$ consecutive composites. Another approach is given in Exercise 2.22.

Exercises for Section 4.1

Note: In all exercises in this chapter p denotes a prime number.

4.1. Modify the proof that there are infinitely many primes to show that there are infinitely many primes that are congruent to 3 (mod 4). *Hint:*

Assume that there are only finitely many such primes: $p_1 = 3$, $p_2 = 7$, ..., p_r and consider $4 p_1 p_2 \cdots p_r - 1$.

4.2. What is wrong with the following proof that there are infinitely many primes that are congruent to 1 (mod 4)? (A correct proof is given in Theorem 6.4.)

> Assume that there are only finitely many primes congruent to 1 modulo 4: $p_1 = 5$, $p_2 = 13$, ..., p_r. Consider $4 p_1 p_2 \cdots p_r + 1$. This is not divisible by any of our primes, but it must be divisible by a prime congruent to $1 \pmod 4$, so the assumption that there are only finitely many such primes must be wrong.

4.3. Prove that li(x) is asymptotic to $x/(\log x)$. *Hint*: Use l'Hôpital's rule twice and the fact that the derivative of a definite integral can be computed by the fundamental theorem of calculus.

4.4. A number of the form $p_1 p_2 \cdots p_r + 1$ is called a **Euclid number**. They can be defined in *Mathematica* as follows.

```
Euclid[n_] := 1 + Apply[Times, Prime[Range[n]]]
```

Use a pseudoprime test to determine which of the first 100 Euclid numbers are (most likely) prime.

4.5. Euclid's proof of Theorem 4.1 can be used to build an infinite list of primes by choosing at each step the *least* prime dividing $P + 1$. The list then begins with 2, 3, 7, 43, 13 because 13 is the least divisor of $2 \cdot 3 \cdot 7 \cdot 43 + 1$. Investigate by computation the following open question of Dan Shanks (1991): Is it true that every prime appears in this list?

4.6. Prove that the infinite product $(1 - \frac{1}{2})(1 - \frac{1}{3})(1 - \frac{1}{5}) \cdots (1 - \frac{1}{p_n}) \cdots = 0$. This should be interpreted as: $\lim_{n \to \infty} (1 - \frac{1}{2})(1 - \frac{1}{3})(1 - \frac{1}{5}) \cdots (1 - \frac{1}{p_n}) = 0$. *Hint*: Use the geometric series formula to show that for each N, there is a k such that $\sum_{n=1}^{k} \frac{1}{n} < \prod_{i=1}^{N} \frac{1}{1 - 1/p_i}$. Show that as N approaches infinity, so does k. Then use the fact that $\sum_{n=1}^{\infty} \frac{1}{n}$ diverges.

4.7. Prove that $\sum_{p \text{ prime}} \frac{1}{p}$ diverges by assuming convergence and getting a contradiction as follows.

(a) $\sum_{p \text{ prime}} \left(-\log[1 - \frac{1}{p}] - \frac{1}{p} \right) = \infty$. (Use Exercise 4.6.)

(b) Use the Maclaurin series for $\log(1 + x)$ to express the general term of the series in (a) is $\frac{1}{2p^2} + \frac{1}{3p^3} + \frac{1}{4p^4} \cdots$.

(c) Use the geometric series formula to see that the general term in (b) is less than $\frac{1}{2p(p-1)}$, and such terms form a convergent series, contradiction.

4.8. Use a pseudoprime test to get the likely primes in the first 10,000 integers beyond a googol (10^{100}). How does the number of such probable primes compare with the heuristic value for the expected number of primes in this interval?

4.9. Generate some graphs of the error when using li(x) to estimate $\pi(x)$. Does li(x) overestimate or underestimate $\pi(x)$? An amazing theorem of Littlewood asserts that the sign of li(x) − $\pi(x)$ changes sign infinitely often. The place where the sign-change first occurs is called the Skewes number; we do not know much about this number, except that it must be less than 10^{371}. Note the amazing moral here: millions of hours of computations can lead one to believe in a conclusion [in this case, that li(x) is always larger than $\pi(x)$] that is false.

4.10. Dirichlet's Theorem (1837) states that any arithmetic progression $\{ak + b : k = 0, 1, 2, ...\}$ where $\gcd(a, b) = 1$ contains infinitely many primes. Therefore, there are infinitely many primes of the form $4k + 1$ and of the form $4k + 3$. Do some computations and create some graphs to investigate the race between these two families of primes. Can you find n such that in the interval $[2, n]$ the primes congruent to 3 (mod 4) are outnumbered by those congruent to 1 (mod 4)? In fact, a famous theorem of Littlewood asserts that in such prime number races, the lead changes hands infinitely often. Investigate some random 50-digit primes. Are they more likely to be congruent to 3 (mod 4) than 1 (mod 4)? Littlewood's theorem was extended recently by M. Rubenstein and P. Sarnak [RS] who showed that if we assume a generalization of the Riemann Hypothesis, then, in the prime number race for congruence classes modulo n, any configuration of the coprime residue classes shows up as the race order infinitely many times. Generate some graphics to illustrate the leader in the mod-5 race; that is, study which congruence class mod-5 is dominant among the first n primes.

4.11. Perform some computations that show that the random sieve over-estimates the number of primes in a large interval by 12%, or, more precisely, $2e^{-\gamma}$.

4.12. Complete the proof that if $x > 200$, then $\frac{2}{3}\frac{x}{\log x} < \pi(x)$ by following steps (a)–(d).

(a) Prove that if p^e divides $\binom{n}{k}$ then $p^e \le n$ (use the formula — see Exercise 1.24 — for the number of primes in a factorial).

(b) Use (a) to deduce that $\binom{n}{k} \le n^{\pi(n)}$.

(c) Use (b) and the fact that $2^n = \sum_{k=0}^{n} \binom{n}{k}$ to show that $2^n \le (n+1)n^{\pi(n)}$.

(d) Take logarithms in (c) to obtain the desired result, using the condition on 200 to get the needed inequality involving logarithms.

4.13. This exercise proves Bertrand's Postulate: If $n \ge 2$, then there is a prime between n and $2n$.

(a) Prove that $\prod_{p \le n} p < 4^n$. This can be proved by induction on n. Verify that it is correct for $n \le 3$. Show that if it is true for n when n is odd, then it is also true for $n + 1$. The only tricky step is from $n = 2m$ to $n + 1 = 2m + 1$. Prove that if the inequality holds for all integers less than or equal to $n = 2m$, then

$$\prod_{p \leq 2m+1} p \leq \binom{2m+1}{m} \prod_{p \leq m+1} p < \binom{2m+1}{m} 4^{m+1} .$$

Use the inequality mentioned in the proof of part (b) in Theorem 4.4 to finish the proof.

(b) Prove that if $n > 2$ and p is a prime such that $\frac{2}{3}n < p \leq n$, then p does not divide $\binom{2n}{n}$. It follows that if there are no primes between n and $2n$, then all prime divisors of $\binom{2n}{n}$ are less than or equal to $\frac{2}{3}n$.

(c) Use Exercise 4.12(a) to show that if $e(p)$, the largest power of p that divides $\binom{2n}{n}$, is greater than 1, then $p < \sqrt{2n}$. Then use part (b) to show that if there are no primes between n and $2n$, then

$$\binom{2n}{n} \leq (2n)^{\lfloor \sqrt{2n} \rfloor} \prod_{p \leq \frac{2}{3}n} p .$$

(d) Use induction to prove that $\binom{2m}{m} > \frac{4^m}{2m+1}$.

(e) Use parts (a), (c), and (d) to prove that if there are no primes between n and $2n$, then

$$\frac{4^n}{2n+1} < (2n)^{\lfloor \sqrt{2n} \rfloor} \prod_{p \leq \frac{2}{3}n} p < 4^{2n/3} (2n)^{\sqrt{2n}} .$$

This implies that

$$4^{n/3} < (2n+1)(2n)^{\sqrt{2n}} < (2n)^{2+\sqrt{2n}} .$$

Show that this inequality cannot hold for any $n \geq 750$, and therefore Bertrand's Postulate is valid for $n \geq 750$.

(f) Show that Bertrand's Postulate holds for all n larger than 1 and less than 750.

4.2 Prime Testing and Certification

▦ Strong Pseudoprimes

We have seen that Fermat's Little Theorem is effective for identifying composite integers and is usually reliable when trying to find primes. Given a random integer n of, say, 100 digits, if n satisfies $2^{n-1} \equiv 1 \pmod{n}$, then the odds are overwhelming that n is prime. But overwhelming likelihood is not certainty. We can improve the odds by checking some other bases: $b^{n-1} \equiv 1 \pmod{n}$. In some cryptographic applications it would be safer to use randomly chosen bases rather than just starting at 2 and moving up. There is no composite integer that is guaranteed to pass the pseudoprime test for all bases. Even Carmichael numbers will fail the pseudoprime test for all bases that are not relatively prime to the number being tested. However, the existence of Carmichael numbers implies that there could be relatively few bases for which a composite integer is revealed to be composite.

A very strong improvement on the pseudoprime test comes out of the simple observation that all primes except 2 are odd. Let n be an odd integer that we suspect might be prime. We can write n as $2m + 1$. If n really is prime, then Fermat's Little Theorem tells us that for any base b between 1 and n, $b^{2m} \equiv 1 \pmod{n}$. This means that b^m is a mod-n square root of 1; but we know (Proposition 2.4) that there are only two such square roots, ± 1. So if it turns out that b^m is congruent to something other than ± 1, we have a proof that n is composite. Moreover, if b^m does turn out to be congruent to $+1$ and if m is even, say $m = 2k$, we can apply the same reasoning to b^k: It must be $\pm 1 \pmod{n}$ if n is prime. And so on down.

Consider 341, the first 2-pseudoprime. Because $341 = 2 \cdot 170 + 1$, we can look at $2^{170} \pmod{341}$.

> **PowerMod[2, 170, 341]**
>
> 1

This doesn't help. But $170 = 2 \cdot 85$ so we next check out 2^{85}.

> **PowerMod[2, 85, 341]**
>
> 32

Jackpot! 32 is not $\pm 1 \pmod{341}$, so this *proves* that 341 is composite.

The general situation is described as follows. Given n, an odd integer, and b, write $n - 1$ as $2^r m$ where m is odd. Then look at the mod-n sequence b^m, b^{2m}, b^{4m}, ..., $b^{2^r m} = b^{n-1}$. Let us call this the **b-sequence** for n. If this sequence fails to end in $+1$, or ever has a $+1$ preceded by an integer that is not ± 1, then n is definitely composite. Otherwise n is exhibiting prime-like behavior; if such an n is not prime, it is called a **b-strong pseudoprime**, or **b-spsp**. Note that the sequence is very fast to compute because each term is just the square of the preceding term. The following table shows the different sorts of behaviors, where $*$ denotes an integer not congruent to ± 1. The observant reader will note that one type of behavior is missing from the table: the case where $b^{n-1} \equiv -1 \pmod{n}$. In fact, if n is odd then for any b, $b^{n-1} \not\equiv -1 \pmod{n}$; see Exercise 4.18.

b^m	b^{2m}	b^{4m}	b^{8m}	$\cdot$	$\cdot$	$\cdot$	$\cdot$	$\cdot$	$\cdot$	$\cdot$	$\cdot$	b^{n-1}	
+1	+1	+1	+1	+1	+1	+1	+1	+1	$\cdot$	$\cdot$	$\cdot$	+1	prime or b-spsp: b^m is $+1$
$*$	$*$	$*$	$\cdot$	$\cdot$	$\cdot$	$*$	-1	$+1$	$\cdot$	$\cdot$	$\cdot$	+1	prime or b-spsp: $b^{m\,2^k}$ is -1 for some $k < r$
$*$	$*$	$*$	$\cdot$	$\cdot$	$\cdot$	$*$	$+1$	$+1$	$\cdot$	$\cdot$	$\cdot$	+1	Composite; an entry different from ± 1 squares to 1
$*$	$*$	$*$	$\cdot$	$\cdot$	$\cdot$	$*$	$*$	$*$	$\cdot$	$\cdot$	$\cdot$	$*$	Composite; b^{n-1} is not ± 1

Strong pseudoprimes exist. If $b = 2$ then 2047, which factors as $23 \cdot 89$, is a 2-spsp. Because $2046 = 2 \cdot 1023 + 1$, we can get the 2-sequence as follows.

PowerMod[2, {1023, 2046}, 2047]

{1, 1}

The behavior of this sequence is primelike, so 2047 is a 2-strong pseudoprime.

One uses the strong pseudoprime idea to gain evidence for primality by checking n using a few bases. If n appears to be prime according to the strong pseudoprime criterion, then it is extremely likely that n is prime. For example, under 10^{16} there are only 52,593 composite integers[1] n that are both a 2-spsp and a 3-spsp. Moreover, this test runs essentially as fast as the ordinary pseudoprime test. Unlike the case for the ordinary pseudoprime test, it makes sense to consider composite values of b that are not squares (see Exercises 4.15 and 4.16); but it is efficient to look only at the prime bases.

Algorithm 4.2 The Strong Pseudoprime Test for Primality

Choose bases b in order from the primes 2, 3, 5, ..., and find r and odd m such that $n - 1 = 2^r m$. Then examine the b-sequence for n to see if the first $+1$ is preceded by a -1; if it is not or if there is no $+1$, n is proved composite. Otherwise n is either prime or a b-strong pseudoprime. Several bases can be used to increase the likelihood of correctness. The following code

---

[1] These integers are provided in the StrongPseudoprimeData package on the disk accompanying this book. When that package is loaded — it is done automatically as needed within the CNT package — one can use StrongPseudoprimes[{2,3}, max] to get all 2- and 3-strong psps under max, where max $\leq 10^{16}$. This data set, computed by Daniel Bleichenbacher [Ble], can be useful in finding special pseudoprimes (as in Exercise 4.15) or in evaluating new primality tests.

includes the `SpspSequence` function so that we can look at some *b*-sequences if we so desire. These functions are in the `CNT` package too.

The `IntegerExponent[n, b]` function (new in version 4.0; included in `CNT` package for version 3 users) returns the highest power of *b* that divides *n*. `NestList` is used to iterate the modular squaring function, and so generates the *b*-sequence. Then `MemberQ` and `Or` (`||`) are used to check whether the *b*-sequence indicates strong pseudoprimality.

```
SpspSequence[b_, n_] := Module[{s = IntegerExponent[n - 1, 2]},
  NestList[Mod[#², n] &,
    PowerMod[b, (n - 1) / 2ˢ, n], s] /. n - 1 → -1]

StrongPseudoprimeQ[b_, n_] := If[PrimeQ[n] || n == 1, False,
  bSeq = SpspSequence[b, n];
  Union[bSeq] == {1} || MemberQ[bSeq, -1]]
```

```
SpspSequence[2, 7919]
```

```
{1, 1}
```

Now we can easily find the 2-strong pseudoprimes among the first 10,000 integers.

```
Select[Range[3, 10000, 2], StrongPseudoprimeQ[2, #] &]
```

```
{2047, 3277, 4033, 4681, 8321}
```

The test is certainly imperfect, for there are five false positives. But the 2-spsp test does have an impressive success rate of 99.95% on this interval. There are several 3-spsps in this interval too, but if we run the test using 2 and 3 together, it is formidable indeed. The first bad integer for this double test is 1,373,653. And below 10^{16} the odds of being fooled when using the bases 2 and 3 together are 1 in 200 billion.

```
StrongPseudoprimeQ[{2, 3}, 1373653]
```

```
{True, True}
```

```
{SpspSequence[2, 1373653], SpspSequence[3, 1373653]}
```

```
{{890592, -1, 1}, {1, 1, 1}}
```

```
FactorInteger[1373653]
```

```
{{829, 1}, {1657, 1}}
```

Indeed, if we made a test that combined the base-*b* strong pseudoprime tests for *b* = 2, 3, 5, 7, 11 then that test would be very efficient, as there are no counterexamples less than $25 \cdot 10^9$. The composite integer 3215031751 is the first integer that is a *b*-spsp for *b* = 2, 3, 5, and 7. It is caught by *b* = 11. In such a case we would say that 11 is the first ***witness*** to the composite-ness of the integer in question.

```
StrongPseudoprimeQ[Range[2, 12], 3215031751]
```

```
{True, True, True, True, True,
 True, True, True, True, False, True}
```

```
FactorInteger[3215031751]
```

```
{{151, 1}, {751, 1}, {28351, 1}}
```

A record of sorts was set by the composite number $n = pqr$ (found by D. Bleichenbacher [Ble]), where p, q, and r are the following primes.

```
p = 18215745452589259639;
q = 4337082250616490391;
r = 867416450123298079;
n = p q r;
```

The first witness for this 56-digit composite number is 101.

```
Select[Range[101], ! StrongPseudoprimeQ[#, n] &, 1]
```

```
{101}
```

There is also a package function that gives the first strong psp witness.

```
FirstSPSPWitness[n]
```

```
101
```

It is strongly suspected (indeed, it follows from the unproved extended Riemann Hypothesis) that checking all bases up to $2(\log n)^2$ is a true test of primality; such a test would run in polynomial time. Exercise 4.15 shows that, below 10^{16}, the first strong pseudoprime witness for n is less than $1.66(\log n)^2$. If we avoid small numbers by starting at the first 3-strong pseudoprime (2047), the witness upper bound below 10^{16} becomes $0.052(\log n)^2$.

Now we have a better understanding of *Mathematica*'s PrimeQ function. It combines a 2-spsp test, a 3-spsp test, and a Lucas pseudoprime test (a detailed discussion of Lucas pseudoprimes is in Chapter 8). There is no known counterexample to the assertion that these three tests pass the primes and only the primes; it has been checked by D. Bleichenbacher that there is no counterexample under 10^{16}. Finally, we note that for any fixed set of bases $\mathcal{B}$ there are infinitely many numbers that are b-strong pseudoprimes for each $b \in \mathcal{B}$ [AGP].

▣ Industrial-Grade Primes

Strong pseudoprimes are much rarer than pseudoprimes. Below one million, there are 2445 2-pseudoprimes but only 46 2-strong pseudoprimes. Some years ago, Pomerance, Selfridge, and Wagstaff [PSW] took the search for 2-pseudoprimes up to 25 billion. They found 21853 pseudoprimes and only 4842 strong pseudoprimes. Even better is the fact that there is no analog of Carmichael numbers for strong pseudoprimes. Following Bach and Shallit [BS], we shall call a base b a ***strong liar*** for the composite integer n if n passes the strong pseudoprime test for the base b. As we shall show in Section 4.3, the set of strong liars for a composite n cannot be larger than $(n-1)/4$, and, in fact, it is usually much smaller than this. Thus the probability of a composite integer passing the strong pseudoprime test using m random bases is at most 4^{-m}. If we run this test 10 times and n passes each time, we can be 99.99999% confident that n is prime, for if it were not, then each of the random choices of b would have have to come from a set of integers contained in $[1, n]$ whose size is no greater than $n/4$. The chance of this happening in a specific case is at most $1/4$, and so the chance of this happening in 10 independent trials is no greater than 4^{-10}, or about 10^{-7}. If an integer passes this test for 100 random choices of b, then the probability that it is not prime is less than 10^{-60}. This probabilistic approach is known as the ***Miller–Rabin test***, and the `MillerRabin-PrimeQ` function from the `CNT` package performs this test any number of times.

```
n = 10^80 + 129;
MillerRabinPrimeQ[n, 100]

True
```

The odds that each of the 100 random integers was a strong liar is so remote that, for any practical purpose, we can be certain that $10^{80} + 129$ is prime.

The issue has been raised whether or not this is good enough. Remember that a positive probability of failure surrounds any project carried out on a computer because no computer program is absolutely 100% guaranteed to run correctly. In practice, several strong pseudoprime tests are good enough for those times — such as constructing an RSA cryptosystem (discussed in Chapter 5) — when one needs to identify large primes. Henri Cohen has called such highly probable primes ***industrial-grade primes***. But it would be nice to be able to prove rigorously (called: ***certifying*** a prime) that a given integer is prime. The remainder of this chapter will look at tests that give us absolute certainty. These tests share the drawback that they do not identify composite integers. They simply run indefinitely when a composite integer is the input. For this reason, you will want to use a pseudoprime test before attempting to certify primality. You need to know that your number is probably prime before you try to prove that it is prime.

▣ Prime Certification Via Primitive Roots

So far our discussion of prime testing has focused on methods that tell us that n is probably prime. But suppose we wish to have an *absolute* proof of primality. Such a proof is often called a ***certificate of primality***, the point being that the certificate should contain enough information so that it can quickly be turned into a rigorous proof of primality. We now discuss one way to get certificates for primes p for which we can find the complete factorization of $p - 1$.

The method that we shall focus on here is based on the observation that n is prime if and only if $\phi(n) = n - 1$. As we have seen in Exercise 3.3, the only practical means of calculating $\phi(n)$ is first to find the complete factorization of n, which seems to defeat our purpose. But if, for an integer n, we can find b such that $\mathrm{ord}_n(b) = n - 1$, then we will know that n is prime, since $\mathrm{ord}_n(b)$ divides $\phi(n)$ and so this proves that $\phi(n) = n - 1$. And if n is prime, there definitely exists such a b: any primitive root will do (see Section 3.5). The primitive roots are fairly common: there are $\phi(n - 1)$ of them by Corollary 3.13. So to prove n is prime all we have to do is find one of these primitive roots and verify that its order really is $n - 1$.

Let b be between 1 and n. We want to see if $n - 1$ is the order of b modulo n. If it is, we have our proof; if not, we keep looking. We could try all integer powers up to and including $n - 2$ and check that none is congruent to 1, but this is neither practical nor necessary. If $b^{n-1} \equiv 1 \pmod{n}$, then the true mod-$n$ order of b must divide $n - 1$. This means that in our search for a smaller order we can restrict ourselves to the divisors of $n - 1$.

We can do much better than this. If we know that the factorization of $n - 1$ is $p_1^{e_1} p_2^{e_2} \cdots p_r^{e_r}$, where $p_1 = 2$, and if the order of b is not $n - 1$, then, because $\mathrm{ord}_n(b)$ divides $n - 1$, there must be at least one p_i such that $(n - 1)/p_i$ is a multiple of the order of b. This means that if b's order is not $n - 1$, then $b^{(n-1)/p_i} \equiv 1 \pmod{n}$ for some i, $1 \le i \le r$. We will sometimes call b a **witness** to n's primality (the term "witness" in the context of strong pseudoprimes is used to refer to a witness for compositeness). The theorem that follows gives a slight improvement to the preceding criterion by reducing the size of the exponents from $(n - 1)/p_i$ to $(n - 1)/(2p_i)$.

Theorem 4.5. Suppose that n is odd and $n - 1$ has the prime factorization $p_1^{e_1} p_2^{e_2} \cdots p_r^{e_r}$, where $p_1 = 2$. Suppose that b has the following property: $b^{(n-1)/2} \equiv -1 \pmod{n}$ but for each $i > 2$, $b^{(n-1)/(2p_i)} \not\equiv -1 \pmod{n}$. Then $\mathrm{ord}_n(b) = n - 1$. It follows that $\phi(n) = n - 1$ and so n is prime.

Proof. The first congruence tells us that $\mathrm{ord}_n(b)$ divides $n - 1$ but does not divide $(n - 1)/2$. It follows that 2^{e_1} divides $\mathrm{ord}_n(b)$. We claim that the remaining congruences imply that $\mathrm{ord}_n(b)$ does not divide any $(n - 1)/p_i$; it will follow that each $p_i^{e_i}$ divides $\mathrm{ord}_n(b)$, and this will mean that $n - 1$ divides $\mathrm{ord}_n(b)$, and therefore $\mathrm{ord}_n(b) = n - 1$, completing the proof. To prove the claim, let $b^{(n-1)/(2p_i)} \equiv a \pmod{n}$, where $a \not\equiv -1$, and suppose that $b^{(n-1)/p_i} \equiv 1 \pmod{n}$. Then $a^2 \equiv 1 \pmod{n}$ and $a^{p_i} \equiv -1 \pmod{n}$; but $a^{p_i} = (a^2)^{(p_i-1)/2} a \equiv a \pmod{n}$, a contradiction because $a \not\equiv -1$. $\square$

The preceding theorem leads to the following algorithm for proving a number is prime.

Algorithm 4.3 Prime Certification via Primitive Roots

This algorithm terminates when it has found an element of order $n-1$ modulo n, which proves that n is prime. It will not terminate if n is not prime. We use oddprimes to denote the set of odd primes that divide $n-1$. StringForm["abcd``efg``", x, y] produces the string where x and y are substituted for the two ``s in order; this can be used to create informative output. The _?OddQ is used to restrict the function to odd arguments, but it is worth modifying the code so that it recognizes (most) composite integers; see Exercise 4.20. Because PowerMod is a "listable" function we can use a list as the second argument, and check whether 1 occurs in the resulting list.

```
PrimeQWithProof[n_?OddQ] := Module[
  {oddprimes = Rest[Map[First, FactorInteger[n - 1]]], b = 2},
  While[PowerMod[b, (n - 1) / 2, n] ≠ n - 1 ||
    MemberQ[PowerMod[b, (n - 1) / (2 oddprimes), n], n - 1], b++];
  StringForm["n = `` is prime; `` has order n - 1.", n, b]]
```

This algorithm requires the factorization of $n-1$. That will be difficult for a random large prime, say the first prime beyond 10^{50}. But for some special classes of primes this presents no problem at all. For example, we can look at the 11th and 75th Euclid numbers, in which case $n-1$ is very easy to factor. The following computations provide witnesses to the primality of these large integers.

```
PrimeQWithProof[Euclid[11]]

n = 200560490131 is prime; 34 has order n - 1.


PrimeQWithProof[Euclid[75]]

n =
  171962010545840643344833405683175430195845756358957425604 38`
  771105058321655238562613083979651479555788009994557822024`
  565226932906295208262756822275663694111
  is prime; 19 has order n - 1.
```

Note that to be a useful certificate, the output should include the prime factorization of $n-1$; for, with that in hand, one could easily check that the claimed witness really has maximal order. This point will be discussed in more detail in the Pratt certificate subsection that follows.

As we saw in Exercise 3.6, the average value of $\phi(m)$ is $6m/\pi^2$. Of course, $m = n-1$ is not arbitrary. It must be even, and for even integers the average value of $\phi(m)$ is $4m/\pi^2$. Experimental evidence suggests that when m is one less than a prime, then $\phi(m)$ is about $0.37m$, still a very high fraction of all the residues. Thus one can expect to find a primitive root quickly. The difficulty is not in finding a primitive root when n is prime. The drawback to this algorithm is that you have to be able to factor $n-1$ in order to *prove* that n is prime. If n has fewer than 50 digits, then this is often possible with the methods that will shall investigate in this book. When n has more than 50 digits, you need to switch to a more sophisticated technique such as those described briefly at the end of this chapter.

An Improvement

It is possible with a little additional insight to make some improvements to Algorithm 4.3. An observation made by Brillhart, Lehmer, and Selfridge in 1975 is that we do not really need to find a primitive root. It is sufficient if for each of our odd prime divisors of $n-1$, p_2 through p_r, we can find a b_i for which $b_i^{(n-1)/2} \equiv -1 \pmod{n}$ and $b_i^{(n-1)/(2p_i)} \not\equiv -1 \pmod{n}$. The first congruence tells us that the order of b_2 divides $n-1$ but not $(n-1)/2$. This implies that 2^{e_1} must divide the order of b_2. Because the order of b_2 divides $\phi(n)$, 2^{e_1} must also divide $\phi(n)$. The second congruence tells us (by the same reasoning as in the proof of Theorem 4.4) that the order of b_i (which, from the first congruence, must divide $n-1$) does not divide $(n-1)/p_i$, and so the order of b_i is divisible by $p_i^{e_i}$. This implies that $\phi(n)$ is divisible by $p_i^{e_i}$. When we have found such a b_i for each odd prime divisor of $n-1$, we will have established that $n-1$ divides $\phi(n)$, which can only happen if n is prime. This argument fails if n has the form $2^m + 1$, for then there is no odd prime p_2. In that case we use the method of Algorithm 4.3.

The algorithm given below is the modified form of Algorithm 4.3 in which the *certificate of primality* is the list of bases that work for each of the odd prime divisors of $n-1$.

Algorithm 4.4 Improved Prime Certification via Primitive Roots

This algorithm terminates when it has found a proof that $\phi(n) = n-1$, which proves that n is prime. It will not terminate if n is not prime. The If[] construction deals with the case that $n-1$ is a power of 2. Scan[Function[k, bvals[k] = b], P1] scans the list P1 and applies the function defined by Function[] to each element p in the list. That function assigns a b-value to bvals[p]. The While construction starts with the full set of odd prime divisors and deletes them as witnesses are found; once that set becomes empty the complete set of witnesses has been found.

```
BLSPrimeProof[n_] := Module[{b = 2, bvals,
  P = oddprimes = Rest[Map[First, FactorInteger[n - 1]]]},
 If[P == {}, While[PowerMod[b, (n - 1) / 2, n] ≠ n - 1, b++];
  StringForm[ "n = `` is prime; the prime divisors of n-1
    and their witnesses are ``", n, Transpose[{{2, b}}]],
  While[P ≠ {}, If[PowerMod[b, (n - 1) / 2, n] == n - 1 ,
   P1 = Select[P, PowerMod[b, (n - 1) / (2 #), n] ≠ n - 1 &];
    Scan[Function[k, bvals[k] = b], P1]; P = Complement[P, P1]];
  b++]; StringForm["n = `` is prime; the odd prime
    divisors of n-1 and their witnesses are ``", n,
 Transpose[{oddprimes, Map[bvals, oddprimes]}]]]]]
```

This variation can be many times faster than Algorithm 4.3.

BLSPrimeProof[131]

```
n = 131 is prime; the odd prime divisors
   of n-1 and their witnesses are {{5, 2}, {13, 2}}
```

BLSPrimeProof[**Euclid**[11]] // **Timing**
PrimeQWithProof[**Euclid**[11]] // **Timing**

```
{0.0166667 Second,
 n = 200560490131 is prime; the odd prime divisors of n-1
   and their witnesses are {{3, 2}, {5, 3}, {7, 2}, {11, 2},
     {13, 2}, {17, 2}, {19, 2}, {23, 2}, {29, 2}, {31, 2}}}

{0.3 Second, n = 200560490131 is prime; 34 has order n - 1.}
```

Pratt Certificates

The primality test of Algorithm 4.3 does not really produce a certificate because the output is not detailed enough for the user to verify that Theorem 4.5 holds. For that, the user would need to know the prime factorization of $n - 1$, and that information was not provided. Moreover, even if that factorization was provided, how would the user know with certainty that the primes in the list are really prime. In 1975, Vaughan Pratt made the nifty observation that the same method can be used on those primes! That is, we can use a recursive technique to certify n, where the same procedure is used to certify the primes in the factorization of $n - 1$. This leads to a tree structure for the certificate. The `PrattCertificate` function from the CNT package can be used to get such trees; here is an example for the 1000th prime.

PrattCertificate[7919, **Tree → True**]

```
7919 → 7
    2
    37 → 2
        2
        3 → 2
            2
    107 → 2
        2
        53 → 2
            2
            13 → 2
                2
                3 → 2
                    2
```

Here is how to read this tree. The primes 2, 37, and 107 are the primes that occur in the factorization of 7918 (some of them could occur to a power higher than 1, but that is not shown as it is not relevant). The number 7 to the right of the first arrow is the witness for the primality of 7919 in that it has full order; that is, it satisfies the congruence conditions of Theorem 4.5. Now the same rules apply recursively: 2 and 53 are the prime factors of 106 and 2 is the value of b showing that 107 is prime. Note that 2 is considered to be self-certifying: everyone knows that 2 is prime.

A slight modification to the code of Algorithm 4.3 yields code that produces these Pratt certificates. Of course, we repeat the drawback that one must be able to factor $n - 1$ for this to work.

Algorithm 4.5. Pratt Certificates of Primality

We call this `PrattCertificateBasic` to avoid a clash with the package function. It is essential to make w and `primes` local variables (via `Module`) so that they are distinct variables at every stage of the recursion. The routine finds a witness w using the citerion of Theorem 4.4, and then calls itself on the primes showing up in the factorization of $n - 1$ to certify them in the same way. We make `PrattCertificateBasic` listable so that we can use it on a list of primes (the last line of the code). We also check the input to see that `PrimeQ` thinks it is prime, for otherwise the search for a witness will never end. The output is a nested list.

```
Attributes[PrattCertificateBasic] = Listable;

PrattCertificateBasic[2] = 2;

PrattCertificateBasic[p_Integer?PrimeQ] :=
  Module[{w = 2, primes = First /@ FactorInteger[p - 1]},
    While[PowerMod[w, (p - 1) / 2, p] ≠ p - 1 || MemberQ[
      PowerMod[w, (p - 1) / (2 Rest[primes]), p], p - 1], w++];
    {p, w, PrattCertificateBasic[primes]}] /; p > 2
```

Here is an example (we give the straight-line form followed by the tree form) that certifies the primality of $10^{30} + 57$.

The version in the CNT package gives the certificate in the form of a tree. Because $p - 1$ is even, there will be a 2 in each list of primes; because 2 is self-certifying we may as well shorten our certificates by suppressing the 2s. The package version has an option to do this.

```
PrattCertificate[10^30 + 57, Show2s → False]
{1000000000000000000000000000057, 5,
  {{3, 2}, {79043, 2,
    {{39521, 3, {{5, 2}, {13, 2, {{3, 2}}}, {19, 2, {{3, 2}}}}}}},
  {3998741, 2, {{5, 2}, {17, 3}, {19, 2, {{3, 2}}},
    {619, 2, {{3, 2}, {103, 5, {{3, 2}, {17, 3}}}}}}},
  {290240017, 10, {{3, 2}, {11, 2, {{5, 2}}},
    {181, 2, {{3, 2}, {5, 2}}}, {3037, 2,
    {{3, 2}, {11, 2, {{5, 2}}}, {23, 5, {{11, 2, {{5, 2}}}}}}}}},
  {454197539, 2, {{331, 3, {{3, 2}, {5, 2}, {11, 2, {{5, 2}}}}},
    {686099, 2, {{7, 3, {{3, 2}}},
      {7001, 3, {{5, 2}, {7, 3, {{3, 2}}}}}}}}}}}
```

PrattCertificate[10^{30} + 57, Tree → True, Show2s → False]

```
1000000000000000000000000000057 → 5
                              3 → 2
                          79043 → 2
                              39521 → 3
                                    5 → 2
                                   13 → 2
                                       3 → 2
                                   19 → 2
                                       3 → 2
                        3998741 → 2
                              5 → 2
                             17 → 3
                             19 → 2
                                 3 → 2
                            619 → 2
                                3 → 2
                              103 → 3
                                    3 → 2
                                   17 → 3
                      290240017 → 5
                              3 → 2
                             11 → 2
                                 5 → 2
                            181 → 2
                                3 → 2
                                5 → 2
                           3037 → 2
                                3 → 2
                               11 → 2
                                    5 → 2
                               23 → 5
                                   11 → 2
                                        5 → 2
                      454197539 → 2
                            331 → 2
                                3 → 2
                                5 → 2
                               11 → 2
                                    5 → 2
                         686099 → 2
                                7 → 3
                                    3 → 2
                             7001 → 3
                                    5 → 2
                                    7 → 3
                                        3 → 2
```

We repeat that the information in this table can be used to, very quickly, get a complete and rigorous proof of the primality of 10^{30} + 57: just check that the primes in the vertical columns do exhaust the various numbers of the form $n - 1$, and then check that the congruences of Theorem 4.5 are satisfied by the various witnesses to the right of the arrows. Note that finding these witnesses is generally quite easy: most of the time 2 works. It is a good exercise in recursive programming to write a routine that takes the straight-line certificate and performs the computations necessary to verify that it behaves as claimed (Exercise 4.23).

Both the length of a certificate and the time needed to verify the claims for the entries in the certificate are polynomial functions of the length of the input (details omitted). This means that Pratt's method proves that the problem of determining primality is in the complexity class $\mathcal{NP}$. Recall that $\mathcal{P}$ denotes the set of problems solvable in polynomial time. The class $\mathcal{NP}$ stands for problems solvable in ***nondeterministic polynomial time***. A set X is in $\mathcal{NP}$ if for each x in X there is a certificate whose length is polynomial in the length of x and that can be checked to be a certificate in time polynomial in the length of x. If the certificate checks out properly, then x lies in X. If x does not lie in X the program does nothing. The certificate is viewed as heaven-sent because the definition makes no provision for finding the certificate. It might seem that this notion is uninteresting, because heaven is not in the business of providing certificates to us. But in fact the notion is very important.

Consider the set of composite integers, for example. It is easy to see that this set is in $\mathcal{NP}$ because a certificate for a composite n can be a pair $\{a, b\}$ where $n = ab$. Of course, such pairs may be very hard to find, but if we have a pair, it is easy to multiply a by b and check that the product is n; so the pair works as a certificate. Now, the class $\mathcal{NP}$ is apparently asymmetric: the fact that X is in $\mathcal{NP}$ gives no information about the complement of X. So composites are in $\mathcal{NP}$ but it might seem unlikely that the primes are in $\mathcal{NP}$, because there is no obvious way to certify a number's primality. But the work of Pratt just described shows that every prime does indeed have a short certificate. The code in Algorithm 4.5 actually searches for the certificate; that algorithm itself is not a polynomial time algorithm because of the factoring step. It is the verification of the Pratt certificate (Exercise 4.23) that will be fast.

Amazingly, it is not known that the class $\mathcal{NP}$ is really different from $\mathcal{P}$. Indeed this problem, called the $\mathcal{P} = \mathcal{NP}$ problem, is the most important unsolved problem of theoretical computer science. Of course, $\mathcal{P} \subseteq \mathcal{NP}$ because a fast algorithm that works without a certificate can be viewed as working with the empty certificate.

For serious certification work we can improve the basic Pratt certificate in several ways, the aim being to make it smaller. As noted, we may as well suppress the 2s. And we can use the knowledge that *Mathematica*'s `PrimeQ` is reliable up to 10^{16} and avoid certifying such small primes. The `PrattCertificate` function in the package uses an `AssumedPrimeBound` that controls this last point. Using the options, the certificate for $10^{30} + 57$ becomes remarkably short.

```
PrattCertificate[10^30 + 57, Tree → True,
  Show2s → False, AssumedPrimeBound → 10^16]
```

```
1000000000000000000000000000057 → 5
                                   3
                                   79043
                                   3998741
                                   290240017
                                   454197539
```

One can shorten the Pratt certificate even more by appealing to Theorem 4.6 that follows. We use `FactorInteger` and if that works, then we have the full factorization of $n-1$. But we don't actually need all the primes and the certificate works so long as we look at enough primes of the factorization of $n-1$ so that they multiply to more than $\sqrt{n}$. This cuts down on the number of conditions the witness has to satisfy. The ideas of Theorem 4.6 go back well into the nineteenth century. The variation that follows was proved in 1914 by Henry Cabourn Pocklington (1870–1952).

Theorem 4.6. Suppose $n-1$ factors into $F \cdot R = (\prod_{i=1}^{s} p_i^{a_i}) \cdot R$, where $R < \sqrt{n}$. And suppose that a witness b exists so that

1. $b^{n-1} \equiv 1 \pmod{n}$, and
2. $\gcd(b^{(n-1)/p_i} - 1, \ n) = 1$ for $i = 1, \ldots, s$.

Then n is prime.

Proof. Let p be the smallest prime divisor of n and let $d = \operatorname{ord}_p(b)$. Then d divides $p-1$. Further, by assumption (1), d divides $n-1$. But d does not divide $(n-1)/p_i$, because doing so would violate assumption (2) (for then p would divide both n and $b^{(n-1)/p_i} - 1$). This means that $p_i^{a_i}$ divides d for each i, so F divides d. But d divides $p-1$, so F divides $p-1$ and therefore $p-1 \geq F \geq \sqrt{n}$, a contradiction unless $p = n$. $\square$

The version of `PrattCertificate` in the `CNT` package incorporates this theorem via the `FullPrattCertificate` option. When that is set to `False`, then we start with the largest primes and continue down until we have captured $\sqrt{n}$. The resulting reduction in certificate size can be impressive. For example:

```
PrattCertificate[10^30 + 57, Tree → True, Show2s → False,
  AssumedPrimeBound → 10^16, FullPrattCertificate → False]
```

```
1000000000000000000000000000057 → 2
                                   290240017
                                   454197539
```

To check such a certificate one would check that the two primes really divide $n-1$ and that, when multiplied together, they are greater than $\sqrt{n}$.

There are four other primes in the factorization of $10^{30} + 56$, but they can be ignored thanks to Theorem 4.6.

One can use Theorem 4.6 to construct large primes that are certifiably prime. The basic idea is due to U. Maurer [Mau].

Algorithm 4.6. Getting Certified Primes

Suppose that we want p, a d-digit prime that is certifiably prime. Select a prime q having $\lceil d/2 \rceil + 1$ many digits (this can be done recursively). Then choose a random integer $r < q$ and adjust it by adding (randomly) one of two small integers so that $1 + rq \equiv \pm 1 \pmod 6$; then check whether $p = rq + 1$ can be declared probably prime by a pseudoprime test. If so then, because we know (with certainty!) that q is prime, we have enough of a factorization of $p - 1$ to apply Theorem 4.6 and check whether $b = 2$ is a witness to p's primality. As discussed in Exercise 4.29, it is almost always the case that 2 will work as the witness b, so that we lose nothing by relying on 2 to be the witness. If p fails to be probably prime we can just add 6 to the random multiplier r and try again (and again and again). Below 10^7 we can simply use `RandomPrime`; this calls `PrimeQ`, which is known to tell the truth in that domain (we could use this under 10^{16}, but it seems faster to use 10^7 as the crossover point).

And now here is a recursive routine to get large certified primes. The `CertifiedPrime` routine returns the prime only. If the `ShowPrimeCertificate` option is set to `True` then it returns a complete certificate. The code makes use of some tricks for speed: the mod-6 residue of the first random prime selected (call it q_0, the base case of the recursion) is stored in the form $adjust = 1 - q_0^{-1} \pmod 6$. Then as the random multipliers r are generated, they are immediately adjusted to be 0 (mod 6), and then $adjust$ is added; this guarantees that each $1 + qr \equiv q_0 \pmod 6$. Because of the size of q and the bounds on r, it follows that $r < q$ and $1 + qr$ has d base-10 digits. We can gain some speed by using a rudimentary prime test instead of the more cautious `PrimeQ`. Thus we use `QuickPrimeQ` which combines a single gcd to check the existence of a small prime divisor in [5, 863] with a 2-pseudoprime test. We are assuming that the unlikely event of randomly getting a 2-pseudoprime for one of the q-values (which would send the routine into an infinite loop since 2 will never be a Pocklington witness) will not occur.

```
Options[CertifiedPrime] = {ShowCertificate → False};

primeProd = Apply[Times, Prime[Range[3, 150]]];

QuickPrimeQ[n_] :=
  GCD[n, primeProd] == 1 && PowerMod[2, n - 1, n] == 1;

CertifiedPrime[d_, opts___] :=
  If[ShowCertificate /. {opts} /. Options[CertifiedPrime],
    Identity, First][PrimeAndCertificate[d]];
```

```
PrimeAndCertificate[d_] :=
  Module[{p, r, q, q1 = PrimeAndCertificate[Ceiling[d / 2] + 1]},
    q = q1[[1]]; r = Random[Integer, {10.^(d-1), 10.^d}/q];
    r = r - Mod[r, 6] + adjust;
    While[r = r + 6; p = 1 + q r;
      ! QuickPrimeQ[p] ∨ GCD[p, PowerMod[2, r, p] - 1] ≠ 1];
    Prepend[q1, p]];

PrimeAndCertificate[d_ /; d < 8] :=
  {q0 = RandomPrime[d]; adjust = 1 - PowerMod[q0, -1, 6]; q0};

CertifiedPrime[50]

34087174545519525901693137627351741082557462228341

CertifiedPrime[50, ShowCertificate → True]

{34087174545519525901693137627351741082557462228341,
  32105645361217806642310537, 68973614186831, 57734449, 70753}

cert = CertifiedPrime[150, ShowCertificate → True]

{20063107730747797152781759046253545482011620675588163525535\
  24946389658364223457842527507221525937258205217348793967683\
  6047041475791607921401700022114227,
  20012157369956105114838427753173062783488875312430466954029\
  95701167569281573, 43570690952651894125575973114632928257 7,
  14600780049002840987 9, 344635088753, 5902903}
```

It is not hard to check that all the primes in `cert` are proved prime by the witness 2 (except for the last prime, which is less than 10^7). The following code checks them all at once.

```
Table[{p, q} = cert[[{i, i + 1}]]; {q > √p , PowerMod[2, p - 1, p],
  GCD[p, PowerMod[2, (p - 1) / q, p] - 1]}, {i, Length[cert] - 1}]

{{True, 1, 1}, {True, 1, 1},
  {True, 1, 1}, {True, 1, 1}, {True, 1, 1}}
```

Now here's a surprise: The generation of certified primes by Maurer's method is, for large integers, faster than using `PrimeQ`, which provides numbers that are only probably prime; see Exercise 4.32.

Exercises for Section 4.2

4.14. For plain pseudoprimes we use only prime bases because of Exercise 2.41, which shows that the first pseudoprime witness must be prime. Show that the analogous statement for strong pseudoprimes is false. More precisely, show that the following is false: If n is a b-spsp and a c-spsp, then n is a bc-spsp. *Hint*: Let $n = 410041$.

4.15. (a) Exercise 4.14 raises the possibility that the first strong pseudoprime witness to the compositeness of an odd n might be composite. A search for such a number should focus on numbers that are known to be 2- and 3-strong pseudoprimes, because we want the first witness to be neither 2 nor 3. The StrongPseudoprimes function (based on computations of D. Bleichenbacher) accesses a list of all such candidates below 10^{16}. The usage is StrongPseudoprimes[{2, 3}, max], where max $\leq 10^{16}$. Use this function to find an odd n whose first witness is composite. Find the first instance of each composite that occurs.

(b) Use the results of part (a) and some further computations to show that between 2047 and 10^{16}, the first strong pseudoprime witness b for n is less than $0.052(\log n)^2$.

4.16. Exercise 4.15 suggests that the first strong pseudoprime witness cannot be 4. Prove that this is so. More generally, prove that the first witness cannot be a square. Prove that the first witness cannot be a perfect power. Find an odd composite n so that neither 2 nor 3 are witnesses, but 12 is. There are no examples below 10^{16} for which 12 is the *first* witness, but such examples do exist. In 1994, two months of computing time led Shuguang Li (University of Hawaii at Hilo) to 1502401849747176241, for which 12 is the first strong psp witness.

4.17. Exercise 2.42 asserts that each composite Fermat number $2^{2^k} + 1$ is a 2-pseudoprime. Show that each composite Fermat number is a 2-strong pseudoprime.

4.18. (a) Suppose that 2^t divides $p - 1$ for each prime divisor p of n. Prove that 2^t divides $n - 1$.

(b) Prove that if n is odd and b is arbitrary, then $b^{n-1} \not\equiv -1 \pmod{n}$. *Hint*: Suppose that the congruence holds and $\{p_i\}$ are the distinct prime divisors of n. Look at the power of 2 in $n - 1$ and the power of 2 in each $\text{ord}_{p_i}(b)$. Use part (a).

4.19. (J. H. Davenport) Here is an improvement to the strong pseudoprime idea in the case that $n \equiv 1 \pmod 4$. Compute the strong pseudoprime sequences corresponding to $b = 2, 3$, and 5. If one of these proves n is composite, we know that n is composite. Otherwise, look at the square roots of -1 that show up in the three strong psp sequences [note that such a square root can occur only if the first entry of the sequence is not ± 1; this requires that the sequence have length at least three, which requires $n \equiv 1 \pmod 4$]. If more than two such square roots occur among the three sequences, n can be declared composite. The odds of success are higher if one allows b to take on more than three values. Implement this strong strong test with $b = 2, 3, 5, 7, 11$. There are 107 integers below 10^{16} that are 2-, 3-, 5-, 7-, and 11-strong pseudoprimes, and they are quickly available as Strong-

`Pseudoprimes[{2, 3, 5, 7, 11}, 10`^{16}`]`. Of these, 79 are congruent to 1 (mod 4). How many of these 79 does this strong strong test catch?

4.20. Modify the code for `PrimeQWithProof` so that it returns a message indicating failure (as opposed to running forever) if the input is composite according to `PrimeQ`.

4.21. Let p be an odd prime divisor of $n - 1$ where n is prime. Prove that if b is selected randomly from $\{1, ..., n - 1\}$ and if $b^{(n-1)/2} \equiv -1 \pmod{n}$, then the probability that $b^{(n-1)/(2p)} \equiv -1 \pmod{n}$ is $1/p$.

4.22. Write a program that takes any integer less than 100,000,000 and proves either primality or compositeness using the list of all primes less than 10,000. Use `Eratosthenes[10000]` to generate the list of primes.

4.23. Write a routine that takes a straight-line Pratt certificate and verifies that all the claims are true.

4.24. Write a routine that takes a straight-line Pratt certificate in shortened form, as discussed following Theorem 4.6, and verifies that all the claims are true.

4.25. Use `PrimeQ` or a pseudoprime test to find some large prime Fibonacci numbers, and then try to certify their primality via Pratt's method.

4.26. Modify `PrattCertificateBasic` so that it makes use of Algorithm 4.5 to avoid finding a witness that works for all the prime divisors, but rather finds a list of witnesses, one for each prime dividing $n - 1$.

4.27. Modify `PrattCertificateBasic` so that it makes use of Theorem 4.6 to avoid using the full factorizations of the numbers $n - 1$ that occur.

4.28. In Theorem 4.6, why is it not sufficient to verify that $b_i^{n-1} \equiv 1 \pmod{n}$ and $b_i^{(n-1)/p_i} \not\equiv 1 \pmod{n}$? To be more specific: Find an example of a composite integer n for which $n - 1 = F \cdot R$ where $R < \sqrt{n} < F$ and $\gcd(F, R) = 1$, such that for every prime p that divides F, there is an integer b that satisfies $b^{n-1} \equiv 1 \pmod{n}$ and $b^{(n-1)/p} \not\equiv 1 \pmod{n}$.

4.29. Explain why we may as well restrict the witness search to $b = 2$ in Algorithm 4.6. More precisely, if p is prime and $p = 1 + rq$ where q is prime and $q > \sqrt{p}$, why are the odds that 2 fails to certify the primality of p via the Pocklington method (Theorem 4.6) very small?

4.30. Prove the following small variation on Theorem 4.6. It requires that we factor more of n (we lose a 2) but gets us a little more efficiency in the search for b since we have, in cases where F is odd, eliminated the need to find a witness b for the case of the prime 2. The proof is almost identical to that of Theorem 4.6.

Suppose n is odd and $n - 1$ factors into $2R \cdot F = 2R(\prod_{i=1}^{s} p_i^{\alpha_i})$, where $F > \sqrt{n}$ or, if F is odd, $F > R$. And suppose a witness b exists so that

1. $b^{n-1} \equiv 1 \pmod{n}$, and
2. $\gcd(b^{(n-1)/p_i} - 1, \ n) = 1$ for $i = 1, \ldots, s$.

Then p is prime.

4.31. Prove the following extension of Pocklington's method due to (U. Maurer [Mau]): Suppose n is odd and $n-1$ factors into $2R \cdot F = 2R(\prod_{i=1}^{s} p_i^{a_i})$, where $F > n^{1/3}$. Suppose a witness b exists satisfying the two conditions of Theorem 4.6. Choose x and y so that $2R = xF + y$ where $0 \le y < F$ (so y is just Mod[$2R$, F]). If $y^2 - 4x$ is not a perfect square, then n is prime. *Hint*: First improve Theorem 4.6 to show that every prime factor of n is congruent to 1 (mod F). This shows that n can have at most two prime factors. This can be used to build a quadratic equation of the form $k^2 - yk + x$ that has an integer solution, which means $y^2 - 4x$ is a square.

4.32. Compare the speed of generating d-digit primes (with d between 70 and 150) using Maurer's method against just generating a random number and then searching upwards. The latter search, for a fair timing comparison, should not use the built-in `PrimeQ`, but should instead mimic it using `MillerRabinPrimeQ` with 2 bases and the `RandomBases` option set to `False` to force it to use 2 and 3. This is an adequate simulation of `PrimeQ`.

4.33. Implement a modification of Maurer's algorithm for certified primes that accepts the digit-size d and a relatively prime pair $\{a, m\}$ as input and returns a certified d-digit prime that is congruent to a (mod m).

4.3 Refinements and Other Directions

▪ Other Primality Tests

In the nineteenth century, Édouard Lucas discovered how to use what we now call Lucas sequences to prove primality when it is possible to factor $n + 1$. This approach will be explained in Chapter 8. It gives us the Lucas–Lehmer test, a very fast test for checking possible Mersenne primes, integers of the form $2^p - 1$. It relies on the fact that $(2^p - 1) + 1$ is very easy to factor. In 1975, Brillhart, Lehmer, and Selfridge [BLS] published modified primality testing algorithms that relied on partial factorizations of $n + 1$, similar to Pocklington's theorem (Theorem 4.6) or that could be implemented with a partial factorization of $n - 1$ and a partial factorization of $n + 1$.

All of these modifications ran up against the difficulty of factoring very large integers. The breakthrough came in 1980 when Adleman, Pomerance, and Rumely showed how to use algebraic characters to test for primality. In 1981, H. Cohen and H. W. Lenstra simplified and improved this algorithm, and then A. K. Lenstra implemented it. Known as the **cyclotomy test**, it is now the algorithm of choice for proving prime those candidates for which we do not expect to be able to factor $n \pm 1$ easily, basically anything over 100 digits. It is effective up to about 1000 digits. This test and the mathematics behind it are explained in [Coh] and [CP].

A different approach to proving primality for large integers was taken by Atkin in the late 1980s. Based on ideas of Goldwasser and Kilian from 1986, it relies on the arithmetic of elliptic curves. Its running time is less predictable than the cyclotomy test. Within the realm of what is now computable, the cyclotomy test is usually faster but the Atkin test has successfully proven integers of more than a thousand digits to be prime. It is known that somewhere between 1000 and 10,000 digits the Atkin test will be consistently faster. This algorithm and the mathematics behind it are also explained in [Coh].

Mathematica has a package that uses elliptic curve techniques to certify large primes. Here is how to load and use it. If you wish to see the form of the elliptic curve certificate, set the `Certificate` option to `True`. For inputs less than 10^{10} the `ProvablePrimeQ` function just uses a Pratt certificate (this changeover point is controlled by the `SmallPrimes` option).

```
Needs["NumberTheory`PrimeQ`"]
```

```
p = 10^100 + 267;
ProvablePrimeQ[p]
```

```
True
```

The positive result means that a certificate was found that proves that p is prime.

▣ Strong Liars Are Scarce

Our next goal is to prove that there are no strong Carmichael numbers. By this we mean that for any composite n one can always find a base b relatively prime to n such that the b-strong pseudoprime test declares that n is composite. In fact, we will show that there are lots of such bases b.

Theorem 4.7. If n is odd and composite, then at most $(n-1)/4$ of the residue classes can be strong liars.

Proof. All unspecified congruences are modulo n. We begin with the case that n is a prime power, $n = p^e$, where $e \geq 2$. Let g be a primitive root for n so that $\text{ord}_{p^e}(g) = (p-1)p^{e-1}$. If n is a g^a-pseudoprime, then $g^{a(p^e-1)} \equiv 1 \pmod{p^e}$, which means that $(p-1)p^{e-1}$ divides $a(p^e - 1)$; this implies that p^{e-1} divides a. In fact, p^e is a pseudoprime for the base g^a if and only if p^{e-1} divides a. This implies that here are exactly $p-1$ bases for which p^e is a pseudoprime. But

$$\frac{p-1}{p^e} \leq \frac{p-1}{p^e - 1} \leq \frac{p-1}{p^2 - 1} = \frac{1}{p+1} \leq \frac{1}{4}.$$

Let $p_1, p_2, \ldots, p_r$ be the primes that divide n, $n = p_1^{e_1} \ldots p_r^{e_r}$. We can now assume that r is at least 2. Let $n - 1 = 2^s t$ where t is odd. We define k to be the largest integer such that $b^{2^k} \equiv -1$ for some b. To show that k is well defined, we first observe that $(-1)^{2^0} \equiv -1$, so there exists an integer that satisfies this congruence. We next note that if b and k satisfy this congruence they must also satisfy $b^{2^k} \equiv -1 \not\equiv 1 \pmod{p_i}$ and $b^{2^{k+1}} \equiv 1 \pmod{p_i}$ for $1 \le i \le r$ and therefore $\mathrm{ord}_{p_i}(b) = 2^{k+1}$. This implies that 2^{k+1} divides $p_i - 1$ or, equivalently, $p_i \equiv 1 \pmod{2^{k+1}}$. Because this is true for all primes dividing n, we see that $n \equiv 1 \pmod{2^{k+1}}$, which means that 2^{k+1} divides $n - 1$. This implies that k must be strictly less than s.

Let k and b be as in the preceding paragraph and let $m = 2^k t$. Note that $b^m \equiv (-1)^t = -1$. We define the set $L = \{a : a^m \equiv \pm 1, 1 \le a < n\}$. We shall show that all strong liars are in L and that L has at most $(n-1)/4$ elements. If a is a strong liar, then either $a^t \equiv \pm 1$ or $a^{2^j t} \equiv -1$ for some j with $1 \le j \le k$. If $a^{2^j t} \equiv \pm 1$, then $a^{2^{j+1} t} \equiv 1$, so a must be in L. We note also that if $a \in L$, then so is ab because $(ab)^m \equiv -a^m$.

For each $a \in L$, we consider the set

$$S(a) = \left\{ x : 1 \le x < n \text{ and for each } i \text{ with } 1 \le i \le r, \ x \equiv a \text{ or } ab \,(\mathrm{mod}\, p_i^{e_i}) \right\}.$$

There are 2^r elements in $S(a)$, exactly two of which — a and ab — are also in L (Exercise 4.34). Each integer appears in at most two of these sets (Exercise 4.35), and thus we have found at least $|L|(2^r - 2)/2$ integers that are not strong liars. If $r \ge 3$, this proves that there are at least three integers that are not strong liars for every integer that is. If $r = 2$, we have shown that there is at least one element in the union of the $S(a)$ that is not a strong liar for every element that is.

We next observe that if x is in the union of the $S(a)$, then x is at least a *liar* in the sense that n is an x-pseudoprime (Exercise 4.35). We claim that if a is not a Carmichael number, then the liars make up at most half the positive integers less than a. If x is a liar and y is relatively prime to a and not a liar, then xy is not a liar. If x_1 and x_2 are distinct liars, then $x_1 y$ and $x_2 y$ are distinct nonliars.

All that is left is to show that if $r = 2$, then n cannot be a Carmichael number. But this was done in Proposition 3.8. □

This theorem yields a method for getting probabilistic proofs of primality, implemented in the CNT package as MillerRabinPrimeQ and discussed earlier on page 117.

▣ Finding the *n*th Prime

It seems very impressive that we can get, say, the billionth prime *exactly* in only a few seconds.

```
Prime[10^9]
```

```
22801763489
```

Of course, there is some cleverness underlying the algorithm for this; it is done without computing millions of primes and counting them. Recall the prime-counting function $\pi(x)$. A good algorithm for $\pi(x)$ quickly leads to a good algorithm for getting p_n, the nth prime: one can just use a root-finding technique to invert π. The code that follows uses the built-in **FindRoot**, but it would be quite simple to program an interval-halving method that would work almost as fast. As with all root-finding problems, we need some starting values that are close to the final answer. It is known that p_n is asymptotically equal to $n \log n$, so we could use multiples of that. But we may as well use the current best-known bounds, which are due to Rosser (1938) and Robin (1983); see [Rib, p. 249; BS, p. 299].

$$n(\log n + \log\log n - 1.0072629) \le p_n < n(\log n + \log\log n + 8), \text{ when } n \ge 2.$$

We can use this trapping interval to start the root-finding program. We find x such that $\pi(x) \approx n$ and use a simple function to perform the fine corrections to get the exact prime we want.

```
CorrectPrime[x_, n_] := (a = Round[x]; s = PrimePi[a]; While[
    ! PrimeQ[a] || PrimePi[a] ≠ n, a -= (Sign[s - n] /. 0 → 1)]; a)
```

```
prime[1] = 2;
prime[n_] := Module[{x},
    CorrectPrime[x /. FindRoot[PrimePi[x] == n,
        {x, {n (Log[n] + Log[ Log[n]] - 1.0072629),
            n (Log[n] + Log[Log[n]] + 8)}}], n]]
```

```
{prime[1234567], Prime[1234567]}
```

```
{19394489, 19394489}
```

So the key is coming up with a fast way to compute $\pi(x)$. Legendre, in 1808, found an excellent way to do this. He used what is now known as a ***Legendre sum***. Define $\phi(n, a)$ to be the number of positive integers m that are less than or equal to n and are not divisible by any of the first a primes $p_1, p_2, ..., p_a$. One can give an explicit, but unwieldy formula for ϕ as follows, where the signs alternate because we first subtract the number of multiples of p_1, p_2, and so on, and then we add back the number that are subtracted doubly, then recorrect by subtracting the number that were added triply, and so on.

$$\phi(n, a) = \lfloor n \rfloor - \sum_{i=1}^{a} \left\lfloor \frac{n}{p_i} \right\rfloor + \sum_{i,j=1}^{a} \left\lfloor \frac{n}{p_i \, p_j} \right\rfloor - \sum_{i,j,k=1}^{a} \left\lfloor \frac{n}{p_i \, p_j \, p_k} \right\rfloor + \cdots$$

But the following succinct recursive rule is much more useful. If $m \leq n$ and m is not divisible by any of the first $a - 1$ primes, then either m is not divisible by p_a and so appears in the count in $\phi(n, a)$, or it is, in which case m / p_a gets counted the right number of times in $\phi(\lfloor n / p_a \rfloor, a - 1)$.

$$\phi(n, a - 1) = \phi(n, a) + \phi\left(\left\lfloor \frac{n}{p_a} \right\rfloor, a - 1\right)$$

Isolating $\phi(n, a)$ in the preceding formula leads to a recursive definition. We cache the values (this is done by the phrase `Legendre`ϕ`[n, a]` = `Legendre`ϕ`[n, a - 1] - ···`, which stores the values *as soon as they are computed*; this is a standard technique and is often essential in recursions to avoid a tremendous slowdown due to repeated computation of the same value). Note that for this to work we need the ath prime. We avoid `Prime[a]`, because the whole point here is to see how `PrimePi` and `Prime` can be computed from scratch. So we first use a sieve to generate the primes below 10,000 This will allow our implementation of $\pi(x)$ to work for x under $10{,}000^2$, or 100 million (though our implementation will be too slow for numbers much beyond ten million).

```
primes = Eratosthenes[10000];
Length[primes]
```

```
1229
```

Now we implement the recursion, with a reset of a system default so that we can do large recursions.

```
$RecursionLimit = 5000;
Legendreϕ[n_Integer, 0] := n
```

```
Legendreϕ[n_Integer, a_Integer] := (Legendreϕ[n, a] =

    Legendreϕ[n, a - 1] - Legendreϕ[Floor[ n / primes[[a]] ], a - 1])
```

Here is an example followed by a check that uses a one-by-one count. The general behavior of the Legendre function for fixed a is shown in Figure 4.4.

```
{Legendreϕ[500, 30], Length[
  Select[Range[500], Intersection[First /@ FactorInteger[#],
    Take[primes, 30]] == {} &]]}
```

```
{66, 66}
```

We can speed up the code by eliminating part of the recursion. Note how the first part of the recursion will decrement a until it gets down to 1. Thus we can replace it by a simple sum. A special case is added for $n < p_{a+1}$, as that is simply 1. We use an `Apply[Plus, …]` construction as that is fastest;

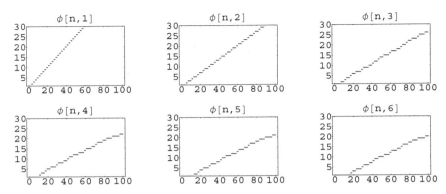

Figure 4.4. The Legendre sums up to 100 with $a = 1, 2, 3, 4, 5$, and 6.

that requires making Legendreϕ listable so it will properly run through the list returned by Quotient[]. This variation gives a worthwhile speedup to the function.

```
Clear[Legendreϕ];
Attributes[Legendreϕ] = Listable;
Legendreϕ[n_Integer, 0] := n
Legendreϕ[n_Integer, a_] := 1 /; n < primes[[a + 1]]
Legendreϕ[n_Integer, a_Integer] :=
  (Legendreϕ[n, a] = n - Apply[Plus,
      Legendreϕ[Quotient[n, Take[primes, a]], Range[0, a - 1]]])
```

We add the following code so we can clear out the cached values for timing tests.

```
initialize := (
  Clear[Legendreϕ];
  Attributes[Legendreϕ] = Listable;
  Legendreϕ[n_Integer, 0] := n;
  Legendreϕ[n_Integer, a_] := 1 /; n < primes[[a + 1]];
  Legendreϕ[n_Integer, a_Integer] :=
    (Legendreϕ[n, a] = n - Apply[Plus, Legendreϕ[
        Quotient[n, Take[primes, a]], Range[0, a - 1]]]))
```

Now Legendre's idea on how to use ϕ to compute $\pi(x)$ is very simple. The following recursive rule will do the job:

$$\pi(n) + 1 = \pi\left(\sqrt{n}\right) + \phi\left[n, \pi\left(\sqrt{n}\right)\right].$$

To prove this observe that if p is a prime and $p \leq n$, then either $p \leq \sqrt{n}$ or $p > \sqrt{n}$; in the latter case p gets counted in the ϕ-term and because the ϕ-count includes the nonprime integer 1, we must add 1 to the left side. This shows that the left side is less than or equal to the right. For the opposite direction, suppose that m is an integer occurring in the count on the right side. Then one of the following hold: $m = 1$; or m is prime and $m \le \sqrt{n}$; or $\sqrt{n} < m \le n$ and m is not divisible by any prime $p \le \sqrt{n}$. But this last case means that m is prime, and so appears in the π-count on the left side. Isolating $\pi(n)$ in this formula leads to a simple recursive implementation. Because the second argument to ϕ is at most $\pi(\sqrt{n})$ and because $\pi(\sqrt{10{,}000^2})$ is the number of primes in the sieve-list that we computed to 10,000, we see that our sieve will be adequate for $n < 10^8$. Here is the implementation; the amount of total code is nicely compact.

```
LegendrePi[x_] := 0 /; x < 2
LegendrePi[x_] := Module[{a = LegendrePi[N[√x]]},
    a + Legendreφ[Floor[x], a] - 1]
```

```
{LegendrePi[10000], PrimePi[10000]}
```

```
{1229, 1229}
```

```
{LegendrePi[1234567], PrimePi[1234567]}
```

```
{95360, 95360}
```

Here is a chart, created by a CNT package function, that shows the recursion in action. The answer, $\pi(1{,}000{,}000)$, is the last entry of the last column. Because iterating the square root function quickly gets us down to 1 — the repeated square roots are shown in the first column — the number of recursive steps is quite small.

```
LegendrePiFull[10⁶]
```

Square roots, m	$\lfloor \sqrt{m} \rfloor$	$\pi(\lfloor \sqrt{m} \rfloor)$	$\phi(m, \pi(\lfloor \sqrt{m} \rfloor))$	$\pi(m) = \pi(\lfloor \sqrt{m} \rfloor) + \phi(m, \pi(\lfloor \sqrt{m} \rfloor)) - 1$
1	1	0	1	0
2	1	0	2	1
5	2	1	3	3
31	5	3	9	11
1000	31	11	158	168
1,000,000	1000	168	78,331	78,498

Note that it is essential to make a a local variable, otherwise confusion would ensue in the recursion. One can easily program this part without recursion (Exercise 4.44), but it is not so easy to eliminate recursion for the definition of ϕ.

This routine is slow, with the main bottleneck being in the ϕ computations. Some important improvements due to Meissel and Lehmer can lead to a significant speedup. But the most modern approach to the problem — the one that led to the current world record of $\pi(10^{20})$ by Déléglise and Rivat [DR] — is due to Lagarias, Miller, and Odlyzko in 1985 (see [BS, Chap. 9]. Indeed, the data of these researchers for powers of 10 has been hard-wired into the CNT package.

$\underline{\pi}[10^{\text{Range}[20]}]$

```
{4, 25, 168, 1229, 9592, 78498, 664579, 5761455,
 50847534, 455052511, 4118054813, 37607912018,
 346065536839, 3204941750802, 29844570422669,
 279238341033925, 2623557157654233, 24739954287740860,
 234057667276344607, 2220819602560918840}
```

An important improvement to Legendre's ideas came from the astronomer E. D. F. Meissel in 1871. In 1885, he used his method to compute $\pi(10^9) = 50{,}847{,}478$. He made some arithmetical errors, as his value is low by 56.

$\underline{\pi}[10^9]$

```
50847534
```

Meissel's formula requires $P_k(n, a)$, the number of positive integers less than or equal to n that are a product of k prime factors, each of which is greater than p_a. The sum of $P_k(n, a)$ for $k = 1, 2, 3, \ldots$ gives a count of all the positive integers less than or equal to n whose smallest prime factor is at least p_{a+1}. This leads to the following formula for $\pi(n)$:

$$\pi(n) = \phi(n, a) + a - 1 - P_2(n, a) - P_3(n, a) - \cdots .$$

The required number of terms $P_k(n, a)$ depends on a. If a is such that $p_{a+1} > \sqrt{n}$, then $P_2(n, a)$ equals 0 and we have the original formula of Legendre. If a is chosen so that $n^{1/3} < p_{a+1} \leq n^{1/2}$, then $P_2(n, a)$ is not 0, but $P_3(n, a)$ and subsequent terms do vanish, because $p_i p_j p_k > n^{1/3} n^{1/3} n^{1/3} = n$ for all i, j, k beyond a.

Thus all that remains is to find a compact formula for $P_2(n, a)$. Recall that $P_2(n, a)$ is the number of integers in the interval $[1, n]$ that are of the form $p_i p_j$ with $a + 1 \leq i \leq j$. Considering one prime at a time we have

$P_2(n, a)$ = the number of integers $p_{a+1} p_j \leq n$ with $j \geq a + 1$

+ the number of integers $p_{a+2} p_j \leq n$ with $j \geq a + 2 + \cdots$

+ the number of integers $p_b p_j \leq n$ with $j \geq b$

where $b = \pi(\sqrt{n})$. Using subtraction to account for the inequalities on j gets the following formula.

$$P_2(n, a) = \pi\left(\frac{n}{p_{a+1}}\right) - a + \pi\left(\frac{n}{p_{a+2}}\right) - (a + 1) + \pi\left(\frac{n}{p_{a+3}}\right) - (a + 2) + \cdots =$$
$$-\frac{(b - a)(b + a - 1)}{2} + \sum_{i=a+1}^{b} \pi\left(\frac{n}{p_i}\right)$$

Here is some code that implements the Meissel formula. For large values it is much faster than Legendre's formula. But if $n \leq 10^4$ we use

Legendre's formula, because that is faster for small values of n. And we use some caching, so reinitialization of `MeisselPi` would be necessary for repeated timing comparisons.

```
Clear[MeisselPi];
Attributes[MeisselPi] = Listable;
MeisselPi[x_] := 0 /; x < 2;
MeisselPi[n_Integer] := Module[{a = LegendrePi[√N[n]]},
    a + Legendreφ[n, a] - 1] /; n ≤ 10⁴

MeisselPi[n_Integer] := MeisselPi[n] =
    Module[{a = MeisselPi[N[n]^(1/3)], b = MeisselPi[√N[n]]},
    Legendreφ[n, a] + (b + a - 2) (b - a + 1) / 2 - Apply[Plus,
        MeisselPi[Quotient[n, Take[primes, {a + 1, b}]]]]]]

MeisselPi[x_] := MeisselPi[Floor[x]] /; ! IntegerQ[x]

initialize; MeisselPi[10⁶] // Timing
```

{2.71667 Second, 78498}

```
initialize; LegendrePi[10⁶] // Timing
```

{7.21667 Second, 78498}

The Meissel improvement is substantial. D. H. Lehmer, and later investigators, pushed Meissel's idea farther to get ever more speed; see Exercise 4.46.

Exercises for Section 4.3

4.34. Referring to the proof of Theorem 4.7, prove that a and ab are the only elements of $S(n)$ that are also in L.

4.35. Referring to the proof of Theorem 4.7, prove that each integer appears in at most two of the sets $S(a)$, $a \in L$, and that each $a \in L$ appears in exactly two of these sets.

4.36. Referring to the proof of Theorem 4.7, prove that if $x \in S(a)$ for some $a \in L$, then $x^{n-1} \equiv 1 \pmod{n}$.

4.37. Let p be an odd prime, $p - 1 = 2^s t$ where t is odd. Let a be a positive odd integer that is not divisible by p. Prove that the number of incongruent solutions of $x^a \equiv 1 \pmod{p^e}$ is $\gcd(a, t)$. *Hint*: Every residue class that satisfies this congruence must have an order that divides a and thus the order must divide $\gcd(a, \phi[p^e])$. Prove that $\gcd(a, \phi[p^e]) = \gcd(a, t)$. If d divides $\gcd(a, t)$, then there are $\phi(d)$ elements of order d. Use Theorem 3.1(h).

4.38. Let n be an odd integer with factorization $p_1^{e_1} p_2^{e_2} \cdots p_r^{e_r}$ where $p_i - 1 = 2^{s_i} t_i$. Use Exercise 4.37 to prove that if a is a positive odd integer relatively prime to n, then the number of incongruent solutions of $x^a \equiv 1 \pmod{n}$ is equal to $\prod_{i=1}^r \gcd(a, t_i)$.

4.39. Let p be an odd prime, $p-1 = 2^s t$ where t is odd. Let a be a positive odd integer that is not divisible by p. Prove that the number of incongruent solutions of $x^{2^j a} \equiv -1 \pmod{p^e}$ is $2^j \gcd(a, t)$ if $j < s$ and 0 otherwise. *Hint*: A residue class satisfies this congruence if and only if its order divides $2^{j+1} a$ but not $2^j a$. There is no element of this order if $j \geq s$. If $j < s$, then the order must be of the form $2^{j+1} d$ where d is a divisor of $\gcd(a, \phi[p^e])$. The number of elements of order $2^{j+1} d$ is $\phi(2^{j+1} d)$, which equals $2^j \phi(d)$.

4.40. Let n be an odd integer with factorization $p_1^{e_1} p_2^{e_2} \cdots p_r^{e_r}$ where $p_i - 1 = 2^{s_i} t_i$. Use Exercises 4.37–39 to prove that if a is a positive odd integer relatively prime to n, then the number of incongruent solutions of $x^{a 2^j} \equiv -1 \pmod{n}$ is equal to $2^{jr} \prod_{i=1}^r \gcd(a, t_i)$ if $j < s_i$ for all i and 0 otherwise.

4.41. Combine the results of Exercises 4.37–4.40 to prove that for odd $n = p_1^{e_1} p_2^{e_2} \cdots p_r^{e_r}$ where $p_i - 1 = 2^{s_i} t_i$, the number of strong liars is equal to

$$\left(\prod_{i=1}^r \gcd[a, t_i] \right) \left(1 + \sum_{j=0}^{s-1} 2^{jr} \right) = \left(\prod_{i=1}^r \gcd[a, t_i] \right) \left(1 + \frac{2^{rs} - 1}{2^r - 1} \right)$$

where a is the largest odd integer that divides $n - 1$ and s is the minimum of $\{s_1, \ldots, s_r\}$.

4.42. What is the probability that $n = 49939 \cdot 99877$ is a strong pseudoprime to the base b if b is chosen at random from $\{1, 2, \ldots, n - 1\}$?

4.43. Suppose q, $2q + 1$, and $4q + 1$ are all prime, and $n = (2q + 1)(4q + 1)$. Prove that the number of strong liars for n is $2q^2$. This means that if we could find arbitrarily large primes q for which $2q + 1$ and $4q + 1$ are prime (this is an open question), then the proportion of strong liars can be made arbitrarily close to $1/4$.

4.44. Implement the `LegendrePi` function in a way that avoids recursion.

4.45. Let $P_m = \prod_{i=1}^m p_i$. The following formulas, which make use of the standard Euler ϕ-function, can be useful in sophisticated computations of Legendre sums. Prove them.

(a) (Symmetry) If $P_a / 2 < n < P_a$, $\phi(n, a) = \phi(P_a) - \phi(P_a - n - 1, a)$

(b) (Reduction) If $P_a \leq n$, $\phi(n, a) = \lfloor P_a / n \rfloor \phi(P_a) + \phi(\mathrm{Mod}[n, P_a], a)$

4.46. Derive and implement Lehmer's improvement to Meissel's formula for $\pi(x)$. Here is an outline.

(a) Show that if $a = \pi(n^{1/4})$, then $\pi(n) = \phi(n, a) + a - 1 - P_2(n, a) - P_3(n, a)$.

(b) Show that

$$P_3(n, a) = \text{the number of integers } p_{a+1} \, p_j p_k \le n \text{ with } a + 1 \le j \le k$$

$$+ \text{ the number of integers } p_{a+2} \, p_i p_k \le n \text{ with } a + 2 \le j \le k + \cdots$$

$$+ \text{ the number of integers } p_c \, p_j p_k \le n \text{ with } c \le j \le k$$

where $c = \lfloor n^{1/3} \rfloor$.

(c) Show that

$$P_3(n, a) = P_2\left(\frac{n}{p_{a+1}}, a\right) + P_2\left(\frac{n}{p_{a+2}}, a\right) + P_2\left(\frac{n}{p_{a+3}}, a\right) + \cdots + P_2\left(\frac{n}{p_c}, a\right) =$$

$$\sum_{i=a+1}^{c} P_2\left(\frac{n}{p_i}, a\right).$$

(d) Use the formula for $P_2(n, a)$ to get the following, where $b_i = \pi\left(\sqrt{n/p_i}\right)$:

$$P_3(n, a) = \sum_{i=a+1}^{c} \sum_{j=i}^{b_i} \left(\pi\left(\frac{n}{p_i p_j}\right) - (j - 1) \right).$$

(e) Replace the $-(j-1)$ terms by finding a closed form using the standard formula for the sum of a sequence of consecutive integers.

(f) Put this all together, using Meissel's formula for $P_2(n, a)$, to get the following, where $a = \lfloor n^{1/4} \rfloor$, $b = \lfloor \sqrt{n} \rfloor$, and $c = \lfloor n^{1/3} \rfloor$.

$$\pi(n) = \phi(n, a) + \frac{(b + a - 2)(b + a - 1)}{2} -$$

$$\sum_{i=a+1}^{b} \pi\left(\frac{n}{p_i}\right) - \sum_{i=a+1}^{c} \sum_{j=i}^{\pi\left(\sqrt{n/p_i}\right)} \left(\pi\left(\frac{n}{p_i p_j}\right) - (j - 1) \right)$$

4.47. Extend Legendre's method to obtain a program that computes $\pi_1(n)$, the number of primes less than or equal to n that are congruent to 1 (mod 4). This can be done by proving the formulas in (a)–(c) and then implementing them recursively in a way similar to the implementation of Legendreϕ. For a more sophisticated approach, see [HB].

Define ϕ_1 and ϕ_3 as follows: $\phi_i(n, a)$ is defined to be the number of positive integers between 1 and n, inclusive, such that m is relatively prime to the first a primes and $m \equiv i$ (mod 4).

(a) Prove that $\phi_1(n, a) = \phi_1(n, a - 1) - \phi_1(\lfloor n / p_a \rfloor, a - 1)$.

(b) Prove that $\phi_3(n, a) = \phi_3(n, a - 1) - \phi_3(\lfloor n / p_a \rfloor, a - 1)$.

(c) Prove that $\phi_3(n, a) = \phi(n, a) - \phi_1(n, a)$. For implementation, one can use the preceding formula together with the implementation of Legendreϕ given in the text.

(d) Establish what the necessary base cases for the recursive definition of ϕ_1 are.

(e) Write a program to compute $\phi_1(n, a)$.

(f) Prove that $\pi_1(n) = \pi_1\left(\sqrt{n}\right) + \phi_1\left[n, \pi\left(\sqrt{n}\right)\right] - 1$, and then implement this formula to get a program that computes $\pi_1(n)$.

4.4 A Dozen Prime Mysteries

There are many unsolved problems about prime numbers. Several have the form: Are there infinitely many primes of a certain form? Here are a dozen prominent open questions about primes. See [Guy] and [Rib] for results related to these, as well as many more questions about primes.

1. Is there a good formula for the primes? Note that formulas do exist. A fascinating formula in terms of a polynomial was discovered in 1970 (see [JSWW]). And, assuming the extended Riemann Hypothesis, one can say that n is prime if and only if n passes the b-strong pseudoprime test for any $b < 2(\log n)^2$; this is easy to compute and so is, in an important sense, a "good" formula. But one wonders if there is a simple combination of a fixed number of pseudoprime tests that characterizes primality. The three tests used by PrimeQ work up to 10^{16} and no counterexample is known. Thus a specific case of the question here is: Is it true that n is prime if and only if n passes the 2- and 3-strong pseudoprime test and a certain Lucas test (discussed in Section 9.2). Experts feel that it is unlikely, but one cannot help but be intrigued by the possibility that a handful of pseudoprime tests might characterize primality. Other possible characterizations are mentioned in Section 9.2.

2. **(Goldbach's Conjecture)** Is every even number beyond four a sum of two primes?

3. Are there infinitely many Mersenne primes?

4. Are there infinitely many twin primes? (A ***twin prime*** is one of a pair of primes whose difference is 2, such as 41 and 43.)

5. Are there infinitely many primes of the form $n^2 + 1$?

6. Are there infinitely many prime Fibonacci numbers?

7. Are there infinitely many primes of the form $n! + 1$?

8. Are there infinitely many ***Sophie Germain primes*** (meaning: a prime p such that $2p + 1$ is prime too; these primes played an important role in 19th-century work on Fermat's Last Theorem).

9. Are there arbitrarily long strings of primes in arithmetic progression? For example, $142{,}072{,}321{,}123 + 1{,}419{,}763{,}024{,}680i$ ($i = 0, 1, \ldots, 20$) is a string of 21 such primes.

10. **(Giuga's Conjecture)** Is it true that $1^{p-1} + 2^{p-1} + \cdots + (p-1)^{p-1} \equiv -1 \pmod{p}$ if and only if p is prime? (one direction is an easy consequence of Fermat's Little Theorem. It is known that a composite n satisfying the equation must have at least 12,000 digits).

11. **(Bertrand's Postulate Extension)** Is there always a prime between n^2 and $(n+1)^2$?

12. **(Gilbreath's Conjecture)** Does the sequence of iterated prime gaps always yield 1s in the first column?

Gilbreath's conjecture requires some explanation. Start with a long list of primes. Form the differences of adjacent primes, writing them below the smaller of the two primes. Form the differences of these differences. Continue. It seems as if the infinite column of numbers in the leftmost column are all 1s. But this has not been proved.

IteratedPrimeGaps[20, 20]

Primes	2	3	5	7	11	13	17	19	23	29	31	37	41	43	47	53	59	61	67	71
Differences 1	1	2	2	4	2	4	2	4	6	2	6	4	2	4	6	6	2	6	4	2
Differences 2	1	0	2	2	2	2	2	2	4	4	2	2	2	2	0	4	4	2	2	4
Differences 3	1	2	0	0	0	0	0	2	0	2	0	0	0	2	4	0	2	0	2	2
Differences 4	1	2	0	0	0	0	2	2	2	2	0	0	2	2	4	2	2	2	0	2
Differences 5	1	2	0	0	0	2	0	0	0	2	0	2	0	2	2	0	0	2	2	2
Differences 6	1	2	0	0	2	2	0	0	2	2	2	2	2	0	2	0	2	0	0	0
Differences 7	1	2	0	2	0	2	0	2	0	0	0	0	2	2	2	2	2	0	0	0
Differences 8	1	2	2	2	2	2	2	2	0	0	0	2	0	0	0	0	2	0	0	0
Differences 9	1	0	0	0	0	0	0	2	0	0	2	2	0	0	0	2	2	0	0	0
Differences 10	1	0	0	0	0	0	2	2	0	2	0	2	0	0	2	0	2	0	0	2
Differences 11	1	0	0	0	0	2	0	2	2	2	2	2	0	2	2	2	2	0	2	2
Differences 12	1	0	0	0	2	2	2	0	0	0	0	2	2	0	0	0	2	2	0	2
Differences 13	1	0	0	2	0	0	2	0	0	0	2	0	2	0	0	2	0	2	2	2
Differences 14	1	0	2	2	0	2	2	0	0	2	2	2	2	0	2	2	2	0	0	2
Differences 15	1	2	0	2	2	0	2	0	2	0	0	0	2	2	0	0	2	0	2	0
Differences 16	1	2	2	0	2	2	2	2	2	0	0	2	0	2	0	2	2	2	2	0
Differences 17	1	0	2	2	0	0	0	0	2	0	2	2	2	2	2	0	0	0	2	0
Differences 18	1	2	0	2	0	0	0	2	2	2	0	0	0	0	2	0	0	2	2	0
Differences 19	1	2	2	2	0	0	2	0	0	2	0	0	0	2	2	0	2	0	2	0
Differences 20	1	0	0	2	0	2	2	0	2	2	0	0	2	0	2	2	2	2	2	0

Exercise for Section 4.4

4.48. Use a computer to explore some of the mysteries in this section.

CHAPTER 5

Some Applications

5.0 Introduction

Number theory started out 3000 years ago as an abstract field of study for Egyptians, Babylonians, and Greeks, but today its applications span an incredibly diverse collection of areas of engineering, science, and communication. Indeed, this is a truly modern development. It is fair to say that until the middle of the twentieth century number theory was still being pursued for its own sake. But many aspects of modern computing, and the development of a secure scheme for coding secret messages based on prime numbers, have guaranteed that the subject will henceforth be known for its beauty *and* its utility.

Two good sources for a general overview of the applications of number theory are [Schr] and [Ros]. In this chapter we discuss two important uses of integers in the real world: public-key cryptography to send secret message and the use of check digits to catch and sometimes correct errors when messages are transmitted. And we also begin a discussion of how to factor large integers. The methods presented here will only work on integers of modest size, but they show the sorts of ideas that can be brought to bear; and current state-of-the-art factoring routines can handle numbers of about 150 digits.

5.1 Coding Secrets

The world of secret codes was turned on its head in the mid-1970s when it was discovered that, by a clever use of large prime numbers, one could send messages using an encryption method that was completely public, but yet entirely secure in that only the intended recipient could decrypt the message. Such a scheme means that senders of sensitive data, from credit card numbers to computer passwords to stolen state secrets, need not worry about keeping their means of communication confidential: All correspondence can be fully open, and the method of coding can be open as well; decryption, though, will be impossible except by the intended recipient.

Of course, this development is only the latest in a long line of important mathematical contributions to code-breaking. Mathematicians and cryptographers played critical roles in almost all wars from the time of Caesar on to the present day. See [Hod], [Kah], and [Tuc] to learn more about code-breaking in World Wars I and II.

At first glance, the claim seems impossible. Recall a simple cipher allegedly used by Julius Caesar. Each letter is shifted three units to the right, so "SEND HELP" becomes "VHQG KHOS". Obviously if someone intercepts the encrypted message and knows the method of encryption, it is a trivial matter to decrypt: Just invert the coding procedure and shift down three letters. Public-key cryptography, however, is a way of encoding for which knowledge of the encryption procedure does not help with decryption.

■ Tossing a Coin into a Well

To see why prime numbers are relevant to keeping secrets, consider a slightly different problem. Alice and Bob wish to flip a coin over the telephone or via e-mail. Alice will flip, Bob will call HEADS or TAILS, and Alice will tell Bob whether he is right in a way that prevents cheating. There is an obvious flaw to the suggestion that Alice just use a real coin: the temptation to cheat might be too strong to resist. Here is how prime numbers can be used to eliminate any possibility of cheating by either party. Alice picks two large primes p and q (100 digits each, say) with $p \equiv 1 \pmod 4$ and $q \equiv 3 \pmod 4$ and tells Bob their product, n. Bob then commits himself to HEADS or TAILS and tells Alice his choice. Then Bob guesses whether the larger of p and q is congruent to 1 (mod 4). If he is right the toss is considered HEADS; if he is wrong it is taken to be TAILS. Alice then reveals the primes and they can quickly see whether Bob's choice regarding the modulo 4 condition was right or wrong. This in turn translates into HEADS or TAILS, and they can see whether Bob's choice of one of these was correct.

Here is how Alice might proceed. `RandomPrime` uses the three pseudoprime tests of `PrimeQ` to vouch for primality; more stringent tests can be used if more certainty is required. `RandomPrime` allows the user to specify a congruence class, so here Alice specifies that the larger of the two primes should be congruent to 1 (modulo 4).

```
p = RandomPrime[40, {1, 4}];
q = RandomPrime[39, {3, 4}];
n = p * q
403946566947857892067028923703965433023964900425896883861713‚
  9916971376800347947
```

Presented with this large number, Bob has no idea what its prime factors are, and current factoring algorithms would not be able to factor this in the few seconds before Bob must commit to a decision. Of course, larger primes can be used for more security.

Note that, after committing to his choice and receiving p and q from Alice, Bob can easily multiply the primes together to make sure there was no funny business: he checks that pq agrees with the n he received earlier. He should also run the Miller–Rabin probabilistic test a few dozen times on p and q to make sure they are prime; otherwise Alice might choose four primes a, b, c, and d and decide, after Bob commits to his choice, to tell Bob that p is ab (or perhaps ac or ad), which would allow her to rig the results.

The protocol is effective because Bob cannot factor n in reasonable time, and so he has no information about p and q. This sort of electronic coin flip is called "coin flipping into a well" because it is as if Alice flips a coin into a well, Bob calls it, and then they both walk over to see how it came up. In the protocol given here Alice learns the answer first, but this is of no consequence. Note that because the coin flip is tied to the correctness of Bob's mod-4 choice, as opposed to directly relating it to the mod-4 class of the larger prime, Alice cannot rig in advance that, say, HEADS come up. She does not know what Bob's choice will be until she is committed to p and q. Exercise 5.1 addresses the situation where Alice wishes to pass a random bit to Bob in such a way that she does not know what it is, and Bob and she are both convinced that it is random. An application of this is to the situation where Alice and Bob wish to play poker by e-mail. Alice can deal a random card to Bob by passing enough bits to determine five numbers in [1, 52]. Bob can then do the same for Alice and they can see who has the better hand. If the same card occurs in both hands, they start over. See [Schn] and the references therein for more details of this sort of electronic communication.

A natural question is: How should Alice choose her primes? The simplest strategy is to generate random integers and use several strong pseudoprime tests to verify primality. If sufficiently many spsp tests are used the probability that p is not really prime can be made so small as to be negligible. But recall that Algorithm 4.6 (see also Exercise 4.33) contained a much simpler algorithm for generating large primes that are certifiably prime.

■ The RSA Cryptosystem

The fact that two primes can be easily multiplied but a product of two primes cannot be easily factored is the basis of the RSA public-key scheme. It is a little more complicated than the coin-flipping idea just described because it requires Euler's theorem. Suppose Alice wishes to send the message "Sell shares now. Alice." to Bob. First she must convert the message to a single integer (or perhaps a sequence of integers in case the message is very long). This can be done in a number of ways. The CNT package function `StringToInteger` uses standard ASCII coding. The ASCII code for "S" is 83, but 31 is subtracted so that all characters will yield 2-digit numbers. Thus, "S" yields "52".

ToCharacterCode["S"]

{83}

```
m = StringToInteger["Sell shares now. Alice."]
```

5270777701847366837084017980881501347774687015

Now, some time ago Bob has chosen two large primes, p and q, and he has published their product, n. He must also choose an encryption exponent, called e, and that gets published too; e must be relatively prime to $(p-1)(q-1)$, which is easy for Bob to arrange. The pair (n, e), called the **public key**, is all anyone needs to send a secret message to Bob. Alice would simply take m, her message in integer form, and compute m^e (mod n), communicating the result to Bob over a completely open line. To see this in action, we first generate two primes for Bob, and use 17 as the encryption exponent.

```
p = RandomPrime[53];
q = RandomPrime[47];
e = 17;
n = p * q
```

21741375953661120431968263608869661714378678371517900154674 0
   4523885726686294933838099843179538711757

```
GCD[e, (p - 1) (q - 1)]
```

1

Alice then uses the public key — the pair (n, e) — to form the eth power of m, call it s, and sends it to Bob. It is important in this basic example that $m < n$; longer messages can be broken into several pieces.

```
s = PowerMod[m, e, n]
```

10107980210108599438385268504282963503123561308351520340853 2
   5632184560386200341411732073719246153448

Now for the good part: Even though the whole world knows n, e, and the encrypted message s, Bob is the only person who knows p and q. Therefore, he knows the explicit value of $\phi(n)$, which is $\phi(pq)$, or $(p-1)(q-1)$. Using this he is able to compute d, the inverse of $e \pmod{\phi(n)}$; d is the decoding key.

```
d = PowerMod[e, -1, (p - 1) (q - 1)]
```

12789044678624188489393096240511565714340399041686626478155 7
   8376118237165244355932939870841110804 01

Using d he can look at s^d (mod n), which is just $(m^e)^d$ (mod n), or m^{ed} (mod n). But $ed \equiv 1 \pmod{\phi(n)}$, so by Euler's Theorem (or the more explicit Corollary 3.8), $m^{ed} \equiv m^1 = m$ (mod n).

```
decrypt = PowerMod[s, d, n]
```

5270777701847366837084017980881501347774687015

IntegerToString[decrypt]

Sell shares now. Alice.

This method, called **_RSA encryption_** after its inventors, R. Rivest, A. Shamir, and L. Adleman, is most remarkable. It is an example of a public-key cryptosystem. One wonders if the great codebreakers of the past, from John Wallis to Alan Turing, would have thought such a thing was possible. The idea of a public-key cryptosystem was put forward by W. Diffie and M. Hellman in 1976. While RSA, which was discovered in 1978, was not the first public-key system, it is the easiest to understand and works well in practice.

Before committing to an RSA system for serious work, one must be aware of several things:

- The choice of p and q should be such that factoring their product is difficult. This depends on the state of factoring algorithms. Current practice is to avoid primes p for which $p-1$ is divisible only by small primes. Part of the reason is that Pollard's $p-1$ factoring algorithm (see Section 5.4) will be able to factor pq if $p-1$ is divisible only by small primes. An even more important reason is that if $p-1$ and $q-1$ have many small prime divisors, then there is a high probability that the encryption exponent e has a small order (mod $\phi(n)$). This can help an eavesdropper (see Exercise 5.5).

- Message senders must be aware of the *chosen-plaintext* attack. If an eavesdropper knows that the message is either YES or NO, then she can just encode these two and see which yields a match. Thus short messages should be padded with random sequences of letters.

- Short messages can be insecure. For if m^e turns out to be less than n, then the modular reduction disappears and an eavesdropper can just take the eth root of what she overhears to recover m.

- Of course, e must be coprime to $(p-1)(q-1)$. But that does not guarantee security. There can be bad values of e that must be avoided like the plague. This is discussed in Exercise 5.5.

- If a message would yield an integer larger than n, then the message must be broken into blocks, each of which gets coded as an integer less than n.

- A single numerical error in transmission is a disaster, so perhaps some error-correcting or redundancy might be useful. See Section 5.5 for the use of check digits to catch some transmission errors.

- No system is immune to the *rubber-hose method* of breaking a code. If Bob can read the secret message, then so can Eve if she is torturing Bob to force him to do her bidding. No encryption scheme can prevent this. Bribery is a less violent alternative.

For a more comprehensive treatment of pitfalls in encryption, see [Schn].

The CNT package has RSAEncode and RSADecode functions that take care of the conversion to an integer and the breaking down into blocks. Here is how these functions would be used to encode the Zimmermann telegram, which was sent in January 1917 from the German foreign minister to the German ambassador in Mexico City. The British were able to decipher this most secret and sensitive message, and when they revealed it to the Americans, it was one of the most important factors causing the United States to enter World War I. For more information on the code-breaking methods involved see [Tuc].

```
telegram =
  "We intend to begin on the first of February unrestricted
   submarine warfare. We shall endeavor in spite of this
   to keep the United States of America neutral. In the
   event of this not succeeding, we make Mexico a proposal
   of alliance on the following terms: make war together,
   make peace together, generous financial support and an
   understanding on our part that Mexico is to reconquer
   the lost territory in Texas, New Mexico, and Arizona.
   The settlement in detail is left to you. You will inform
   the President of Mexico of the above most secretly as
   soon as the outbreak of war with the United States
   of America is certain and add the suggestion that he
   should, on his own initiative, invite Japan to immediate
   adherence and at the same time mediate between Japan
   and ourselves. Please call the President's attention
   to the fact that the ruthless employment of our
   submarines now offers the prospect of compelling England
   in a few months to make peace. Signed, ZIMMERMANN.";
```

CodedTelegram = RSAEncode[telegram, e, n]

```
{1980611911027020311734026043044791758558548665816957378873₆⸱
   18105693177533546351760701846034817208,
 820224860315250480329415359180743634127572100334977868552₃₃⸱
   1125155986933988054966009180349942276357,
 1274493733528638969492591702323881462941327168814845644165₉⸱
   9021633104231085043571220257925481337928,
 156398221266119300925102043383861575006762106231895142993₆⸱
   0813283147433397713795025693416866892906₇,
 1144125038092472833402089962864642424546127942719435119969₇⸱
   271286105977527488780136202874722203225₉,
 1652955868985299154936222846306684921820950731763217823545₂⸱
   453463752343313473940138717558022987002₈₈,
 1345513065757744074404584126395914578683167763127881098401₈⸱
   0180988856225676765359736845598924340743₅,
 1687506763531756515567385127216079013371814207013056151135⸱
   0445887928530580263882165834476807348858₉,
 1899195886874731785377116644038525406624238790865318278990₇⸱
   652995061497634791632261386972628717104₄₈,
 1321505895541486306042541914504316469701146476314424456439₇⸱
   5364435531872824406122593580603093201904₅,
 1478220340251464306877201085788152333680716689182210294359₇⸱
   0285326226781716924482289964351413300871₈,
```

```
350880939406119469259936540535365192653521723573592608995114:
  99296144395239478488872186977213477966З,
785755260863064498237379746855264452191670089763374338598Zl:
  78008358946003305102983783862954797883Я,
213176312856043103282166661845810740135621891523835684372б9:
  046991153515542409296883746134295915078Я0,
209824163457722398598387872805141144852396206437009174414кO:
  43388196021862132581737858353017661469817,
188178731597158907636096306945181065175452240919502783581к8:
  96331169540102508055834807385105929604760,
122214509562090766935124208078241954374044188158465022208кЗ:
  59695667133308908845867226769644952715770,
646431652290089163554590642659169838984497646177713676100O2:
  74970840337214262722389021743784575079Z6,
169934036066435753496228298621217357202064042349098135459O:
  73035659120139103810494822506886820192314,
197218668918607114850098265878079585687981320754808337039O5:
  71862368078241589187855119765439924505124}
```

Short[RSADecode[CodedTelegram, d, n], 8]

We intend to begin on the first of February unrestricted
   submarine warfare. We shall endeavor in spite of
   this to keep the United States of America neutral.
   In the event of this not succeeding, we make Mexico
   a proposal of alliance on the following terms: …
  at the same time mediate between Japan and ourselves.
   Please call the President's attention to the fact
   that the ruthless employment of our submarines now
   offers the prospect of compelling England in a few
   months to make peace. Signed, ZIMMERMANN.

▣ Digital Signatures

Suppose Bob, a stockbroker, receives a message from his client, Alice, saying: "Sell all shares in XY Co. Alice. Feb. 17, 1999". Because the key to send messages to Bob is fully public, he has no way of knowing whether this message is really from Alice; and he needs certification of Alice's signature to avoid a future claim by Alice that she sent no such message. The public-key scheme allows Alice's signature to be encoded in such a way that it would stand up in a court of law. Here is how it is done.

Recall that the directory of public keys contains Bob's key (n, e) as well as one for Alice, (n_A, e_A). Alice takes the message and turns it into a number as usual, but before sending it to Bob, she encrypts it using n_A and her personal decoding key, d_A, which only she knows. She then takes this and encrypts it using n and e. Now, Bob, upon receipt, decrypts as usual. What he sees is not English, but gibberish. But it is very special gibberish! If he pretends to encode it using Alice's public key, it will turn into the original message. This means that the sender had to know d_A, which proves that the sender was Alice. The reason this works is that the main rule of RSA, $m^{ed} \equiv m^1 = m \pmod n$, is independent of the order in which e and d are multiplied; $m^{de} \equiv m^1 \pmod n$ too.

Here are the details, where we use Bob's keys defined earlier and create a new key for Alice, n_A, e_A, and d_A.

```
pₐ = RandomPrime[53];
qₐ = RandomPrime[47];
nₐ = pₐ qₐ;
eₐ = 17;
dₐ = PowerMod[eₐ, -1, (pₐ - 1) (qₐ - 1)];

message = StringToInteger[
   "Sell all shares in XY Co. Alice. Feb. 17, 1999"];

signedmessage = PowerMod[message, dₐ, nₐ];
sentmessage = PowerMod[signedmessage, e, n];
```

Now when Bob decrypts what he receives, the integer he gets turns into nonsense if he tries to turn the integer into a string.

```
decryptedgibberish = PowerMod[sentmessage, d, n]
```

```
14617201268407865822858286180176319702177445584677764955948 3
88210087857722151728650415654017034857 2
```

```
IntegerToString[decryptedgibberish]
```

```
M03+cGm`q;Y;\o0^2e4lKVsbl_~ZOEq)'mXg4RguQHWU eAtg
```

This ostensibly meaningless string is in fact very special. Encoding it with Alice's public key turns it into proper English.

```
IntegerToString[PowerMod[decryptedgibberish, eₐ, nₐ]]
```

```
Sell all shares in XY Co. Alice. Feb. 17, 1999
```

Only someone who knew d_A could have produced such a string, and this proves that Alice sent the message.

The ideas of public-key cryptosystems have diverse applications in the world of electronic commerce and communication. See [Schn] for an encyclopedic treatment by an expert in the field.

Exercises for Section 5.1

5.1. Show how a modification of the coin-flipping protocol in the text leads to a system by means of which Alice can pass a random bit (0 or 1) to Bob in such a way that:

- Bob learns what the bit is and has faith that it is random (in no way controlled by Alice's choices).

- Alice does not learn what the bit is (but has faith that it is truly random).

- Alice receives some information from Bob so that, if Bob must prove at a future date that he used the proper bit, he can do so in a way that satisifies Alice that he did not cheat.

We do not know any applications of this idea, but closely related ideas of oblivious transfer of messages have a variety of uses (see [Schn]).

5.2. Why is the following coin-flipping protocol seriously flawed? Alice picks two large primes p and q and reveals their product to Bob. Bob then guesses whether $p \equiv q \pmod 4$ or not. If he is correct, the toss is deemed to be HEADS, otherwise TAILS.

5.3. Explain why the following protocol for coin-flipping electronically works; in particular explain why it is important that the same modulus n be used by Alice and Bob.

Alice and Bob use the same primes p and q, with $n = pq$ as usual. But they choose large random exponents e_A and e_B and keep them secret. They can each find the mod-$\phi(n)$ inverse of their exponent, d_A and d_B. Alice forms two integers m_0 and m_1 so that m_i has an i encoded near the beginning and has random digits near the end. She encodes them using e_A and n and sends the results, M_0 and M_1, to Bob. Bob picks one, encodes it using e_B and n, and returns it to Alice. Alice "decodes" this using d_A and n and sends the result back to Bob. Bob "decodes" this using d_B and looks in the specified location for a 0 or 1, which becomes his random bit. He sends this last message, which is one of the m_i, back to Alice, thus proving to her, because of the random digits she included, that he has played by the rules. Alice then tells Bob e_A so he knows she really encoded one 0 and one 1.

5.4. Show how the protocol of Exercise 5.3 can be used to allow Alice and Bob to play poker by e-mail. Hint: Instead of 2 messages, Alice will use 52 messages $M_1, \ldots, M_{52}$.

5.5. Suppose that Bob uses the following primes to get n for his RSA public key.

```
p = 1258112374445127142093961703741802651152228819947;
q = 7622734338175093326899031636952217670023;
n = p q;
PrimeQ[{p, q}]
```

```
{True, True}
```

And because he is concerned about security, he generates a large random number to use as the encryption exponent e. And, of course, he checks that $\gcd[e, (p-1)(q-1)] = 1$.

```
e = 368240228970308920223050662489643355058600584012993920997\
3982968664717046427297322817 25;
```

Soon after, Alice then sends a message to Bob using his public key.

```
message =
  RSAEncode["Look out. Charlie is a double agent.", e, n]
```

294421888496598651046306114748003816118716007363195680744164`
907503961379712974238948242`8

Eve, Charlie's handler, obtains the coded message and tries something she once heard about at a cryptographers' convention. She re-encodes the coded message using Bob's public key, and then repeats this several times. Lo and behold (see table), on the twenty-second iteration the original message pops out! What is special about *e* that makes it vulnerable to this method of attack? What could Bob have done to prevent it?

The following *Mathematica* code takes the message and runs 22 modular powers on it using the exponent *e*. (NestList[f, a, m] applies *f* to *a* repeatedly *m* times and shows all the intermediate results.) The Integer-ToString function shows us the characters corresponding to the integer at each stage.

```
Map[IntegerToString,
   NestList[PowerMod[#, e, n] &, message, 22]] // TableForm
```

1	<K4ws☐ZuR#^%*NO☐E/*v/☐h^2WoiH_yj"☐,nf<i6xO7;	Eavesdrop in string form
2	[5*5{H■~FkP|Uj_%|;!■6■\Ug{Vq1z;h@Aav6=/1|}X<	Encode of preceding
3	k>zc4Rzwxav`Lhx☐S0_O☐"2q{&7.z\3D@☐.cdT92{!0☐	
4	A'fwZASz1\■v☐e4E{c(;.IgVG[aZh*fw~^!Yk☐I#)VU"	
5	[%wW>4GwfV=>#Q|a%%56<rNKOHDO-W■7X>m Xatcy!'$	
6	dn,tU.2JA|7+)■oZ2>r[=|_☐1?t+jroSyA*9cM3■g`&8	
7	b(n.Rw@~]I☐'0L]UM[>U V5%,eky☐U{i~L^r,D!2Z_4`	
8	dn&\X■6B^■jCO#JE(]D5WiNc■SiBppV!b~S#0|8fFDwQ	
9	IRbOs^K):C'6'x1y.Kyb■HHN#zs'SpXKq^O#L_,T)B■O	
10	B7■1=[W[zw6E_Ai*☐ob5tzEg>{☐2D.@)1mup/C5 }al■	
11	4':S/qJ[8(fr"TKD3hR2Us`BkcqD■■?w>[NOXA4ovI=8	
12	5■oR4a22Y=*jSXZw),Van@"(uhd0z'1C(zYi<ZgLScC■	
13	Zn]n;2dUutE6UJYMqV]?☐C-D☐HH8E91&Bgo[c?<$1YDD	
14	?P;ovdF,e&y;e☐#I`V='$%]@`■_Ke'Z☐.te|Q.;F■(K:	
15	.a■W/C$qBm■Z☐hp+>RevBmuPDIrUj]R$4^B6*(m2<.+	
16	&RC#SiiFuo3Uz08j!BTH:^#~v9:A -G :1f■q=^6[9Nv	
17	m■"an&%L2nGH7]_&WBQtKr■[wr;pTIFJU>&Si~f^2"a3	
18	3■TsEK3CxJ7}h[VP`f>pgH@■U2xH[Z;)"UN■"i■tW_Xt	
19	5tu11g?]rNKYV620tO3m.e?!dvp>8`ZFIgCj1x9Vj;"F	
20	XyC=31R-4KDB` ?FOb~/uyq'7☐g=vI:VY$v1Y~rR|[xR	
21	!G!OOJ)4FXqA}Q☐@5q27e#a4I<9z)-JN 4By>S'.ZIQO	
22	1>&~x2d'hg"w{]4☐32■ou>C@M>?$z!jjgiB0] P=khj:	
23	Look out. Charlie is a double agent.	Surprise!

5.6. In the RSA keys one uses large primes, but typically such large numbers are only probably prime. Investigate the consequences of using a 2-pseudoprime for one of the primes used to form *n*. What happens if the prime is a Carmichael number?

5.7. Investigate what happens in the RSA scheme if the message *m* (assumed to be less than *n*) happens to be divisible by one of the two primes that divides n (which is pq), say $m = p^t k$, where $\gcd(k, n) = 1$. Prove that $(p^t k)^{de} \equiv p^t k \pmod{n}$.

5.8. Suppose that Bob is using the RSA public key (n, e) of Exercise 5.5 and Alice is using the public key (n_A, e_A) from the Digital Signatures subsection. Suppose further that Alice and Bob always use digital signatures to validate their messages. Show how Eve can send a message to Alice saying: "Good news. Eve has just come over to our side. Follow her instructions. Bob." in such a way that Alice will believe the signature is correct and the message is from Bob.

5.9. Suppose that Alice and Bob are both using RSA with the same public-key n, but different values of e which are relatively prime to each other. Suppose Charlie sends the same message to each of them, and Eve captures it. Show that Eve can read the message. *Hint*: Eve will want to use r and s so that $re_1 + se_2 = 1$.

5.2 The Yao Millionaire Problem

Suppose Alice and Bob wish to learn who has the higher salary, but without revealing his or her own salary. This problem, known as the Yao millionaire problem, has a nice solution using the ideas of the RSA protocol. Suppose that Alice's salary is $65,000 and Bob's is $68,000, they will work only to the nearest thousand dollars, and they each know that all salaries are between $60,000 and $70,000. These last two assumptions simplify the presentation, but are completely inessential.

First, Alice generates a large random number, less than but with the same number of digits as the n of Bob's public key. To help us keep track of it, we will use the number that follows, which is not quite random.

```
AliceRandom =
    13854655523423924323527835325325284124853123355225123726\
    8272008618196534141077777777777777777892545;
```

Alice then encrypts this number using Bob's public key, but she subtracts 65 before sending it to him. Bob is using the same key he used in Section 5.1, which we regenerate here.

```
p = 65063480765285204250156273721704681149145199258873459;
q = 33415636080234567231657824070952917550391471959;
{n, e} = {p q, 17};
d = PowerMod[e, -1, (p - 1) (q - 1)];

AliceToBob = PowerMod[AliceRandom, e, n] - 65

2041326450943018374426884780509802057404910392812327849721\98\
    39708547634173076657667675042026105199916
```

Bob could recover Alice's random number if he knew to add 65 from the received number, but he does not know this. So the best he can do is add each of 60, 61, 62, ..., 70 before decoding and be confident that one of these is Alice's random number. Of course, it sticks out as the sixth entry in the list that follows.

```
BobData = PowerMod[AliceToBob + Range[60, 70], d, n]
```

{16652312518080253707408620592899241697603768323125463804780﹂
7209943196793213957407236419517766033180,
12660238505917474372175562333358882760318398271645605376345﹂
5993127783626277951799806298727176657114 6,
12384231810501053831138829430528043235596974091070869047915﹂
5508615922292884948948294021831230520492,
11403812568706304086784431030486534253566865074888194006246﹂
13026639410730241405249414692551908495530,
46802358244852620914024211682170091737469330360766185821563﹂
705419206586847053949009898750039210902 2,
13854655523423924323527835325325284124853123352552251237268﹂
272008618196534141077777777777777777892545,
20411255168352111500331020944316845152938973223724846136376﹂
49118756352456182470887801622268885486959,
27338261546263590998817396315880885334001685827878392213087﹂
8994376572826504690628971159637002800762,
15490007677602859374640170867876665408094444245396604846838﹂
0550409247518667842023324425466310101611,
33095869783617854595502829129079548476494920958908660396313﹂
4692448858519816196051362575512974975767,
26960620391981830993283647072176322678468649717808421484117﹂
8505588766078335110902379502237124388756}

Bob now has an 11-number data set. What he would like to do is add 1 to the last two entries only, which correspond to offsets of 69 and 70, the two numbers greater than his salary.

```
BobDataModified = BobData + {0, 0, 0, 0, 0, 0, 0, 0, 0, 1, 1}
```

{16652312518080253707408620592899241697603768323125463804780﹂
7209943196793213957407236419517766033180,
12660238505917474372175562333358882760318398271645605376345﹂
5993127783626277951799806298727176657114 6,
12384231810501053831138829430528043235596974091070869047915﹂
5508615922292884948948294021831230520492,
11403812568706304086784431030486534253566865074888194006246﹂
13026639410730241405249414692551908495530,
46802358244852620914024211682170091737469330360766185821563﹂
705419206586847053949009898750039210902 2,
13854655523423924323527835325325284124853123352552251237268﹂
272008618196534141077777777777777777892545,
20411255168352111500331020944316845152938973223724846136376﹂
49118756352456182470887801622268885486959,
27338261546263590998817396315880885334001685827878392213087﹂
8994376572826504690628971159637002800762,
15490007677602859374640170867876665408094444245396604846838﹂
0550409247518667842023324425466310101611,

```
33095869783617854595502829129079548476494920958908660396313
6924488585198161960513625755129749757678,
26960620391981830993283647072176322678468649717808421484117
850558876607833511090237950223712438875}
```

Bob wishes to sends this modified data set back to Alice, but he cannot do so for she could then just encrypt the data set and learn Bob's salary. Such an encryption would lead to numbers near the encrypted value of `Alice-Random`, except in the cases that Bob changed them. Thus, Bob's salary of $68,000 would become evident to Alice.

PowerMod[BobDataModified, e, n]

```
{20413264509430183744268847805098020574049103928123278497219
83970854763417307665766767504202610519976,
20413264509430183744268847805098020574049103928123278497219
83970854763417307665766767504202610519977,
20413264509430183744268847805098020574049103928123278497219
83970854763417307665766767504202610519978,
20413264509430183744268847805098020574049103928123278497219
83970854763417307665766767504202610519979,
20413264509430183744268847805098020574049103928123278497219
83970854763417307665766767504202610519980,
20413264509430183744268847805098020574049103928123278497219
83970854763417307665766767504202610519981,
20413264509430183744268847805098020574049103928123278497219
83970854763417307665766767504202610519982,
20413264509430183744268847805098020574049103928123278497219
83970854763417307665766767504202610519983,
20413264509430183744268847805098020574049103928123278497219
83970854763417307665766767504202610519984,
11261035848518617361769797764437112325137790399697851098742
51801133875063784883713738247310540467860,
67480850079621970016799955932751870183926830641504108372218
8536511412249960828990547995181384888654}
```

To avoid this problem, Bob chooses some moderately large prime p and reduces everything modulo p before adding 1 to the last three entries and sending the results to Alice, together with p tacked on to the end.

p = RandomPrime[30]

```
703025443247135298024011470387
```

BobToAlice =
Append[Mod[BobData, p] + {0, 0, 0, 0, 0, 0, 0, 0, 0, 1, 1}, p]

```
{443727103860513954399139389249, 700456752974657268766432683382,
501835344456823215685644597614, 603455865559663131623304512220,
335593111480808590103296624630, 637624851316423804466941016527,
440768953412042766089618100463, 137078156437021906032343694032,
448842527549441342499994057680, 685015425640788629294594348741,
601792027335385441551522358063, 703025443247135298024011470387}
```

Mod[AliceRandom, p]

```
637624851316423804466941016527
```

Alice checks to see if her original random number, modulo p, is on this list. If it is, she knows that Bob did not add 1 to the corresponding encrypted version, and therefore her salary is less than or equal to Bob's; otherwise her salary is higher. In our example, the 6376...527 is on the list Alice sees, so she knows Bob's salary is \$65,000 or greater.

There are some flaws with this protocol!

- Alice or Bob could claim a false salary. There is no way around this problem; we simply assume that the intent to discover the truth is sincere or that an attempt to deceive would be found out at some future date.

- Alice learns the answer first. She can then refuse to tell Bob the result, or lie about it.

- If the protocol should lead to two identical numbers (or a zero) in the list that Bob returns to Alice, Alice can get some information she is not supposed to. Bob can easily avoid this problem. See Exercise 5.10.

Protocols similar to this have diverse applications, to electronic auctions, secure voting, contract signing by e-mail, and the like; see [Schn].

Exercise for Section 5.2

5.10. In the Yao millionaire protocol, it could happen that two of the numbers Bob produces and is planning to ship back to Alice differ by one. Why is this bad? What can be done to avoid this problem?

5.3 Check Digits

▪ Basic Check Digit Schemes

The most common typing errors are single-digit errors, where an incorrect character is entered, and transposition errors, where a pair of adjacent, characters are switched. Thus a classical problem is to attach an additional character to a data string, called a ***check digit*** and denoted c in this chapter, in the hope that its presence can help a computer detect errors, with special emphasis on the two most common errors.

A very simple scheme, which is currently used by the U.S. Postal Service on its money orders, is to simply take the raw number and let c be its mod-9 remainder. For example, if the raw number is 19023687, then $c = 0$ and the official number of the object would be taken to be 190236870. If one erroneously enters 190246870, the check digit won't check, because 0 is no longer the mod-9 value of the rest, and a computer can detect this. The origin of this method is probably the traditional folk method, called "casting out nines," where one repeatedly sums digits until a single digit remains, replacing 9s by 0 (these are the same modulo 9) to check arithmetic operations (see Exercise 5.11).

This method is certainly simple enough, but it has several problems. It fails to catch the error caused by the typing of a 9 instead of a 0, or vice versa, because such errors leave the mod-9 value unchanged. Worse, transpositions have no effect on the mod-9 value at all, so the method fails to detect any transposition errors (with the exception of transpositions involving c; these are detected). Of course, the reader may have already noticed that the mod-9 method is inefficient in the sense that c can never be 9, and so a potential check digit is wasted. But it turns out that even using a mod-7 system (in which c is never 7, 8, or 9) is far better than the mod-9 system, as we now show. This method is used on airline tickets.

Proposition 5.1. If a raw identification number has nine digits and a check digit c is taken to be the mod-7 residue of the raw number, then the single-digit-error correction rate is 93.81% and the transposition error correction rate is 93.87%.

Proof. The number of possible errors in the nine leftmost positions are (9 positions)(10 possible digits)(9 possible wrong digits), or 810. Of these, errors of the form $7 \leftrightarrow 0$, $8 \leftrightarrow 1$, or $9 \leftrightarrow 2$ will not be caught. Thus the success rate is $(9 \cdot 84)/810$. For the check position, there are no failures because the correct digit can never by 7, 8, or 9. Thus, the overall single-digit correction rate is $(9 \cdot 84 + 63)/(810 + 63)$, or $(756 + 63)/(810 + 63)$, or 93.81%.

Transposition errors of the form $70 \leftrightarrow 07$, $81 \leftrightarrow 18$, or $92 \leftrightarrow 29$ are undetected, but not all of these can occur in situations involving the check digit. There are nine possible transposition positions. For each of the leftmost eight, there are 90 possible transpositions, and 84 are caught; thus the rate is $(8 \cdot 84)/(8 \cdot 90)$. For the rightmost two digits there are $3 \cdot 7 + 7 \cdot 6$ possible errors and all are caught. Thus, the overall rate is $(8 \cdot 84 + 63)/(8 \cdot 90 + 63)$, or 93.87%. □

Of course, we can perform much more complex caclulations than a single modular reduction, and there are methods that take advantage of that fact. The ***3-weight method*** is used by banks: if the raw number is $a_1 a_2 a_3 a_4 a_5 a_6 a_7 a_8$, then the check digit is the dot product:

$$7a_1 + 3a_2 + 9a_3 + 7a_4 + 3a_5 + 9a_6 + 7a_7 + 3a_8 \pmod{10}.$$

We leave the computation of the success rate to Exercise 5.12. A widely used scheme that was first developed by IBM comes tantalizingly close to being perfect: it catches all the single-digit errors and misses only the transposition error $9 \leftrightarrow 0$. Here are the details. The method takes, say, a 13-digit number $a_1 a_2 \ldots a_{13}$ and defines the check digit to be

$$-(a_1, a_2, \ldots, a_{13}) \cdot (2, 1, 2, 1, 2, 1, \ldots, 1, 2) - r \pmod{10}$$

where r is the number of the odd-indexed digits $a_1, a_3, a_5, ..., a_{13}$ that are greater than 5. Complicated! But it is in some sense the best possible because, as we now prove, no scheme that uses ordinary modular arithmetic can catch all single-digit errors and all transposition errors. There is a perfect scheme in base 11 (Exercise 5.13), but the fact that "10" can show up as a check digit is a big drawback.

Theorem 5.2. There is no perfect check-digit scheme based on mod-10 addition.

Proof. Suppose that we had such a system, where c was computed by a mod-10 sum of values derived from each of the digits. Because different functions can be used in different positions, let $\sigma_i : \mathbb{Z}_{10} \to \mathbb{Z}_{10}$ be the function for the ith digit. Then digit d in position i contributes $\sigma_i(d)$ to the determination of c. To catch all single-digit errors, it must be that each σ_i is a permutation of {0, 1, 2, 3, 4, 5, 6, 7, 8, 9}. To catch all transpositions in the first two positions, we must have that $\sigma_1(a) + \sigma_2(b) \not\equiv \sigma_2(a) + \sigma_1(b) \pmod{10}$, which means that $x \mapsto \mathrm{Mod}[\sigma_2(x) - \sigma_1(x), 10]$ is a permutation of $\mathbb{Z}_{10}$. This leads to a contradiction as follows, where $\equiv$ is modulo 10.

$$5 \equiv 45$$
$$= 0 + 1 + 2 + \cdots + 9 \equiv \sigma_2(0) - \sigma_1(0) + \sigma_2(1) - \sigma_1(1) + \cdots + \sigma_2(9) - \sigma_1(9)$$
$$= \sigma_2(0) + \sigma_2(1) + \cdots + \sigma_2(9) - \sigma_1(0) - \sigma_1(1) - \cdots - \sigma_1(9)$$
$$= 45 - 45 = 0 \quad \square$$

A Perfect Check Digit Method

A ***group*** is a set of objects together with an associative operation, often called multiplication but sometimes called addition, that takes pairs of objects to a third object in the set. The group axioms require that there should be an identity (an element that leaves other elements unchanged under multiplication) and that each element should have an inverse, multiplication by which yields the identity. For example, {0, 1, 2, 3, 4, 5, 6, 7, 8, 9} forms a group under modular addition: 0 is the identity, and the inverse of a is $\mathrm{Mod}[-a, 10]$; this group is called $\mathbb{Z}_{10}$.

There is only one other group with ten elements: the group of symmetries of a regular pentagon, with two such symmetries yielding a third by first applying one and then the other. To be more precise, multiplication $\sigma \circ \tau$ is the symmetry of the pentagon defined by taking each vertex A to $\tau(A)$ and this, in turn, to $\sigma(\tau(A))$. It is easy to check that this is a group; in fact, there are two types of symmetries: the five rotations (0°, 72°, 144°, 216°, 288°) and the five reflections about perpendiculars from vertices to opposite sides. This group is an example of a dihedral group, and is denoted $\boldsymbol{D_5}$.

Now — and this fairly brilliant observation is due to J. Verhoeff in 1969 — we can use the ten symmetries to encode the ten digits by using the identity for 0, the rotations, in order, for 1, 2, 3, and 4, and the reflections for 5–9 (see Figure 5.1).

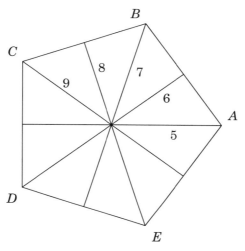

Figure 5.1. The five diagonals of a pentagon and their corresponding digits.

Here is the full multiplication table. Note that, unlike $\mathbb{Z}_{10}$, D_5 is not commutative: for example, $8 \circ 3 = 5$ while $3 \circ 8 = 6$. However, the rotations are commutaive so $i \circ j = j \circ i$ if $0 \le i, j \le 4$.

∘	0	1	2	3	4	5	6	7	8	9
0	0	1	2	3	4	5	6	7	8	9
1	1	2	3	4	0	6	7	8	9	5
2	2	3	4	0	1	7	8	9	5	6
3	3	4	0	1	2	8	9	5	6	7
4	4	0	1	2	3	9	5	6	7	8
5	5	9	8	7	6	0	4	3	2	1
6	6	5	9	8	7	1	0	4	3	2
7	7	6	5	9	8	2	1	0	4	3
8	8	7	6	5	9	3	2	1	0	4
9	9	8	7	6	5	4	3	2	1	0

Now let σ be the permutation of digits that acts as follows: σ switches 1 and 4, and also 2 and 3, and σ cyclically permutes 5, 8, 6, 9, 7; 0 is unmoved. One can summarize σ's behavior by

$$1 \leftrightarrow 4 \qquad 2 \leftrightarrow 3 \qquad 5 \to 8 \to 6 \to 9 \to 7 \to 5 \ .$$

A more concise notation is to just list the cycles: $\sigma = (0)(14)(23)(58697)$.

Given an integer n, such as 1793, Verhoeff's method performs the following operation:

$$\sigma^4(1) \circ \sigma^3(7) \circ \sigma^2(9) \circ \sigma(3)\,.$$

Use the multiplication table and the definition of σ and you will see the result is $1 \circ 6 \circ 5 \circ 2$, or 4. But we are not finished. The final step is to take the inverse in D_5 of the previous result. The inverse of 4 is 1 (because $1 \circ 4 = 0$, or because four rotations followed by one rotation is the same as five rotations, or doing nothing), so the check digit is 1. In general, if $abc...z$ denotes an n-digit number, c is chosen to be the inverse of $\sigma^n(a) \circ \sigma^{n-1}(b) \circ \cdots \circ \sigma(z)$.

As we shall show in a moment, this check digit scheme is perfect: it catches all single-digit errors and all transposition errors. Some investigators had predicted that such a perfect scheme could not exist! Well, it does, and it is used to add a check-digit to the serial numbers of German currency.

Theorem 5.3.   The check-digit scheme based on the symmetries of the pentagon catches all single-digit errors and all transposition errors.

Proof.   We will need the following property of σ, which Exercise 5.16 asks you to verify: $\sigma(a) \circ b \neq \sigma(b) \circ a$ whenever $a \neq b$. Now observe that, because of the inverse used in the last step of the check digit's definition, if c is the check digit of an n-digit number $abc...z$, then $\sigma^n(a) \circ \sigma^{n-1}(b) \circ \cdots \circ \sigma(z) \circ c = 0$. Now suppose —$a$—$c$ is mistyped as —b—c, where — represents any digit sequence, and the possibility that the error involves c itself is allowed. If the check digit test is passed, then $w \circ \sigma^i(a) \circ z$ would equal $w \circ \sigma^i(b) \circ z$, and cancellation in the group yields $a = b$, contradicting the assumption that an actual error was made.

For transpositions, suppose —ab—c is mistyped —ba—c, and the check digit test is passed. Then $w \circ \sigma^{i+1}(a) \circ \sigma^i(b) \circ z$ would equal $w \circ \sigma^{i+1}(b) \circ \sigma^i(a) \circ z$, where w and z are the contributions left and right of the error, respectively. Cancellation yields $\sigma^{i+1}(a) \circ \sigma^i(b) = \sigma^{i+1}(b) \circ \sigma^i(a)$, which can be rewritten as $\sigma(\sigma^i(a)) \circ \sigma(\sigma^i(b)) = \sigma(\sigma^i(b)) \circ \sigma(\sigma^i(a))$. This contradicts the property of σ that started this proof, unless $a = b$. $\square$

▥ Beyond Perfection: Correcting Errors

In a remarkable 1978 paper, A. S. Sethi, V. Rajaraman, and P. S. Kenjale of India [SRK] described a scheme by which check digits can be used to *correct* any single-digit error or transposition error. This is much stronger than the dihedral and other schemes, which merely detect errors. The theory of error-correcting codes has a long history, and, in fact, some schemes that allowed a certain amount of error-correction via check digits have been in use for some time. For example, the abstracts published by the American Chemical Society had used such a scheme for several years, though it now uses a different method. But the SRK scheme is noteworthy in that it detects and corrects *all* single-digit and transposition errors. Such an improvement comes at a cost: their schemes work only for certain bases, and require the use of two check characters. In this section we will describe their base-37 scheme, which allows the use of two check characters for strings made up of digits and the letters A, B, C, ..., Z.

Let p be 37, the first prime beyond 36. Then the SRK method would use the elements of $\mathbb{Z}_{37}$ to denote, in order, the characters "0", "1", "2", ... , "9", "A", "B", "C", "D", ... ,"X", "Y", "Z", "–". So the scheme can be used to add check digits to strings such as "123A–459WZ". It is a little weird to allow the hyphen to occur as a check digit, and a computer implementation might wish to disallow numbers that lead to such a thing.

Their system uses modular arithmetic in a familiar way, but with two check digits. Thus a raw string $a_1 a_2 \cdots a_k$ will be enhanced to $A = a_1 a_2 \cdots a_k a_{k+1} a_{k+2}$. It turns out that there are restrictions on k. We shall see why, in the case at hand, the SRK method leads to $k \le 11$.

The basic definition of the check digits is simple. We will find two sets of weights $W = \{w_i\}$ and $V = \{v_i\}$, and then choose a_{k+1} and a_{k+2} to satisfy the following two congruences, where $n = k + 2$.

$$\sum_{i=1}^{n} w_i a_i \equiv 0 \pmod{p}$$

$$\sum_{i=1}^{n} v_i a_i \equiv 0 \pmod{p} \tag{1}$$

Finding the two check digits will be simple because they satisfy two linear equations in two unknowns: $w_{k+1} a_{k+1} + w_{k+2} a_{k+2} \equiv r$ and $v_{k+1} a_{k+1} + v_{k+2} a_{k+2} \equiv r_1$. The weights will be chosen so that the corresponding 2×2 determinant is not divisible by p — conditon C_3 in the list that follows — and so the system will have a unique rational solution with denominators not divisible by p. Such rationals yield a solution in integers between 0 and p by using the mod-p inverse of the denominators.

Note that condition (1) can also be interpreted as $W \cdot A \equiv V \cdot A \equiv 0 \pmod{p}$, where $\cdot$ is the standard dot product.

The key to the method is to find weights W and V that satisfy the following five conditions, where i and j are assumed to be distinct.

C_1: $1 \le v_i, w_i < p$ for $i = 1, 2, ..., n$.

C_2: The W-values are distinct, as are the V-values.

C_3: $v_i w_j \not\equiv w_i v_j \pmod{p}$ for $i = 1, 2, ..., n$ and $j = 1, 2, ..., n$.

C_4: $(v_{i+1} - v_i)(w_{j+1} - w_j) \not\equiv (w_{i+1} - w_i)(v_{j+1} - v_j) \pmod{p}$ for $i = 1, 2, ..., n - 1$ and $j = 1, 2, ..., n - 1$.

C_5: $(w_i - w_{i+1})v_j \not\equiv (v_i - v_{i+1})w_j \pmod{p}$ for $i = 1, 2, ..., n$ and $j = 1, 2, ..., n$.

How can we find appropriate weights? One of the two methods in [SRK] is to let a be a positive integer less than $p - n$ and let $w_i = i + a$ and $v_i = i(i + a)$ (both reduced modulo p) for $i = 1, ..., n$. C_1 is then clear, as is C_2 for the case of W. For V, we want $(i + a)i \not\equiv (j + a)j$ which simplifies to $(i - j)(i + j + a) \not\equiv 0$. The first factor is not divisible by p because of the bounds on i and j. The second factor will be nonzero (mod p) provided $i + j + a < p$, which holds if $n + (n - 1) + a < p$, or

$$2n + a < p + 1. \tag{2}$$

Conditions C_3 and C_4 follow directly from the definitions: C_3 reduces to $(a + i)(i - j)(a + j) \not\equiv 0$ which follows from the boundary conditions on a, i, and j, and C_4 is similar. And C_5 reduces to

$$(j + a)(2i - j + a + 1) \not\equiv 0 \;(\text{mod } p).$$

But note that $j + a$ is never divisible by p because $1 \le a < p - n$. So the condition reduces to

$$2i - j + a + 1 \not\equiv 0 \;(\text{mod } p).$$

To get this condition, it suffices to choose a and n so that $0 < 2i - j + a + 1 < p$. Because $i \le n$, $j \le n$, and $i \ne j$, the upper bound will hold if $2n + a < p + 2$. But this will follow from condition (2). For the lower bound, let $i = 1$ and $j = n$, the extreme values. We want $0 < 2 - n + a + 1$ or $n < a + 3$. This will hold, and n will be maximized, if we let $a = n - 2$.

So now we see how to get the weights and appropriate values of a and n. For n, condition (2) leads to $2n + a \le p$, which becomes, using $a = n - 2$, $3n - 2 \le p$ or $n \le \lfloor (p + 2)/3 \rfloor$. In our case p is 37, so n can be 13 and a is 11. Then the weights are as follows.

```
p = 37; n = 13; a = n - 2;
W = Table[Mod[i + a, p], {i, 1, n}]
V = Table[Mod[i (i + a), p], {i, 1, n}]
```

{12, 13, 14, 15, 16, 17, 18, 19, 20, 21, 22, 23, 24}

{12, 26, 5, 23, 6, 28, 15, 4, 32, 25, 20, 17, 16}

Now let us see how the five conditions lead to error detection and correction. Given a string A, let S_1 and S_2 denote the two "check-sums" for A, as given by formula (1). All congruences in the paragraphs that follow are modulo 37. First observe that, because of C_1 and C_2, any single-digit error or transposition error leads to values of S_1 and S_2 that are both nonzero (Exercise 5.19).

Let A denote the string entered and suppose a single-digit error was made, say d became e in position i. Then the two check-sums for A would satisfy $S_1 \equiv (e - d)w_i$ and $S_2 \equiv (e - d)v_i$, respectively (since each check-sum is 0 for the errorless string). Cross-multiplying and using the primality of

37 to cancel $e - d$ leads to $S_1 v_i \equiv S_2 w_i$. Now, for any $j \neq i$, $S_1 v_j \not\equiv S_2 w_j$, for otherwise $S_1 v_i S_2 w_j \equiv S_i v_j S_2 w_i$ in which we can cancel the S's (they are both nonzero modulo p) and get a contradiction to condition (3). So in this case there is a unique value of i such that $S_1 v_i \equiv S_2 w_i$. A computer can find such an i — that is, locate the position of the error — and then compute the correct digit by finding which value of d leads to $S_1 \equiv (e - d) w_i$. The correct digit is simply $e - S_1 w_i^{-1}$, using the mod-37 inverse.

A similar argument using C_4 shows how transposition errors can be detected; the critical equation is $S_1(v_{i+1} - v_i) \equiv S_2(w_{i+1} - w_i)$. Note that for transposition errors correction is trivial once the location of the error is known. Finally, we must be sure that the computation never confuses a single-digit error with a transposition error, and that is what C_5 accomplishes.

The implementation of the SRK error-correction scheme is left as an exercise (Exercises 5.22 and 23). All of us are quite used to having computers beep when an error of some sort is made at the keyboard; having the computer correct the error is much more impressive. An implementation is included in the CNT package. Here is an example, where the input string has the right number of letters, 11, for the parameters of this section

SRKProtect["**PAINTTHECAR**"]

PAINTTHECARRN

SRKCorrect["**PAINTTHECATRN**"]

PAINTTHECARRN

SRKCorrect["**PANITTHECARRN**"]

PAINTTHECARRN

A second method of finding the weights is given in [SRK]; it is a little better than the method presented here because it allows the protected string can be as long as 18 characters, as opposed to 13.

Exercises for Section 5.3

5.11. Show that summing the digits of an integer, summing the digits of the result, and continuing until only a single digit remains leaves the remainder modulo 9 of the original number (provided one replaces a final 9 with a 0).

5.12. What are the single-digit and transposition success rates for the 3-weight method that uses the mod-10 dot product $(7, 3, 9, 7, 3, 9, 7, 3) \cdot (a_1, a_2, a_3, a_4, a_5, a_6, a_7, a_8)$?

5.13. The scheme used in ISBN numbers (International Standard Book Numbers, which are used to register every book with a number) is perfect in the sense that all single-digit and transposition errors are caught. It uses base-11 arithmetic, with an X used to represent 10 in base 11. Thus, it is not a purely numeric scheme, which can cause programming and other difficulties. Show that the method is perfect. The raw ID number has 9 digits (the leading digit codes the language, the next four code the publisher, and the next four identify the book). Given such an ID number $a_1 a_2 \ldots a_9$, the check digit c is computed so that $10a_1 + 9a_2 + 8a_3 + \cdots + 3a_8 + 2a_9 + c \equiv 0 \pmod{11}$; if c ends up being 10, then X is used as the check digit.

5.14. Prove that the IBM check digit scheme catches all single-digit errors and all transposition errors except $9 \leftrightarrow 0$.

5.15. Implement a routine that creates random single-digit or random transposition errors and then run a simulation to see if the success rate of a particular check-digit scheme matches the rates predicted in this section and earlier exercises.

5.16. Verify that the permutation σ of Theorem 5.3 satisfies the critical property: $\sigma(a) \circ b \neq \sigma(b) \circ a$ whenever $a \neq b$.

5.17. Find some other permutations of 0, 1, 2, 3, ..., 9 that have the same critical property of σ? Why is it preferable to use a permutation that leaves 0 unchanged?

5.18. Write a computer program that takes any integer and appends the check digit as defined by Verhoeff's pentagonal scheme.

5.19. Show that a single-digit or transposition error in a string that is protected by the SRK scheme leads to nonzero values for both check sums in (1).

5.20. Show how condition C_4 in the SRK scheme allows the method to detect and correct a transposition error.

5.21. Show how condition C_5 in the SRK scheme guarantees that the equations governing the detection of a single-digit or transposition error will never clash with each other. In other words, show that the SRK protocol will always be successful in detecting the type of an error, provided only one error is made.

5.22. Implement a routine that accepts an 11-character string using the characters "0", "1", "2", ...,"9", "A", "B", "C", "D", ...,"X", "Y", "Z", "−" and computes the two check digits according to the SRK scheme. Short inputs should be padded with 0s on the left. The *Mathematica* code that follows shows how to turn a string into a list of codes between 0 and 37, and vice versa.

```
ToSRKInteger[s_String] := (c = First[ToCharacterCode[s]] - 55;
    Which[c > 9, c, -8 < c < 3, c + 7, c == -10, 36]) /;
        StringLength[s] == 1

ToSRKInteger[s_String] := Map[ToSRKInteger, Characters[s]]
```

```
FromSRKInteger[n_Integer] := (Which[
    n < 10, ToString[n],
    9 < n < 36, FromCharacterCode[n + 55],
    n == 36, "-"])

FromSRKInteger[s_List] := Apply[StringJoin,
    Map[FromSRKInteger, s]]

ToSRKInteger["ABC-0123X-2Z"]

{10, 11, 12, 36, 0, 1, 2, 3, 33, 36, 2, 35}

FromSRKInteger[{10, 11, 12, 36, 0, 1, 2, 3, 33, 36, 2, 35}]

ABC-0123X-2Z
```

5.23. Implement a routine that accepts a 13-character string using "0", "1", "2", ..., "9", "A", "B", "C", "D", ...,"X", "Y", "Z", "–" and checks to see if the two rightmost characters are the proper SRK check digits. If they are not, then the scheme checks to see if one single-digit error or one transposition error has been made, in which case it returns the correct string.

5.4 Factoring Algorithms

The use of large numbers in cryptography brought the problem of factoring large integers to the forefront of computational number theory. Faced with a large integer to be factored, our first task is to determine whether or not it is probably prime. If, in fact, it is composite, this will almost certainly be revealed by a pseudoprime or strong pseudoprime test. Knowing that our integer is composite, what comes next? Indeed, the factoring problem is difficult in part because, aside from trial division, there is no obvious way to proceed. The reader may find it surprising that there is no single algorithm that one always uses for factoring whatever integer is presented. The choice of algorithm is heavily dependent on the size of the integer and on what else is known about it.

▪ Trial Division

The best place to start is with simple trial division. Using a list of primes generated by the sieve of Eratosthenes, one simply checks if they divide the integer in question.

Algorithm 5.1. Factorization by Trial Division

This algorithm attempts to factor n by trial division by all primes less than or equal to m. The parameter u tracks the unfactored portion that remains.

If at any point, u is less than the square of the prime we are about to trial divide, then we know that u is prime. If we finish trial division with an unfactored u larger than 1, then this algorithm runs a 2-strong pseudoprime test to see if the unfactored piece is probably prime. It returns the value of u and the truth value of the strong pseudoprime test.

```
TrialDivide[n_, m_] := Module[{u = n, ans = {}},
   Scan[
     (If[u < #^2, If[u > 1, AppendTo[ans, {u, 1}]]; u = 1; Return[]];
       If[Mod[u, #] == 0, j = 0; While[Mod[u, #] == 0, u /= #; j++];
       AppendTo[ans, {#, j}]]) &, Prime[Range[π[m]]]];
   If[u == 1, ans,
     Append[ans, {u, StrongPseudoprimeQ[2, u] /.
       {False → "Composite", True → "Probably Prime"}}]]]]
```

```
TrialDivide[268663268514419, 10000]
```

```
{{541, 1}, {7919, 3}}
```

```
TrialDivide[268663268514419, 1000]
```

```
{{541, 1}, {496604932559, Composite}}
```

No algorithm is as efficient as trial division for finding small prime factors (though one can improve efficiency by using gcds of the input with the precomputed products of several small primes at once). On the other hand, even with fast computers, this algorithm is impractical for finding factors larger than 10^7.

The beauty of this algorithm is that if you know you have a prime divisor of a given size, you know exactly how long it will take to find that divisor. This is what we mean by "completely deterministic". All the powerful factoring algorithms rely on a certain random element so that luck begins to play a role. If there is a prime divisor of a given size, we can predict how long the algorithm will run on average, but we cannot be certain that we will not run considerably over that time. These "probabilistic" algorithms also share the characteristic that they are not always guaranteed to find a prime divisor. All they will do for you is to break your integer into a product of two smaller integers. You then have to test each of these factors for probable primality and, if they are shown to be composite, run them through again until they have been reduced to a product of probable primes.

The probabilistic algorithms come in two broad categories. The *first category* consists of those algorithms that are most likely to find the smallest prime divisor. Their running time is more dependent on the size of the smallest prime divisor than on the size of the integer that one wants to factor. Such algorithms include Pollard rho, which was used to factor the Fermat number F_8, Pollard $p - 1$, Williams $p + 1$, and the elliptic curve method, which was used on F_{10} and F_{11}. These are simple, fast algorithms that can be effective when there is a prime divisor having between 7 and 40 digits. They are extremely effective at finding divisors of less than 20 digits. They should always be used as the second line of attack.

If you want to factor an integer of 100 or more digits and you are not getting anywhere with first category algorithms, then you need to move to the **second category**, also known as the **Kraitchik family**, after the number theorist Maurice Kraitchik (1882–1957). Let n be the integer that you want to factor. All of these second category algorithms rest on the search for "randomly generated" pairs of integers (a, b) such that $a^2 \equiv b^2$ (mod n). This implies that n divides $a^2 - b^2 = (a - b)(a + b)$. If p and q are distinct primes that divide n and if (a, b) is really chosen at random, then we have a 50% chance that gcd$(n, a - b)$ is divisible by one of these primes and not the other. In other words, there is at least a 50% chance that gcd$(n, a - b)$ is a nontrivial divisor of n.

All the algorithms of the Kraitchik family are complicated, and one draws no benefit from the existence of small prime divisors of n. Their running time is essentially independent of the size of the smallest prime factor and depends only on the size of n. This is why they should not be used until after trial division and first category algorithms have been pushed as far as possible. On the other hand, they draw no disadvantage from the fact that all of the prime divisors are large and therefore these algorithms are ideally suited to finding factorizations when there are no small prime divisors. From this family, the continued fraction algorithm was used to crack F_7. The Lenstra and Manasse factorization of a 100-digit integer used the multiple polynomial quadratic sieve. F_9 fell to the number field sieve. At present, these methods reach their practical limitations for general integers that are difficult to factor when n has about 130 digits. The method of this type that we shall discuss is the continued fraction method, but that must wait until Section 7.6.

▣ Fermat's Algorithm

Trial division is not the only completely deterministic algorithm available. Fermat discovered an algorithm that he prized which, instead of starting with small primes and working up, begins by looking for divisors near the square root and works down. One of its great advantages is that it involves no division, except for division by 2 at the end.

The idea is as follows. If n is odd and composite, $n = ab$, then $x = (a + b)/2$ and $y = (a - b)/2$ are integers and

$$n = ab = (x + y)(x - y) = x^2 - y^2 .$$

On the other hand, if we can write n as the difference of two squares, $n = x^2 - y^2$, then this yields a factorization of n as $(x - y)(x + y)$.

The algorithm proceeds by starting with $x = \lceil \sqrt{n} \rceil$ and $y = 0$. A fast algorithm for finding $\lfloor \sqrt{n} \rfloor$ is given in Exercise 5.24. Then each cycle increments x or y by 1 until we find the first pair of values for which $n = x^2 - y^2$. If $x^2 - y^2 > n$, then we want to increase y by 1. If $x^2 - y^2 < n$, then we increase x by 1. We can further streamline this algorithm if we just keep track of

$$r = x^2 - y^2 - n,$$
$$u = 2x + 1, \text{ and}$$
$$v = 2y + 1.$$

The variable u tracks the amount by which we increase r when x is increased; v does the same for an increase in y. When $r = 0$, we know that

$$n = (x - y)(x + y) = \left(\frac{u - v}{2}\right)\left(\frac{u + v - 2}{2}\right).$$

Algorithm 5.2. Fermat's Factorization Algorithm

This algorithm takes an odd composite n and returns a pair of integers whose product is n and that are as close as possible to $\sqrt{n}$. The SqrtFloor function is discussed in Exercise 5.24, though one could simply use Floor[Sqrt[ ]]. The If statement near the end is adequate, as opposed to While, because it will always take only one upward u-step to make r positive (because $u > v$).

```
SqrtFloor[n_] := Module[{a = n, b = Quotient[n + 1, 2]},
  While[b < a, a = b; b = Quotient[a² + n, 2 a]]; a]

FermatFactor[n_ ? OddQ] :=
 Module[{s = SqrtFloor[n] + 1, r, u, v = 1},
   If[(s - 1)² == n, Return[{s - 1, s - 1}]];
   {r, u} = {s² - n, 2 s + 1};
   While[r ≠ 0,
     While[r > 0, r -= v; v += 2]; If[r < 0, r += u; u += 2]];
   {u - v, u + v - 2} / 2]

FermatFactor[2027651281]

{44021, 46061}
```

◼ Pollard Rho

In 1974–75, J. M. Pollard published the first two category-1 factorization algorithms, what today we know as **Pollard $p - 1$** and **Pollard rho**. Despite its age, the second is still popular for finding prime divisors in the range of 7 to 20 digits. Pollard $p - 1$ is seldom used anymore, but it illuminates the theoretical foundation for the elliptic curve method.

Pollard called his second algorithm the Monte Carlo method. It is based on the birthday paradox, that given 23 randomly selected people, there is better than a 50–50 chance that at least two of them share the same birthday. In general, if we perform slightly more than $\sqrt{n}$ selections with replacement from a set of n elements, we have better than a 50–50 chance of picking some element twice (Exercise 5.29).

Let n be the composite integer for which we want to find a proper divisor. In the Pollard rho algorithm, we begin with a randomly chosen seed s, and generate the sequence

$$s_0 = s$$
$$s_i = s_{i-1}^2 + 1 \pmod{n}, \ 1 \le s_i < n \,.$$

As an example, let n be 59153 and let the seed s be 24712. The sequence that we get is

```
n = 59153;
s[0] = 24712;
s[i_] := Mod[s[i - 1]^2 + 1, n]
```

```
Map[s, Range[15]]
```

```
{46526, 23795, 48663, 15521, 30426, 56180, 24933,
 15613, 55410, 49942, 17120, 50439, 40498, 11927, 49518}
```

This is an apparently random sequence of integers in [0, 59152].

Unbeknownst to us, 149 is a prime divisor of 59153. Modulo 149, this sequence looks like

```
FunctionTable[Mod[s[#], 149] &, 15,
 FunctionHeading → "s(n) (mod 149)", Columns → 4]
```

n	$s(n)$ (mod 149)	n	$s(n)$ (mod 149)	n	$s(n)$ (mod 149)	n	$s(n)$ (mod 149)
1	38	5	30	9	131	13	119
2	104	6	7	10	27	14	7
3	89	7	50	11	134	15	50
4	25	8	117	12	77	16	117

We see that s_{14} is the same as s_6. Because each s_i is completely determined by the previous value, once we get a repetition we enter a never-ending cycle, in this case a cycle of length 8. We can visualize what is happening as a circle with a tail: s_1–s_5 are on the tail. In other words, it looks like the Greek letter ρ, spelled rho (see Figure 5.1).

If we now go back to what we do know, the sequence as it appears modulo 59153, and we take any $i > 5$, then 149 divides $s_{i+8} - s_i$. Fortunately 397, which is the other factor of 59153 does not divide $s_{i+8} - s_i$. This means that $\gcd(s_{14} - s_6, 59153) = \gcd(11927 - 56180, 59153) = 149$.

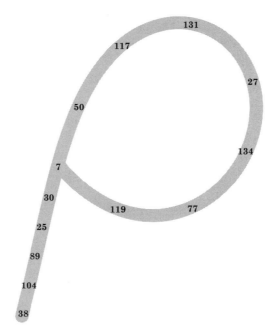

Figure 5.1. The Pollard sequence has a tail and a loop, and so may be viewed as a Greek rho.

Of course, we do not know that the correct cycle length is 8, nor that we must take $i > 5$. But we can expect that if there is a divisor around size m then the cycle length and tail length will be around $0.59\sqrt{m}$. Once we have roughly $1.18\sqrt{m}$ values of s_i, the probability that they are distinct modulo m drops below 50% (see Exercise 5.30), so we can expect that it will take about $1.18\sqrt{m}$ iterations before we get a residue, modulo m, that has appeared before. Since all of the residues that have appeared are, in theory, equally likely to be the first to repeat, the expected position of the first residue to repeat is at the midpoint. We try all possible cycle lengths until we find one that works and take the tail length close to the cycle length. It does not really matter if we try what turns out to be the correct cycle length while we are still on the tail because any multiple of that cycle length will also work.

The version that we describe below is due to Brent. It considers the differences:

$$s_1 - s_3,$$
$$s_3 - s_6, \; s_3 - s_7,$$
$$s_7 - s_{12}, \; s_7 - s_{13}, \; s_7 - s_{14}, \; s_7 - s_{15},$$
$$\vdots$$
$$s_{2^n-1} - s_j, \; 2^{n+1} - 2^{n-1} \le j \le 2^{n+1} - 1.$$

Warning. You should only use category-1 or category-2 algorithms such as Pollard rho after a pseudoprime test has confirmed that the integer in question is composite. In general, a prime input will either prevent the algorithm from terminating or will produce an otherwise inconclusive response.

Algorithm 5.3. Pollard Rho

This version of Pollard rho computes the gcd after each step. Exercise 5.31 asks you to speed it up by computing the gcd only after ten differences have been collected. The use of max_ : 10000 is a way of assigning a default value to max, which is the maximum number of terms of the sequence. $Failed is a system variable that is often used to indicate failure, so we use it here.

```
PollardRho::maxtrm =
  "The number of terms of the Pollard sequence
    has exceeded its maximum with no repeats.";

PollardRho::allfac = "The potential gcd equals
  `` , and so a factor has not been identified.";

PollardRho[n_, max_: 10000] :=
 Module[{s = Random[Integer, {1, n - 1}], g = 1, t, range = 1,
   prod = 1, terms = 0, hf = HoldForm[PollardRho[n, max]]},
  t = Mod[s^2 + 1, n];
  While[terms ≤ max && g == 1,
   Do[t = Mod[t^2 + 1, n]; prod = Mod[prod * (s - t), n];
    terms ++; g = GCD[n, prod]; If[g > 1, Break[]], {range}];
   s = t; range *= 2; Do[t = Mod[t^2 + 1, n], {range}]];
  Which[
   g == 1, Message[PollardRho::maxtrm]; hf,
   g == n, Message[PollardRho::allfac, n]; hf,
   True, {g, n / g} ]]
```

PollardRho[20584996606709]

{1316717, 15633577}

Of course, it can happen that the sequence will not run long enough to find a repeat. As an exercise, you can reprogram it so that old results are saved and if the user decides to attempt a longer sequence, that can be done without starting over from the beginning (Exercise 5.33)

PollardRho[137703491, 10]

```
PollardRho::maxtrm :
  The number of terms of the Pollard sequence
    has exceeded its maximum with no repeats.
```

PollardRho[137703491, 10]

```
PollardRho[137703491, 100]
```

{7919, 17389}

Pollard p − 1

Pollard's earlier category-1 algorithm — the $p-1$ method — is based on Fermat's Little Theorem. Let p be a prime divisor of n, the number that we want to factor. We know that if a, a random integer, is not divisible by p, then $a^{p-1} \equiv 1 \pmod{p}$. This means that p divides $\gcd(a^{p-1} - 1, n)$. Since it is extremely unlikely that $p-1$ is a multiple of the order of a for other prime divisors of n, this greatest common divisor will be p.

Of course, we do not know p, and so we do not know $p-1$. But any multiple of $p-1$ will work just as well. In fact, all we need is a multiple of $\operatorname{ord}_p(a)$. If k is at least as large as the largest prime divisor of $\operatorname{ord}_p(a)$, then $k!$ should be a multiple of $\operatorname{ord}_p(a)$. An advanced result states that on average, the largest prime divisor of m is approximated by $m^{1-1/e}$, or $m^{0.632}$ (see Exercise 5.26). This means that if p is prime divisor of n, then p should divide $\gcd(a^{k!} - 1, n)$ provided $k \geq p^{0.64}$. We do our exponentiation modulo n so that it goes quickly. The drawback is that if p is large, we have a lot of exponentiations to carry out.

Algorithm 5.4. Pollard p − 1

If the algorithm returns $\{n, 1\}$, then you have picked up all of the factors at once. The code then returns a message suggesting that d, the number of iterations before calculating the gcd, be reduced.

```
PollardpMinus1::maxtrm =
  "The number of terms of the Pollard sequence
    has exceeded its maximum with no repeats.";

PollardpMinus1::allfac = "The potential gcd equals
    ``, and so a factor has not been identified.";

PollardpMinus1[n_, max_: 10000] :=
 Module[{a = Random[Integer, {2, n - 2}], g = 1,
   i = 0, hf = HoldForm[PollardpMinus1[n, max]]},
  While[i ≤ max && g == 1, i++; a = PowerMod[a, i, n];
   g = GCD[n, a - 1]];
  Which[
   g == 1, Message[PollardpMinus1::maxtrm]; hf,
   g == n, Message[PollardpMinus1::allfac, n]; hf,
   True, {g, n / g}]]
```

```
PollardpMinus1[2252004283651]
```

{1303831, 1727221}

```
PollardpMinus1[2^{2^5} + 1]
```

{641, 6700417}

This algorithm, of course, works best if $p-1$ has only small divisors. Here's an example, where the smaller prime is $547 \cdot 2 \cdot 3 \cdot 5 \cdot 7 \cdot 11 \cdot 13 \cdot 17 \cdot 19 \cdot 23 \cdot 29 \cdot 31 \cdot 37 \cdot 41 \cdot 43 \cdot 47 + 1$.

```
PollardpMinus1[
  2037613333718005172796442787984180064325842449569596945256256256;
  97]
```

{336344711075904801271,
  6058110226261787479078952914361290709407}

■ The Current Scene

The Pollard $p-1$ algorithm is almost always less efficient than Pollard rho, but it contains an idea that has led to other category-1 algorithms. The reduced residue system modulo p is an example of a finite group: It is closed under multiplication and has an identity and multiplicative inverses. Each element of a finite group has an order, the smallest power of that element that is equal to the identity, and this order must divide the size of the group. Pollard $p-1$ uses a group of size $p-1$ that is based on arithmetic modulo p.

There are other finite groups that use arithmetic modulo p in their definition. Lucas sequences give rise to one of these, and we shall study them in more detail in Chapter 8. The arithmetic structure of Lucas sequences can be used to find those prime factors p for which $p+1$ has only small prime divisors. This gives rise to a $p+1$ factoring algorithm developed by Hugh Williams.

We get real flexibility when we use the arithmetic of elliptic curves, an important area of number theory that is beyond the scope of this book. Elliptic curves arise in many contexts, most famously in Andrew Wiles's proof of Fermat's Last Theorem. One can use them to construct finite groups based on arithmetic modulo p, groups that have all possible numbers of elements from $\lceil p+1-2\sqrt{p} \rceil$ to $\lfloor p+1+2\sqrt{p} \rfloor$. As we choose groups at random, we have a greatly improved chance that we will select one whose order is a divisor of $k!$ for relatively small k. This is the fundamental idea behind Lenstra's elliptic curve algorithm that has been used to factor F_{10} and F_{11}.

The Fermat number $F_{10} = 2^{2^{10}} + 1$ is a 309-digit integer. It has two relatively small prime factors: 45592577 and 6487031809 that were discovered, respectively, by John Selfridge in 1953 and John Brillhart in 1962, but that left a 291-digit integer that was known to be composite. In 1995 R. P. Brent used the elliptic curve method to find a 40-digit prime that divides F_{10}, and then proved that the remaining piece, a 252-digit integer, is prime. As this book goes to press, the largest prime factor found by the elliptic curve method is the 53-digit factor of $2^{677} - 1$ that was discovered by

Conrad Curry in 1998. The latest record for the elliptic curve method can be found at www.loria.fr/~zimmerma/records/ecmnet.html. A description of this algorithm can be found in [CP].

Category-1 factorization methods are useful for finding a prime divisor of around 50 digits or fewer, and they are relatively insensitive to the size of the number you are trying to factor, but the real workhorses of factorization are the category-2 methods. These algorithms have running times that are direct functions of the size of the number you wish to factor and are insensitive to the size of the smallest prime divisor. The first category-2 method was a variation of what today is called the quadratic sieve. It was proposed by Maurice Kraitchik in the 1920s but was not implemented until the advent of massive and inexpensive computer memory in the early 1980s. The first of the category-2 methods to be programmed into an electronic computer was the continued fraction algorithm that we shall see in Chapter 7. It was used by Brillhart and Morrison in 1975 to factor the 39-digit integer F_7. The category-2 algorithms in use today are the quadratic sieve and the number field sieve, each with its own variations. The quadratic sieve was used to factor RSA-129, a 129-digit behemoth created by Rivest, Shamir, and Adleman in 1977 and publicized by Martin Gardner in his *Scientific American* column as a challenge problem to those who would try to crack the RSA public-key cryptosystem. In 1994, the quadratic sieve produced the factorization of RSA-129 into a 64-digit and a 65-digit prime. This was a herculean effort that required eight months on 1600 computers yoked in parallel. But the record for the factorization with the largest least prime divisor goes to the number field sieve, which The Cabal, coordinated by Peter Montgomery, used to factor $(10^{211} - 1)/9$ (a repunit with 211 1s) into a 93-digit and a 118-digit prime. Unfortunately, this version of the number field sieve works only on integers such as $10^{211} - 1$ that are very close to a power of a very small integer. A variation of it, the general number field sieve, works on any integers, but so far has only been successful in factoring integers of less than 120 digits. The quadratic and number field sieves, together with their most useful variations, are described in detail by Crandall and Pomerance [CP].

Most factorization records have arisen from the Cunningham Project, a dispersed effort to factor all integers of the form $b^n \pm 1$ for $2 \le b \le 12$. The current status of this project, the smallest composite integers that have not been factored, and the records for the two largest prime divisors found by each of the category 1 methods and for the two largest composite integers factored by each of the category 2 methods can be found at www.cerias.purdue.edu/homes/ssw/cun/index.html.

Exercises for Section 5.4

5.24. Explain why the following algorithm returns $\lfloor \sqrt{n} \rfloor$. *Hint*: It is based on the Newton–Raphson method for finding the root of a polynomial, in this case a root of $x^2 - n$. Note that Quotient[a, b] is faster than Floor[a, b].

```
SqrtFloor[n_] := Module[{a = n, b = Quotient[n + 1, 2]},
    While[b < a, a = b; b = Quotient[a^2 + n, 2 a]]; a]
```

5.25. Use Algorithm 5.2 to factor 2027651281.

5.26. Perform some numerical computations to support the theorem that, on average, if $p(n)$ is the largest prime divisor of n, then $\log_n p(n)$ is asymptotic to $1 - 1/e$ (**asymptotic** means that the limit of the quotient is 1; the use of *average* means that you must study the average value of $\log_n p(n)$ over a large number of integers).

5.27. Consider the following factorization algorithm: Given a composite integer n, we choose a random integer r between 1 and n and compute $g = \gcd(r, n)$. If g is 1 or n, then we choose another random integer and repeat. If g is not 1 or n, then we have found a proper divisor of n. How does this algorithm differ from Algorithm 5.1 and, in particular, how many tries would you expect this algorithm to take if the smallest prime divisor of n is of size m?

5.28. The following program selects a prime p below one million and a random seed s. Starting with $s_0 = s$, it calculates the sequence $s_{i+1} = s_i^2 + 1 \pmod{p}$ until it finds i and j such that $s_i = s_{i+j}$. It then returns the values of p, i (the length of the tail), and j (the length of the cycle). Do multiple trials and apply whatever statistical tools you have available to see how well the outcomes match the predicted value of slightly more than $\sqrt{p}/2$ for i and j.

```
RhoTrials := Module[
    {p = Prime[Random[Integer, {1, PrimePi[10^6]}]], s, seq = {}},
    s = Random[Integer, {2, p - 1}];
    While[FreeQ[seq, s], seq = {seq, s}; s = Mod[s^2 + 1, p]];
    seq = Flatten[seq];
    Flatten[{p, Position[seq, s] - 1,
      Length[seq] - Position[seq, s] + 1}]]
```

5.29. For each of the following values of m — 25, 100, 365, 10,000, 1,000,000 — find the smallest integer t for which

$$\frac{m-1}{m} \cdot \frac{m-2}{m} \cdots \frac{m-(t-1)}{m} < \frac{1}{2}.$$

5.30. We can get a rough approximation of the smallest value of t (as a function of m) for which $\frac{m-1}{m} \cdot \frac{m-2}{m} \cdots \frac{m-(t-1)}{m} < \frac{1}{2}$ by taking the logarithm of each side, replacing the logarithm of $1 - \frac{j}{m}$ by $-\frac{j}{m}$, and then replace $t^2 - t$ by t^2. Show that this approximation gives t to be roughly $1.18\sqrt{m}$. Test this rough approximation against the values calculated in Exercise 5.29.

5.31. Speed up the Pollard rho algorithm by doing the gcds 10 at a time. Or d at a time.

5.32. If you are willing to run the Pollard rho algorithm for 100,000 iterations, how large a prime factor can you expect to be able to find?

5.33. Improve the code for Pollard rho so that, perhaps via an option, partial results are stored so that the user can continue and use these already computed values if he or she wishes to try again with a longer sequence. Perhaps the output, in case of failure, could have the form: "No success yet. Shall I continue for another 10,000 terms?"

CHAPTER **6**

Quadratic Residues

6.0 Introduction

We learned in Chapter 2 how to solve linear congruences $ax + b \equiv 0$ (mod m). What about quadratic congruences of the form $ax^2 + bx + c \equiv 0$ (mod m)? The problem is both harder and richer. In solving it, we shall develop tools that have implications for primality testing (as in Pépin's test) and factorization (as in the quadratic sieve).

When the mathematicians of the eighteenth century tackled this problem, they soon realized that it boiled down to the simple case $x^2 \equiv a \pmod{p}$, where p is an odd prime. When does this have a solution? If a solution exists, how does one find it? There are two approaches to studying this congruence:

- One can fix p and ask for the values of a that have a square root modulo p.
- One can fix a and ask for the moduli p for which a square root of a exists.

The greatest breakthrough comes from the realization that these questions are connected. Knowing the answer to one will yield an answer to the other. This is the content of the celebrated quadratic reciprocity theorem.

6.1 Pépin's Test

▣ Quadratic Residues

As we saw in the last chapter, if we can factor $n - 1$, then we have a reliable and fast algorithm for proving n to be prime. For most values of n over 10^{100} this is not a practical approach, but if n is of a special form, say $n =$

$2^m + 1$, then we can verify primality quickly, even for very large n. As Fermat realized, if $2^m + 1$ is prime, then m must be a power of 2 (see Exercise 2.42).

Fermat also observed that each of the following numbers is prime.

$$2 + 1 = 3$$
$$2^2 + 1 = 5$$
$$2^4 + 1 = 17$$
$$2^8 + 1 = 257$$
$$2^{16} + 1 = 65537$$

At one time, he believed that $2^m + 1$ would be prime if and only if m is a power of 2, but it is not that simple. In fact, $2^{32} + 1 = 4294967297 = 641 \cdot 6700417$. We define F_k, the kth **Fermat number**, to be $2^{2^k} + 1$ and ask when F_k is prime.

We know from Theorem 4.4 that if b is an integer such that $b^{(F_k-1)/2} \equiv -1 \pmod{F_k}$, then b has order $F_k - 1$ (modulo F_k), and so F_k is prime; indeed, b is a primitive root for F_k. In 1877, Fr. Jean François Théophile Pépin (1826–1904) pointed out that there are values of b that are guaranteed to work for any k for which F_k is prime. This implies that if $b^{(F_k-1)/2} \not\equiv -1 \pmod{F_k}$, then F_k cannot be prime. The key to finding such values of b is the notion of quadratic residues.

We say that b is a **quadratic residue modulo m** if b is relatively prime to m and there exists an integer t such that $b \equiv t^2 \pmod{m}$; if there is no such t, b is called a **nonresidue**. For small numbers we can see the quadratic residues by simply squaring all nonzero residues. Here are the quadratic residues for 7 and 11.

Mod[Range[6]^2, 7]

{1, 4, 2, 2, 4, 1}

Mod[Range[10]^2, 11]

{1, 4, 9, 5, 3, 3, 5, 9, 4, 1}

Some of the basic properties of quadratic residues for prime moduli are summarized in the following theorem.

Theorem 6.1. For any prime modulus $p > 2$, exactly $(p-1)/2$ elements of the reduced residue system modulo p are quadratic residues. The integer b is a quadratic residue modulo p if and only if $b^{(p-1)/2} \equiv 1 \pmod{p}$ [this is called **Euler's criterion**]. If p does not divide $b = cd$, then b is a quadratic residue if and only either c and d are both quadratic residues or neither is a quadratic residue.

Proof. Because $i^2 \equiv (p-i)^2 \pmod{p}$ and i cannot be congruent to $p-i$ modulo p, each quadratic residue must be the square of at least two distinct elements of $\mathbb{Z}_p^*$. On the other hand, if $i^2 \equiv j^2 \pmod p$, then p divides $i^2 - j^2 = (i-j)(i+j)$. Because p is prime, either p divides $i-j$, and so $i \equiv j \pmod p$, or p divides $i+j$ and so $i \equiv -j \pmod p$. This implies that each quadratic residue corresponds to exactly two elements of the reduced residue system modulo p, so the number of quadratic residues is $(p-1)/2$.

Let g be a primitive root modulo p. Because any even power of g is a quadratic residue and because there are $(p-1)/2$ distinct even powers of g (modulo p), no odd power of g can be a quadratic residue (and these are all the nonresidues). If $b \equiv g^e \pmod p$ where e is odd, then $b^{(p-1)/2} \equiv g^{e(p-1)/2} \pmod p$. Because $e(p-1)/2$ is not a multiple of $p-1$, $g^{e(p-1)/2}$ cannot be congruent to 1 (mod p); in fact it must be $-1 \pmod p$ because it squares to 1. Euler's criterion now follows from $g^{e(p-1)/2} \equiv (g^{p-1})^{e/2} \equiv 1 \pmod p$ if e is even and $g^{e(p-1)/2} \equiv (g^{(p-1)/2})^e \equiv (-1)^e = -1 \pmod p$ if e is odd.

Finally, if $b = c \cdot d$, then b is an even power of g if and only if either both c and d are even powers of g (and hence quadratic residues) or both are odd powers of g. $\square$

Given a prime p, it is easy to use Euler's criterion to test whether or not a given integer b is a quadratic residue.

```
QuadraticResidueQ[b_, p_] := (PowerMod[b, (p - 1) / 2, p] == 1)
```

Assuming that p is prime, this returns `True` when b is a quadratic residue and `False` when it is not.

But there are many instances — Pépin's Test is one of them — when we want a simple characterization of all primes for which a given b is a quadratic residue. By Theorem 6.1, if we know the status of all prime factors of b, we can determine whether or not b is a quadratic residue, so we shall restrict our attention to prime values of b. The following program finds all odd primes below `nmax` for which b is a quadratic residue.

```
QuadraticModuli[b_, nmax_] :=
  Select[Prime[Range[2, π[nmax]]], QuadraticResidueQ[b, #] &]

QuadraticModuli[7, 100]

{3, 19, 29, 31, 37, 47, 53, 59, 83}
```

Some of the results are summarized in the following table.

b	Primes Less Than 100
2	7, 17, 23, 31, 41, 47, 71, 73, 79, 89, 97
3	11, 13, 23, 37, 47, 59, 61, 71, 73, 83, 97
5	11, 19, 29, 31, 41, 59, 61, 71, 79, 89
7	3, 19, 29, 31, 37, 47, 53, 59, 83
11	5, 7, 19, 37, 43, 53, 79, 83, 89, 97
13	3, 17, 23, 29, 43, 53, 61, 79
17	13, 19, 43, 47, 53, 59, 67, 83, 89

For $b = 2$, we get precisely the primes congruent to ± 1 (modulo 8). For $b = 3$, we get the primes congruent to ± 1 (modulo 12). For $b = 5$ we get the primes congruent to ± 1 (modulo 5). For $b = 7$, it is a bit more complicated. We get the primes congruent to ± 1, ± 3, or ± 9 (modulo 28). For $b = 13$ we get the primes congruent to ± 1, ± 3, or ± 4 (modulo 13).

Euler, Legendre, and other eighteenth-century mathematicians studied these residue classes for patterns, and eventually made the following remarkable observation.

Theorem 6.2. Quadratic Reciprocity. For distinct odd primes p and q, p is a quadratic residue modulo q if and only if q is a quadratic residue modulo p, *except* in the case where p and q are both congruent to 3 (mod 4), in which case p is a quadratic residue modulo q if and only if q is *not* a quadratic residue modulo p.

In the late 1700s, this was one of the great unproven conjectures in mathematics. Notice what it will do for us. To find the primes for which 5 is a quadratic residue, we only need to find the residue classes that are quadratic residues modulo 5. There are two of them and they are 1 and 4, so the primes congruent to 1 and 4 (modulo 5) are the ones sought. There are six quadratic residues modulo 13. They are 1, 3, 4, 9, 10, and 12. The number 7 is more complicated because it is congruent to 3 (modulo 4). The quadratic residues modulo 7 are 1, 2, and 4. Any prime for which 7 is a quadratic residue must either be congruent to 1 (modulo 4) and congruent to 1, 2, or 4 (modulo 7) [in other words, congruent to 1, 9, or 25 (modulo 28)] or it must be congruent to 3 (modulo 4) and to 3, 5, or 6 (modulo 7) [in other words, congruent to 3, 19, or 27 (modulo 28)].

This theorem was finally proven by the brilliant young mathematician Carl Friedrich Gauss (1777–1855) who published the first proof in his book *Disquisitiones Arithmeticæ* in 1801. We shall postpone this proof until the next section.

▪ Pépin's Test

Returning to the testing of Fermat numbers, $F_k = 2^{2^k} + 1$, for primality, recall that we wanted to find a b for which $b^{(F_k-1)/2} \equiv -1 \pmod{F_k}$ whenever F_k is prime. This will be true if b is not a quadratic residue for any prime F_k. If we look at our table, we see that 3 and 5 look promising (they are not quadratic residues for 3, 5, or 17) while 2, 7, 11, and 13 are out of the question.

Quadratic reciprocity (see Exercise 6.1) tells us that 3 is a quadratic residue modulo p if and only if $p \equiv 1$ or 11 (mod 12). It is clear that for $k > 0$, $F_k \equiv 1 \pmod 4$. Can F_k be congruent to 1 (mod 3)? For $k > 0$, 2^k is even and therefore $2^{2^k} \equiv 1 \pmod 3$. This means that $F_k = 2^{2^k} + 1 \equiv 2 \pmod 3$, and so $F_k \equiv 5 \pmod{12}$. So 3 is not a quadratic residue for any prime Fermat number; this means that if F_k is prime, 3 will be a primitive root. This proves Pépin's test.

Theorem 6.3. Pépin's Test. The Fermat number $F_k = 2^{2^k} + 1$ is prime if and only if $3^{(F_k-1)/2} \equiv -1 \pmod{F_k}$.

Composite Fermat numbers are always 2-strong pseudoprimes (Exercise 4.16). Pépin's test can be interpreted as saying that a composite Fermat number is never a 3-strong pseudoprime. The following computation shows that among the first twelve Fermat numbers, only the first four are prime. FermatNumber is a CNT package function that gives the Fermat numbers.

```
Select[Range[12],
 (f = FermatNumber[#]; PowerMod[3, f - 1 / 2, f] == f - 1) &]
{1, 2, 3, 4}
```

Because of this simple test, it has been possible to prove that F_5 through F_{23} are all composite. At this moment, the two smallest Fermat numbers whose status is unknown are F_{24} and F_{31}. The former has 5,050,446 digits, the latter 646,456,994. Other methods have been used to find large families of Fermat numbers that must be composite. It is generally believed that there are at most a finite number of prime Fermat numbers. Probably the current list — $n = 1, 2, 3, 4$ — is a complete list of Fermat primes, but no proof of this is known.

■ Primes Congruent to 1 (Mod 4)

There is one value of b for which it is easy to determine the primes for which b is a quadratic residue without using quadratic reciprocity. That is $b = -1$. From Theorem 6.1 we know that -1 is a quadratic residue if and only if $(-1)^{(p-1)/2} \equiv 1 \pmod{p}$. If $p \equiv 3 \pmod 4$, then $(-1)^{(p-1)/2} = -1 \not\equiv 1 \pmod p$, while if $p \equiv 1 \pmod 4$ then $(-1)^{(p-1)/2} = 1$. We have proven the following theorem.

Theorem 6.4. For odd primes p, -1 is a quadratic residue modulo p if and only if $p \equiv 1 \pmod 4$.

This simple idea gives a very fast way of computing a square root of $-1 \pmod p$. If $p \equiv 1 \pmod 4$, then half of the integers in $\mathbb{Z}_p$ are nonresidues. If c is such, then $c^{(p-1)/2} \equiv -1 \pmod p$, so $c^{(p-1)/4}$ squares to $-1 \pmod p$. It will be easy to find c by trial and error. Here is an example that shows a typical case.

```
p = RandomPrime[50, {1, 4}]
6379227815651635830681097734139265309373303433 9537
```

Below 20 there are five good values of c.

```
Select[Range[20], PowerMod[#, p - 1/2, p] == p - 1 &]
```

{5, 10, 13, 15, 20}

And here is a value of $\sqrt{-1}$. The other mod-p square root of -1 is just its negative.

```
PowerMod[5, (p - 1) / 4, p]
```

630709564183304445949281384015959778305220983330381

In Exercise 4.1, you were asked to prove that there are infinitely many primes that are congruent to 3 (mod 4). In the next exercise, you were asked to explain why that approach could not succeed in proving that there are infinitely many primes congruent to 1 (mod 4). Equipped with Theorem 6.4, we can now prove that we also have infinitely many primes congruent to 1 (mod 4).

Theorem 6.5. There are infinitely many primes congruent to 1 (mod 4).

Proof. Assume that there are only finitely many such primes: $p_1 = 5$, $p_2 = 13$, ..., p_r, and let P be their product, $P = p_1 p_2 \cdots p_r$. Let q be any prime that divides $P^2 + 1$, which means that $P^2 \equiv -1 \pmod{q}$. Because -1 is congruent to a perfect square modulo q, q must be congruent to 1 (mod 4). But because P is the product of all primes congruent to 1 (mod 4), q must also divide P^2, a contradiction.   □

Exercises for Section 6.1

6.1. Use Theorem 6.2 to prove that 3 is a quadratic residue modulo p, where p is an odd prime, if and only if $p \equiv \pm 1 \pmod{12}$.

6.2. What are the residue classes for which 11 is a quadratic residue modulo the prime p if and only if p is in one of these residue classes?

6.3. Can 5 be a quadratic residue for a prime of the form $2^{2^k} + 1$?

6.4. Can 17 be a quadratic residue for a prime of the form $2^{2^k} + 1$?

6.5. Prove that -3 is a quadratic residue modulo p if and only if p is congruent to 1 (mod 6).

6.6. Prove the following formula for a square root of $-1 \pmod{p}$ where p is prime and congruent to 1 (mod 4).

$$\left(\left[\frac{p-1}{2} \right]! \right)^2 \equiv -1 \pmod{p}$$

Hint: You will need Wilson's Theorem (Exercise 2.12).

6.7. Prove that if p is a prime congruent to 3 (mod 4), then $[(p-1)/2]!$ must be congruent to ± 1 (mod p).

6.8. Prove that there are infinitely many primes congruent to 1 (mod 6). *Hint*: Assume that there are only finitely many such primes and let P be their product. Show that if q is a prime divisor of $12P^2 + 1$, then -3 is a quadratic residue (mod q). Use the result of Exercise 6.5 to show that q must be congruent to 1 (mod 6) and so q divides $12P^2$.

6.9. If $p \equiv 1$ (mod 4), then there are an even number of quadratic residues (mod p). Prove that exactly half of them are between 1 and $(p-1)/2$ (inclusive).

6.10. Show that if $p \equiv 3$ (mod 4) and $1 \le x \le p-1$, then x is a quadratic residue (mod p) if and only if x is a fourth power (mod p). *Hint*: Use the fact that -1 is not a quadratic residue.

6.2 Proof of Quadratic Reciprocity

Gauss's Lemma

Although Adrien-Marie Legendre could not prove quadratic reciprocity, he did come up with a convenient notation for expressing it, the **Legendre symbol**, written $\left(\frac{b}{p}\right)$. It is defined for any integer b and any odd prime p and is equal to $+1$ when b is a quadratic residue modulo p, -1 when b is not divisible by p and is not a quadratic residue, and 0 when b is divisible by p. It follows from Theorem 6.1 that

$$\left(\tfrac{b}{p}\right) \equiv b^{(p-1)/2} \pmod{p} \qquad \text{(Euler's criterion)}$$
$$\left(\tfrac{b+kp}{p}\right) = \left(\tfrac{b}{p}\right) \qquad \text{for any integer } k$$
$$\left(\tfrac{ab}{p}\right) \equiv \left(\tfrac{a}{p}\right)\left(\tfrac{b}{p}\right).$$

The code that follows computes the Legendre symbol by appealing to Euler's criterion via the package function QuadraticResidueQ. The built-in way of getting Legendre symbols uses the fact that it is a special case of the Jacobi symbol, discussed later in this section.

```
LegendreSymbol[b_, p_?PrimeQ] := (qr = QuadraticResidueQ[b, p];
    Which[qr, 1, Mod[b, p] ≠ 0 && !qr, -1, True, 0])
```

Quadratic reciprocity is the statement that $\left(\frac{p}{q}\right)$ and $\left(\frac{q}{p}\right)$ are the same unless $p \equiv q \equiv 3$ (mod 4), in which case $\left(\frac{p}{q}\right) = -\left(\frac{q}{p}\right)$. A slick way of stating this in a single statement is that if p and q are distinct odd primes, then

$$\left(\frac{p}{q}\right) \cdot \left(\frac{q}{p}\right) = (-1)^{(p-1)(q-1)/4}.$$

The first step in proving quadratic reciprocity is Gauss's criterion.

Theorem 6.6. Gauss's Criterion Let p be an odd prime and b a positive integer not divisible by p. For each positive odd integer $2i - 1$ less than p, let r_i be the residue of $b(2i - 1)$ (mod p):

$$r_i \equiv b(2i - 1) \ (\mathrm{mod}\ p), \quad 0 < r_i < p.$$

Let t be the number of r_i that are even. Then $\left(\frac{b}{p}\right) = (-1)^t$.

As an example, let $p = 23$ and $b = 3$ [$7^2 = 49 \equiv 3$ (mod 23) so 3 is a quadratic residue modulo 7]. Then

$$
\begin{aligned}
&r_1 \equiv 3 \cdot 1 \equiv 3\,(\mathrm{mod}\ 23) &\quad &r_5 \equiv 3 \cdot 9 \equiv \mathbf{4}\,(\mathrm{mod}\ 23) &\quad &r_9 \equiv 3 \cdot 17 \equiv 5\,(\mathrm{mod}\ 23) \\
&r_2 \equiv 3 \cdot 3 \equiv 9\,(\mathrm{mod}\ 23) &\quad &r_6 \equiv 3 \cdot 11 \equiv \mathbf{10}\,(\mathrm{mod}\ 23) &\quad &r_{10} \equiv 3 \cdot 19 \equiv 11\,(\mathrm{mod}\ 23) \\
&r_3 \equiv 3 \cdot 5 \equiv 15\,(\mathrm{mod}\ 23) &\quad &r_7 \equiv 3 \cdot 13 \equiv \mathbf{16}\,(\mathrm{mod}\ 23) &\quad &r_{11} \equiv 3 \cdot 21 \equiv 17\,(\mathrm{mod}\ 23) \\
&r_4 \equiv 3 \cdot 7 \equiv 21\,(\mathrm{mod}\ 23) &\quad &r_8 \equiv 3 \cdot 15 \equiv \mathbf{22}\,(\mathrm{mod}\ 23) &\quad &
\end{aligned}
$$

It follows that $t = 4$ — the bold entries in the display — and $(\frac{3}{23}) = (-1)^4 = 1$. If $b = 5$, then there are five even residues in the list of multiples — {5, 15, **2**, **12**, **22**, 9, 19, **6**, **16**, 3, 13} — and 5 is indeed not a quadratic residue modulo 23.

Proof. Write $p = 2m + 1$. There are m positive odd integers less than p. Relabel the residues so that $r_1, r_2, \ldots, r_t$ are all even and $r_{t+1}, r_{t+2}, \ldots, r_m$ are all odd. Let $a_1, a_2, \ldots, a_m$ be the positive odd integers less than p ordered so that $r_i \equiv b a_i$ (mod p).

Consider the integers $p - r_1, p - r_2, \ldots, p - r_t, r_{t+1}, \ldots, r_m$. These are all odd positive integers less than p. We claim that they are distinct. Because the a_i are distinct modulo p, there are no repetitions among the first t or among the last $m - t$. It suffices to prove that we cannot have $p - r_i = r_j$ where i is at most t and j is larger than t. If we did, then we would have that

$$p = r_i + r_j \equiv b a_i + b a_j = b(a_i + a_j) \ (\mathrm{mod}\ p).$$

Because p does not divide b, it must divide $a_i + a_j$. But $0 < a_i + a_j < 2p$, so $p = a_i + a_j$, which contradicts the fact that a_i and a_j are both odd and therefore their sum is even.

Because $p - r_1, p - r_2, \ldots, p - r_t, r_{t+1}, \ldots, r_m$ are m distinct odd positive integers less than p, they must be all of them, and therefore

$$a_1 a_2 \cdots a_m = (p - r_1) \cdots (p - r_t) r_{t+1} \cdots r_m = (-1)^t r_1 r_2 \cdots r_m \ (\mathrm{mod}\ p).$$

But p does not divide any of the odd integers between 0 and p, so we can divide both sides of this congruence by $a_1 a_2 \cdots a_r$. Using Euler's criterion this implies that

$$1 \equiv (-1)^t \, b^{(p-1)/2} \equiv (-1)^t \left(\frac{b}{p} \right) (\text{mod } p). \quad \square$$

Corollary 6.7. For prime modulus p, 2 is a quadratic residue if and only if $p \equiv \pm 1 \ (\text{mod } 8)$.

Proof. For $1 \le i \le (p-1)/2$, $r_i \equiv 2(2i-1) \ (\text{mod } p)$ is even if and only if $2i-1 < p/2$, or $i < p/4 + 1/2$. If $p = 4m+1$, (using the notation of Gauss's criterion) this means that $t = m$, and so 2 is a quadratic residue when $p \equiv 1 \ (\text{mod } 8)$ and not when $p \equiv 5 \ (\text{mod } 8)$. If $p = 4m+3$, then $t = m+1$ and so 2 is a quadratic residue when $p \equiv 7 \ (\text{mod } 8)$ and not when $p \equiv 3 \ (\text{mod } 8)$. $\quad \square$

■ Proof of Quadratic Reciprocity

Theorem 6.2 requires p and q to be distinct primes, so the Legendre symbols $\left(\frac{p}{q} \right)$ and $\left(\frac{q}{p} \right)$ are not zero. We let s be the number of positive odd integers $2i-1 < p$ such that if $r_i \equiv q(2i-1) \ (\text{mod } p)$ with $0 < r_i < p$, then r_i is even. We let t be the number of positive odd integers $2j-1 < p$ such that if $r'_j \equiv p(2j-1) \ (\text{mod } q)$ with $0 < r'_j < q$, then r'_j. From Gauss's criterion (Theorem 6.6), we know that

$$\left(\frac{p}{q} \right) \left(\frac{q}{p} \right) = (-1)^{s+t}.$$

We shall prove that $s + t$ is odd if and only if $p \equiv q \equiv 3 \ (\text{mod } 4)$.

Consider the set S of all integers of the form $qa - pb$ where a runs over the positive odd integers less than p and b runs over the positive odd integers less than q. As an example, if $p = 5$ and $q = 7$, then a is 1 or 3 and b is 1, 3, or 5. The set S can be computed using `Outer`, which takes all combinations from the arguments, as follows.

```
Outer[f, {1, 2, 3}, {7, 8}]

{{f[1, 7], f[1, 8]}, {f[2, 7], f[2, 8]}, {f[3, 7], f[3, 8]}}
```

Now here is the set S.

```
p = 5; q = 7; a = {1, 3}; b = {1, 3, 5};
S = Flatten[Outer[Subtract, q a, p b]]

{2, -8, -18, 16, 6, -4}
```

We leave it as an exercise (6.11) to verify that the numbers in S are always even, nonzero, and distinct.

Consider those pairs (a, b) for which $r = qa - pb$ is positive and less than p. This means that

$$qa = pb + r \equiv r \pmod{p}$$

and so r is one of the residues counted by s. Furthermore, every residue counted by s arises in this way, for if $qa \equiv r \pmod{p}$ where a is positive, odd, and less than p and if r is positive, even, and less than p, then $qa - r = pb$ where b is positive and odd and pb is strictly less than $qp - r$, and therefore b is strictly less than q.

Similarly, if $qa - pb$ is negative and larger than $-q$, then it equals $-r$, where r is one of the residues counted by t. Furthermore, every residue counted by t corresponds to a pair (a, b) for which $qa - pb$ is negative and larger than $-q$. Therefore $s + t$ is the number of elements of S that lie between $-q$ and p.

We have reduced our problem to showing that the number of elements of S between $-q$ and p is even except when $p \equiv q \equiv 3 \pmod{4}$, in which case it is odd. We shall investigate the parity of this set by identifying natural pairs of elements.

Let $qa - pb$ be an element of S in the desired range. Let

$$c = p - 1 - a$$
$$d = q - 1 - b.$$

We see that $qc - pd$ is also in S and that $qc - pd = -q + p - (qa - pb)$. Using the bounds on $qa - pb$, we observe that

$$-q = -q + p - p < qc - pd < -q + p - (-q) = p$$

and so $qc - pd$ is also in the desired range. This means that we can pair up the elements of S in the desired range via

$$(a, b) \longleftrightarrow (p - 1 - a, q - 1 - b).$$

The number of such elements must be even unless some elements pair with themselves. But if $a = p - 1 - a$ and $b = q - 1 - b$, then $a = (p-1)/2$ and $b = (q-1)/2$. This means that there is at most one element of S in the desired range that pairs with itself, and this element exists if and only if $(p-1)/2$ and $(q-1)/2$ are both odd. In other words, the number of elements of S in the desired range is odd if and only if $p \equiv q \equiv 3 \pmod{4}$.   □

🖥 Jacobi's Extension

The notion of quadratic residue is defined only for moduli that are prime and odd. It certainly makes sense to ask whether, say, 2 is or is not a square modulo n where n is composite, but in fact that is not the "right" generalization of the Legendre symbol to composites. It is much more useful to extend the Legendre symbol to what is called the Jacobi symbol $(\frac{a}{n})$ as follows. Given n odd, $n = p_1^{e_1} p_2^{e_2} \cdots p_r^{e_r}$, and a relatively prime to n, the **Jacobi symbol** $(\frac{a}{n})$ is defined by

$$\left(\frac{a}{n}\right) = \left(\frac{a}{p_1}\right)^{e_1} \left(\frac{a}{p_2}\right)^{e_2} \cdots \left(\frac{a}{p_n}\right)^{e_n}$$

where each $\left(\frac{a}{p_i}\right)$ is a Legendre symbol (and so indicates the quadratic character of $a \pmod{p_i}$). Note the special case of the empty product: $(\frac{a}{1}) = 1$. This function is built into *Mathematica* as `JacobiSymbol`, and, because of a special algorithm that we will describe presently, this provides a faster way to get the Legendre symbol $\left(\frac{a}{p}\right)$ than the more straightforward use of Euler's criterion. Here is a simple example to show that the Jacobi symbol does *not* correspond to the quadratic character of a modulo n.

```
JacobiSymbol[8, 15]
```

    1

But 8 is not a square modulo 15.

```
Union[PowerMod[Range[14], 2, 15]]
```

    {1, 4, 6, 9, 10}

The next result summarizes the main properties of the Jacobi symbol. The proofs are straightforward applications of the results about the Legendre symbol presented earlier in this chapters, including quadratic reciprocity, and are left as exercises.

Proposition 6.8. The Jacobi symbol has the following properties.

(a) If $a \equiv b \pmod{n}$, then $(\frac{a}{n}) = (\frac{b}{n})$.

(b) $\left(\frac{ab}{n}\right) = (\frac{a}{n})(\frac{b}{n})$

(c) $(\frac{-1}{n}) = (-1)^{(n-1)/2}$

(d) $(\frac{2}{n}) = (-1)^{(n^2-1)/8}$

(e) $(\frac{n}{m})(\frac{m}{n}) = (-1)^{\left(\frac{m-1}{2}\right)\left(\frac{n-1}{2}\right)}$

(f) $(\frac{a}{n}) = (\frac{n}{r})(-1)^{s(n^2-1)/8}(-1)^{(n-1)(r-1)/4}$, where s and r are defined as follows: let $R = \mathrm{Mod}[a, n]$, let s be the largest e such that 2^e divides R, and let $r = R/2^s$.

Algorithm 6.1. The Jacobi Symbol

To compute $(\frac{a}{n})$ one can use a recursion based on part (f) of the preceding proposition. That leads to an algorithm that is very similar to the standard Euclidean algorithm. We raise the built-in recursion limit within a `Block` so that it reverts to its default outside of the program.

```
jacobi[a_, n_?OddQ] := 0 /; GCD[a, n] ≠ 1
jacobi[_, 1] := 1
jacobi[a_, n_?OddQ] :=
 Block[{$RecursionLimit = ∞, R = Mod[a, n], s, r},
  s = IntegerExponent[R, 2]; r = R / 2^s;
  jacobi[n, r] * (-1)^((n-1)(r-1)/4) * ((-1)^s)^((n^2-1)/8) ]
```

The recursive version just given is slower than the simple `PowerMod` approach for getting Legendre symbols. For a fair comparison one should use the manual exponentiation code from Section 1.4 as opposed to the built-in version. We repeat that code here for convenience.

```
powermodLR[a_, n_, m_] :=
 Fold[Mod[Mod[#1 #1, m] * If[#2 == 1, a, 1], m] &,
  a, Rest[IntegerDigits[n, 2]]]

n = Random[Integer, {1, 10^200}];

p = RandomPrime[200];

Do[jacobi[n, p], {5}] // Timing
Do[powermodLR[n, (p - 1) / 2, p] /. p - 1 → -1, {5}] // Timing

{4.31667 Second, Null}

{1.96667 Second, Null}
```

But the implementation can be improved quite a bit. First, one can eliminate the recursion and get a thoroughly iterative method, much as was done for the Euclidean algorithm in Exercise 1.12. Second, one can eliminate the arithmetic in the $n^2 - 1$ computation by using instead the mod-8 value of n. The table in the code that follows summarizes the mod-8 behavior so that this part of the work can be accomplished by a table lookup. This code also avoids a gcd computation, because that comes for free; the value of `aa` at the completion of the loop will be the gcd of a and n.

```
jacobiIter[a_, n_] :=
 (table = {0, 1, 0, -1, 0, -1, 0, 1}; {aa, bb, c} = {a, n, 1};
  While[bb ≠ 0, {aa, bb, c} = {bb, If[(R = Mod[aa, bb]) == 0,
     0, r = R / 2 ^ (s = IntegerExponent[R, 2])],
     c * If[R == 0, 1, ((-1)^(bb-1/2))^(r-1/2) * table[[Mod[bb, 8] + 1]]^s]}];
  If[aa ≠ 1, 0, c])
```

We first check correctness.

```
And @@ Table[JacobiSymbol[r = Random[Integer, {1, 100}],
      m = Random[Integer, {1, 10}] 2 + 1] == jacobiIter[r, m], {50}]
True
```

And now we do a timing comparison,

```
Do[jacobiIter[n, p], {5}] // Timing
Do[powermodLR[n, (p - 1) / 2, p] /. p - 1 → -1, {5}] // Timing

{0. Second, Null}

{0.0166667 Second, Null}
```

We can also compare the built-in `PowerMod` approach with the built-in `JacobiSymbol` algorithm, which is similar to our code just given. There is no contest.

```
Do[PowerMod[n, (p - 1) / 2, p] /. p - 1 → -1, {5}] // Timing //
  First
Do[JacobiSymbol[n, p], {5}] // Timing // First

1.18333 Second

0.0166667 Second
```

A historically important pseudoprime test — one of the first probabilistic prime tests, sometimes called the Solovay–Strassen algorithm — can be developed from the Jacobi symbol (Exercise 6.13).

■ An Application to Factoring

Gauss came up with a clever way of using quadratic residues to reduce the set of primes that one would try when using trial division to factor a number. His idea is related to the quadratic sieve, a powerful factorization algorithm in use today.

If n is the number to be factored and if we know that a is a quadratic residue modulo n, then it must be a quadratic residue for any prime p that divides n. We can use quadratic reciprocity to find the possible residue classes in which p must lie. This cuts out approximately half of the primes that we need to test. If we find a second integer that is also a quadratic residue modulo n, we can expect to cut the number of primes by another factor of 2. Ten quadratic residues cut our trial divisions by a factor of 1000. Twenty quadratic residues reduce it by a factor of 1,000,000.

To see how this works in practice, we shall factor $n = 21,307,703$. We take a small value of k, say $1 \le k \le 10$, and choose integers i near $\sqrt{kn}$, in this case $\lfloor \sqrt{kn} \rfloor - 20 \le i \le \lfloor \sqrt{kn} \rfloor + 20$. We calculate the 410 values of $i^2 - kn$ and see which ones we can factor completely using only primes below 100. We get 41 of these values that have all of their prime divisors less than 100. We list here nine of them that will turn out to be useful:

$$4604^2 - 21307703 = -1 \cdot 7^2 \cdot 31 \cdot 73$$
$$4616^2 - 21307703 = -1 \cdot 13 \cdot 19$$
$$4627^2 - 21307703 = 2 \cdot 13 \cdot 47 \cdot 83$$
$$7978^2 - 3 \cdot 21307703 = -1 \cdot 5^3 \cdot 13^3$$
$$12205^2 - 7 \cdot 21307703 = -1 \cdot 2^3 \cdot 17^2 \cdot 83$$
$$122019^2 - 7 \cdot 21307703 = 2^3 \cdot 5 \cdot 11^2 \cdot 31$$
$$12223^2 - 7 \cdot 21307703 = 2^{11} \cdot 11^2$$
$$12225^2 - 7 \cdot 21307703 = 2^8 \cdot 19 \cdot 61$$
$$13043^2 - 8 \cdot 21307703 = -1 \cdot 3^2 \cdot 5^2 \cdot 7^2 \cdot 31 \,.$$

Let p be any prime that divides 21,307,703. The third equation from the bottom implies that

$$12223^2 \equiv 2^{11} \cdot 11^2 \pmod{p}$$

and therefore 2 must be a quadratic residue mod p. Because $21,307,703 \equiv 3 \pmod 4$, it must be divisible by a prime congruent to 3 (mod 4), so that we can assume that $p \equiv 3 \pmod 4$. This implies that -1 is not a quadratic residue modulo p, and so by the last equation, 31 cannot be a quadratic residue modulo p.

We now look at the first equation. Because -31 is a quadratic residue, 73 must also be a quadratic residue. From the fifth equation, because -1 is not a quadratic residue and 2 is, 83 cannot be. In the sixth equation, 2 is and 31 is not, so 5 cannot be a quadratic residue. In the fourth equation, because -5 is a quadratic residue, so is 13. From the second equation, 19 is not a quadratic residue. From the third equation, 47 is not a quadratic residue. From the second last equation, 61 is not a quadratic residue.

In summary, we have established that 21,307,703 is divisible by a prime p for which 2, 13, and 73 are quadratic residues and -1, 5, 19, 31, 47, 61, and 83 are not. We can now use the fast method of computing Jacobi symbols to determine the possible primes. There turn out to be two of them, 8167 and 9463.

```
Select[Prime[Range[2, PrimePi[10000]]],
   Union[JacobiSymbol[{2, 13, 73}, #]] == {1} &&
   Union[JacobiSymbol[{-1, 5, 19, 31, 47, 61, 83}, #]] == {-1} &]
{8167, 9463}
```

And, indeed, $21{,}307{,}703$ factors into $8167 \cdot 2609$.

21307703 / 8167

2609

It may seem that we have done an awful lot of work for a very small reward. After all, there are only 1229 primes below 10,000. The point here is that until we identified 8167 and 9463, we had avoided division by integers larger than 100. When we get to 40 and 50-digit integers, being able to avoid divisions will be critically important. Gauss's algorithm is not useful today, but the ideas that it embodies are.

Exercises for Section 6.2

6.11. Let p and q be distinct odd primes and consider the integers of the form $qa - pb$ where a runs over the positive odd integers less than p and b runs over the positive odd integers less than q. Prove that the numbers generated in this fashion are even, nonzero, and distinct.

6.12. Prove Proposition 6.8.

6.13. (a) Explain why the following pseudoprime test is valid. Given n odd, let $a = 2$ and look at $a^{(n-1)/2} \pmod{n}$. If this is not ± 1, n is proved composite. If this does not equal $\left(\frac{a}{n}\right)$, n is proved composite. Otherwise, n is very probably prime. Composite numbers that pass this test are called **Euler pseudoprimes to the base** a. When the test is repeated many times with random choices of a, it can be interpreted as saying that n is prime with very high probability (just as in the Miller–Rabin test). This probabilisiitc form is called the **Solovay–Strassen algorithm**.

(b) Implement the 2-Euler pseudoprime test and find the first 2-Euler pseudoprime. Find the first 2-Euler pseudoprime that is not a Carmichael number.

6.14. Show that if $p \equiv 1 \pmod{4}$ and d is an odd integer that divides $p - 1$, then d is a quadratic residue modulo p.

6.15. Show that for every prime p there is a solution to the congruence

$$x^6 - 11x^4 + 36x^2 - 36 \equiv 0 \pmod{p}.$$

How many solutions are there?

6.16. Prove that if 3 does not divide m, then there is at least one prime $p \equiv 7 \pmod{12}$ that divides $4m^2 + 3$. Use this to prove that there are infinitely many primes that are congruent to 7 (mod 12).

6.17. Show that if p is prime and divides an integer of the form $n^4 - n^2 + 1$, then $p \equiv 1 \pmod{12}$. *Hint*: First show that if p divides $n^4 - n^2 + 1$, then $(2n^2 - 1)^2 \equiv -3 \pmod{p}$ and $(n - n^{-1})^2 \equiv -1 \pmod{p}$.

6.18. Show that if p is an odd prime that divides an integer of the form $n^4 + 4$, then $p \equiv 1 \pmod{4}$.

6.19. Prove that $x^4 \equiv -4 \pmod{p}$ has a solution if and only if $p \equiv 1 \pmod{4}$.

6.20. Prove that for any $\alpha > 2$ and any prime p, the congruence $x^{2^\alpha} \equiv 2^{2^{\alpha-1}} \pmod{p}$ has a solution.

6.3 Quadratic Equations

In this section we will show how to find square roots modulo m. In other words, we will solve the quadratic congruence $x^2 \equiv a \pmod{m}$; Exercise 6.29 will show how to extend this to solve all quadratic congruences. We will discuss the case of a prime modulus in detail; the extension to the general case is handled in the exercises. Of course, a solution to $x^2 \equiv a \pmod{p}$ exists only when a is a quadratic residue modulo p. Some cases are quite easy. Suppose that $p \equiv 3 \pmod{4}$; then it is easy to find a solution to this congruence. Just choose x so that

$$x \equiv a^{(p+1)/4} \pmod{p}.$$

This works because

$$\left(a^{(p+1)/4}\right)^2 \equiv a^{(p+1)/2} \equiv a\,a^{(p-1)/2} \equiv a \pmod{p}$$

where the last congruence follows from the fact that a is a quadratic residue modulo p. Here's an example. The package function `NextPrime` allows a congruence condition to be given as the second argument, so we can easily get a large prime congruent to 3 (mod 4).

```
p = NextPrime[10^30, {3, 4}]
1000000000000000000000000000099
```

Then we get a large square by squaring an integer modulo p.

```
a = Mod[123123123456456456789789789^2, p];
```

And we can recover the square root very easily.

$$\texttt{PowerMod}\Big[\texttt{a},\ \frac{\textbf{p + 1}}{\textbf{4}},\ \texttt{p}\Big]$$

123123123456456456789789789

There are several algorithms for the remaining case. We present Tonelli's algorithm here.

Algorithm 6.2 Tonelli's Algorithm for $\sqrt{a}$ (mod p)

Suppose that $p \equiv 1$ (mod 4). If k is odd, say $k = 2m + 1$, and $a^k \equiv 1$ (mod p), then a^{m+1} is a mod-p square root of a. If we have a nonresidue h handy, then we can also deal with the case that $a^k \equiv -1$ (mod p), because Euler's criterion tells us that $a^k h^{(p-1)/2} \equiv 1$ (mod p) and the desired square root is $a^{m+1} h^{(p-1)/4}$. This leads to the following algorithm, known as Tonelli's algorithm. We take two exponents e_1 and e_2, which start off as $(p-1)/2$ and $p-1$, respectively. Then e_1 will be repeatedly halved until it yields the odd integer k. After each halving, e_2 will be modified so that it remains an even integer and so that the following congruence holds.

$$a^{e_1} h^{e_2} \equiv 1 \text{ (mod } p) \tag{1}$$

Note that (1) holds for the initial values of the exponents: Fermat's Little Theorem tells us that $h^{p-1} \equiv 1$ (mod p) and Euler's criterion yields $a^{(p-1)/2} \equiv 1$ (mod p) because a must be a quadratic residue if a square root is to exist. If we can preserve (1) and get e_1 to have the form $2m + 1$, then we will have the desired square root as $a^{m+1} h^{e_2/2}$, because its square is $a^{2m+2} h^{e_2}$, or $a(a^{2m+1} h^{e_2})$, which is congruent to a.

Now, here is how to modify e_2. First it is halved and then the new value of e_2 is inserted into (1) together with the new value of e_1 (which is half the old). The product is then ± 1, because it squares to 1 (mod p). If it is $+1$, then nothing else is necessary. If it is -1, add $(p-1)/2$ to (the new value of) e_2. This has the effect of turning the expression in (1) into the desired $+1$. Because the final, odd, value of e_1 has not yet been reached and because e_2 starts out with one more 2 in it than e_1 starts out with, e_2 stays even; this is essential because the protocol calls for it to be halved at each step. Note that, at the last step when e_1 finally becomes odd, we must perform the final transformation on e_2 to preserve (1).

The following table illustrates Tonelli's method to get $\sqrt{49}$ (mod 3329). Because $3328 = 2^8 \, 13$, the initial values of (e_1, e_2) are (1664, 3328) and the algorithm requires eight steps before the special final step. The e_1 values are cut in two at each step. The e_2 values are also cut in two, but they get 1664 added to them in the boldface cases, which correspond to the cases in which the Tonelli equation on the pre-e_2 value is 3328. The final answer is 3322, which is -7 (mod 3329).

e_1	a^{e_1}	$a^{e_1} \pmod{p}$	pre-e_2	e_2	Tonelli equation	$a^{e_1}\, h^{\text{pre-}e_2} \pmod{p}$
1664	$49^{128\,13}$	1	3328	3328	1	1
832	$49^{64\,13}$	1	**1664**	**3328**	1	3328
416	$49^{32\,13}$	3328	1664	1664	1	1
208	$49^{16\,13}$	1729	**832**	**2496**	1	3328
104	$49^{8\,13}$	2580	1248	1248	1	1
52	$49^{4\,13}$	2642	**624**	**2288**	1	3328
26	$49^{2\,13}$	2267	**1144**	**2808**	1	3328
13	49^{13}	2447	1404	1404	1	1
7	49^{7}	532	702		532	**3322**

The code that follows assumes that the second argument is prime. It uses Nest and Reverse to form the list of powers of a in advance, which it then scans to get the e_2-sequence. It also shows how to include a message to handle bad calls; in this case such a case arises when the input is not a square modulo p.

```
SqrtModPrimeTonelli::noroot =
  "`` has no square root modulo ``.";

SqrtModPrimeTonelli[a_, p_] :=
  (Message[SqrtModPrimeTonelli::noroot, a, p]; HoldForm[
     SqrtModPrimeTonelli[a, p]]) /; JacobiSymbol[a, p] == -1;

SqrtModPrimeTonelli[a_, p_] := Mod[a, p] /; Mod[a, p] == 0 || p == 2

SqrtModPrimeTonelli[a_, p_] :=
  (r = PowerMod[a, (p + 1) / 4, p]; Min[p - r, r]) /; Mod[p, 4] == 3

SqrtModPrimeTonelli[a_, p_] :=
 Module[{h = Nonresidue[p], e2 = p - 1,
    s = IntegerExponent[(p - 1)/4, 2]},
   k = (p - 1)/2^(s+2);
   Scan[(e2 /= 2; If[Mod[# PowerMod[h, e2, p], p] != 1,
       e2 += (p - 1)/2]) &,
     Reverse[NestList[Mod[#^2, p] &, PowerMod[a, k, p], s]]];
   r = Mod[(PowerMod[a, (k + 1)/2, p]) * PowerMod[h, e2/2, p], p];
   Min[{p - r, r}]] /; Mod[p, 4] == 1
```

```
SqrtModPrimeTonelli[3, 5]
```

SqrtModPrimeTonelli::noroot : 3 has no square root modulo 5.

```
SqrtModPrimeTonelli[3, 5]
```

```
p = NextPrime[10⁴⁰, {1, 4}];
a = Mod[33123², p];
SqrtModPrimeTonelli[a, p]
```

33123

An algorithm known as Shanks's algorithm is slightly faster; it is discussed in Exercise 6.31.

Of course, one really wants a general modular square root function that will return all the square roots of a modulo n even when n is composite. This can be done (Exercise 6.27) by first considering the case that n is a prime power (Exercises 6.23–25) and then, in the general case, pasting together the results for prime powers by using the Chinese Remainder Theorem (Exercise 6.26). If n is a product of many primes, this will be impossible; if n has 30 different prime factors, then the number of square roots will be 2^{30}, or over one billion. *Mathematica*'s standard NumberTheoryFunctions package comes with a general SqrtMod function, but it does not return all the square roots and is not as general as the function you are asked to program in Exercise 6.27. A SqrtModAll function is included in the CNT package. Here's an example.

SqrtModAll[4, 653989852238983167**5790**]

```
{2, 1058840713148829890368,
 1121125460981114001562, 1557118695807102779948,
 1743972939303955113542, 2179966174129943891932,
 2615959408955932670318, 2802813652452785003912,
 3737084869937046671878, 3923939113433899005472,
 4359932348259887783858, 4795925583085876562248,
 4982779826582728895842, 5418773061408717674228,
 5481057809241001785422, 6539898522389831675788}
```

Mod[%², 6539898522389831675790]

```
{4, 4, 4, 4, 4, 4, 4, 4, 4, 4, 4, 4, 4, 4, 4, 4}
```

While it does not always work, in many cases the built-in Solve function is capable of solving polynomial congruences.

x /. Solve[{x¹³ + 3 x² == 4, Modulus == 497985451829}]

```
{1, 61769530393, 129605760352, 188340248546,
 191375290744, 208840723012, 250109778938, 270610253404}
```

Exercises for Section 6.3

6.21. Show that if the sequence a_n is defined recursively by $a_n = c_2 a_{n-2} + c_1 a_{n-1}$ with initial conditions a_0 and a_1, and M is the 2×2 matrix $\begin{pmatrix} 0 & 1 \\ c_2 & c_1 \end{pmatrix}$, then $M^r \begin{pmatrix} a_0 \\ a_1 \end{pmatrix} = \begin{pmatrix} a_r \\ a_{r+1} \end{pmatrix}$.

6.22. Write a program that, on input a and n, uses a brute force search to return all the mod-n square roots of a between 0 and n. Such a program can be used as a check on the more sophisticated and much faster routine discussed in Exercise 6.27.

6.23. Show that if p is an odd prime that does not divide a and if $x^2 \equiv a$ (mod p^n) has two solutions — $x \equiv \pm t$ (mod p^n) — then $x^2 \equiv a$ (mod p^{n+1}) has exactly two solutions: $x \equiv \pm(t + kp^n)$ (mod p^{n+1}), where k satisfies $2kt \equiv (a - t^2)/p^n$ (mod p). It follows that for any positive integer n, $x^2 \equiv a$ (mod p^n) has two solutions if and only if a is a quadratic residue modulo p.

6.24. In this exercise we solve the congruence $x^2 \equiv a$ (mod 2^r) when a is odd.

(a) Show that if a is odd, then $x^2 \equiv a$ (mod 4) has a solution if and only if $a \equiv 1$ (mod 4), and the solutions in that case are 1 and 3.

(b) Show that if a is odd, $r \geq 3$, and $x^2 \equiv a$ (mod 2^r), then $x^2 \equiv a$ (mod 8) [which occurs if and only if $a \equiv 1$ (mod 8)].

(c) Show that if a is odd, $r \geq 3$, and $x^2 \equiv a$ (mod 2^r), then there are precisely four solutions: $\pm x$ and $2^{r-1} \pm x$.

(d) Show that if a is odd and $r > 3$ then a particular solution to $x^2 \equiv a$ (mod 2^r) may be built up from a mod-8 solution using the rule: If $x^2 \equiv a$ (mod 2^{r-1}), then $[(x^3 + (2 - a)x)/2]^2 \equiv a$ (mod 2^r).

(e) Explain how (a)–(d) give a complete solution to $x^2 \equiv a$ (mod 2^r) when a

6.25. This exercise concludes the square root algorithm for the case that the modulus is a prime power by considering the case that the square is divisible by the prime in the prime power modulus. Let p be any prime and assume $e \geq 1$, and $1 \leq s \leq e$. Show that $x^2 \equiv m p^s$ (mod p^e), where p does not divide m, can be solved as follows. If $s = e$, then the set of roots is just $\{0, p^{\lceil e/2 \rceil}, 2 p^{\lceil e/2 \rceil}, \ldots, p^e - p^{\lceil e/2 \rceil}\}$. Otherwise, if s is odd there are no solutions and if s is even a solution exists if and only if there is a solution to $x^2 \equiv m$ (mod p^{e-s}); the solution set is then obtained by taking, for each solution y to $y^2 \equiv m$ (mod p^{e-s}), the set $\{p^{s/2}(y + j p^{e-s}) : j = 0, 1, \ldots, p^{e/2-1}\}$.

6.26. Use the Chinese Remainder Theorem to show that if m and n are relatively prime then there is a one-one correspondence between solutions to $x^2 \equiv a$ (mod mn) and pairs of solutions $\{y, z\}$ where $y^2 \equiv a$ (mod m) and

$z^2 \equiv a \pmod{n}$. So, if a has s square roots modulo m and t square roots modulo n, then there are st square roots of a modulo mn.

6.27. Write a program that will find all square roots of $a \pmod{m}$ for any modulus m. Use Tonelli's method if m is prime and Exercises 6.23-6.25 for prime power moduli. Then use Exercise 6.26 to paste together the results for prime powers.

6.28. Find all solutions of each of the following congruences:

$$x^2 \equiv 2 \pmod{457^2}$$
$$x^2 \equiv 3 \pmod{37^2 457^2}$$
$$x^2 \equiv 5 \pmod{29^3 53^2}.$$

6.29. Fill in the details in the following outline of an algorithm that solves $ax^2 + bx + c \equiv 0 \pmod{m}$ for any integers a, b, and c and modulus $m \geq 2$. Then implement the algorithm.

Let $f(x)$ denote the quadratic polynomial. For each prime p that divides m, let k_p be the largest integer such that p^{k_p} divides m. If x is a solution to $f(x) \equiv 0 \pmod{m}$, then x is a solution to $f(x) \equiv 0 \pmod{p^{k_p}}$. And if we have a solution to $f(x) \equiv 0 \pmod{p^{k_p}}$ for each prime p that divides m, then we can use the Chinese Remainder Theorem to construct a unique solution to $f(x) \equiv 0 \pmod{m}$. Thus we can reduce the problem to solving $f(x) \equiv 0 \pmod{p^k}$ for any prime p and power k.

If x is a solution to $f(x) \equiv 0 \pmod{p^{j+1}}$, then it must be a solution modulo p^j. If x is a solution to $f(x) \equiv 0 \pmod{p^j}$, then $x + tp^j$ is a solution modulo p^{j+1} if and only if

$$a(x + tp^j)^2 + b(x + tp^j) + c =$$
$$(ax^2 + bx + c) + p^j(2axt + bt) + p^{2j}(at^2) \equiv 0 \pmod{p^{j+1}}.$$

Because p^j divides $ax^2 + bx + c = f(x)$, this congruence is satisfied if and only if

$$\frac{f(x)}{p^j} + t(2ax + b) \equiv 0 \pmod{p}.$$

This is a linear congruence that has either 0, 1, or p solutions. Given all solutions modulo p^j, the corresponding solutions to this linear congruence can be used to construct all solutions modulo p^{j+1}. So we can reduce our problem to solving $f(x) \equiv 0 \pmod{p}$.

If $p = 2$, $x^2 \equiv x \pmod{2}$, so the original problem is equivalent to $(a + b)x + c \equiv 0 \pmod{2}$, a linear congruence with 0 or 1 solutions.

If p divides a, the original problem is equivalent to $bx + c \equiv 0 \pmod{p}$, a linear congruence with 0, 1, or p solutions.

If $p > 2$ and p does not divide a, then the original problem is equivalent to $x^2 + a^{-1}bx \equiv -a^{-1}c \pmod{p}$. We can complete the square on the left:

$$(x + 2^{-1}a^{-1}b)^2 \equiv 4^{-1}a^{-2}b^2 - a^{-1}c \pmod{p}.$$

All solutions to this congruence are found by finding the square roots (if any) of $4^{-1}a^{-2}b^2 - a^{-1}c \pmod{p}$ and then subtracting $2^{-1}a^{-1}b$. It follows that this congruence has 0, 1, or 2 solutions.

6.30. Find a characterization of the set of integers n for which the only solutions to $x^2 \equiv 1 \pmod{n}$ are ± 1. Recall from Proposition 2.4 that this fact is true when n is prime.

6.31. Prove that the following algorithm for solving $x^2 \equiv a \pmod{p}$ works and implement it. It is called **Shanks's algorithm**, and timing tests should show you that it is faster than Algorithm 6.2. This is the algorithm used by the CNT package.

Given p prime and a such that $a^{(p-1)/2} \equiv 1 \pmod{p}$, write $p - 1 = 2^t q$ with q odd. Let $z = h^q$ (reduced mod p), where h is a nonresidue modulo p; then $z^{2^{t-1}} \equiv -1 \pmod{p}$ by Euler's criterion. All congruences are modulo p, and all powers are reduced modulo p.

Step 1:   Define v and w by: $v = a^{(q+1)/2}$, and $w = a^q$.

Step 2a: If $w = 1$, then v is the desired square root of a.

Step 2b: Find the least k such that $w^{2^k} \equiv 1$. We know $w^{2^{t-1}} \equiv 1$, so $k \le t - 1$. Update z, t, v, and w to $z^{2^{t-k}}$, k, $vz^{2^{t-k-1}}$, and $wz^{2^{t-k}}$, respectively, and repeat step 2.

The algorithm halts because the value of t is decreasing each time. The key point in the working of the algorithm is that the following congruences are preserved at each step: $aw \equiv v^2$, $z^{2^{t-1}} \equiv -1$, $w^{2^{t-1}} \equiv 1$. Prove this.

CHAPTER 7

Continued Fractions

7.0 Introduction

We began this book with the example of an ancient algorithm for finding rational approximations of $\sqrt{2}$. We start with $a = b = 1$, and simply iterate, resetting $b \to a + b$, then $a \to 2b - a$. The fraction a/b gives an approximation to $\sqrt{2}$ that is off by less than $1/b^2$. The following code generates several of these approximations.

```
nextterm[r_] :=
```
$$\left(\{a, b\} = \{\text{Numerator}[r], \text{Denominator}[r]\}; \; \frac{a + 2b}{a + b} \right)$$

```
fracs = NestList[nextterm, 1, 8]
```

$$\left\{ 1, \frac{3}{2}, \frac{7}{5}, \frac{17}{12}, \frac{41}{29}, \frac{99}{70}, \frac{239}{169}, \frac{577}{408}, \frac{1393}{985} \right\}$$

```
fracs - √2.
```

$$\{-0.414214, 0.0857864, -0.0142136, 0.0024531, -0.000420459,$$
$$0.0000721519, -0.0000123789, 2.1239 \times 10^{-6}, -3.64404 \times 10^{-7}\}$$

The key to what is happening is to rewrite these fractions as

$$1, 1 + \frac{1}{2}, 1 + \cfrac{1}{2 + \frac{1}{2}}, 1 + \cfrac{1}{2 + \cfrac{1}{2 + \frac{1}{2}}}, 1 + \cfrac{1}{2 + \cfrac{1}{2 + \cfrac{1}{2 + \frac{1}{2}}}}, \dots.$$

These are called ***continued fractions***. They have a very rich structure, which we will exploit to solve many number-theoretic questions. They arose in the study of Diophantine equations, the Archimedes cattle problem (Section 7.5) being one of the most famous examples. These Diophantine equations were studied in great depth during the period of classical Indian mathematics, roughly CE 500 to 1200, and continue to be of interest to researchers in mathematics. In Section 7.6 we will see how continued fractions are used to create one of the powerful category-2 algorithms for factorizations: CFRAC, or the Continued Fraction Algorithm.

7.1 Finite Continued Fractions

A *finite simple continued fraction* (we will use simply **CF** to refer to these) has the form

$$a_0 + \cfrac{1}{a_1 + \cfrac{1}{a_2 + \cfrac{1}{a_3 + \cdots}}}$$

where each a_i is a positive integer and the process terminates. By loading a standard package we can generate continued fractions in typeset form.

```
Needs["NumberTheory`ContinuedFractions`"];
```

```
ContinuedFractionForm[{a, b, c, d, e, f}]
```

$$a + \cfrac{1}{b + \cfrac{1}{c + \cfrac{1}{d + \cfrac{1}{e + \cfrac{1}{f}}}}}$$

It is easy to write a rational, say $\frac{33}{14}$, as a CF:

$$\frac{33}{14} = 2 + \frac{5}{14} = 2 + \cfrac{1}{\frac{14}{5}} = 2 + \cfrac{1}{2 + \frac{4}{5}} = 2 + \cfrac{1}{2 + \cfrac{1}{\frac{5}{4}}} = 2 + \cfrac{1}{2 + \cfrac{1}{1 + \frac{1}{4}}}.$$

Definition. The notation $[a_0; a_1, a_2, ..., a_n]$ is used to represent the CF

$$a_0 + \cfrac{1}{a_1 + \cfrac{1}{a_2 + \cfrac{\ddots}{ + \cfrac{1}{a_{n-1} + \frac{1}{a_n}}}}}$$

Proposition 7.1. Every finite CF represents a rational number. Every rational number can be represented by a finite CF.

Proof. The forward direction is obvious, because we can just simplify the fraction. A proper proof uses induction based on the number of terms: $[a_0; a_1, a_2, ..., a_n] = a_0 + \frac{1}{[a_1; a_2, ..., a_n]}$, which is a sum of an integer and, by an inductive assumption, a rational. For the reverse direction use the

Euclidean algorithm and induction on the length of the remainder sequence. Given a rational a/b, let q and r be the quotient and remainder in the first step of the Euclidean algorithm. Then $a = qb + r$, so $a/b = q + r/b = q + \frac{1}{b/r}$, and the pair (b, r) has a shorter Euclidean algorithm sequence than (a, b), and therefore, by an inductive assumption, b/r has a representation as a finite CF, and hence so does a/b. □

Note that every rational number has two representations. This is because any integer a can be written as $(a - 1) + \frac{1}{1}$. Thus, $[a_0; a_1, a_2, ..., a_n]$ is the same as either $[a_0; a_1, a_2, ..., a_n - 1, 1]$ (if $a_n \neq 1$) or $[a_0; a_1, a_2, ..., a_{n-1} + 1]$ (if $a_n = 1$).

```
FromContinuedFraction /@ {{2, 1, 3}, {2, 1, 2, 1}}
```

$\left\{ \dfrac{11}{4}, \dfrac{11}{4} \right\}$

The construction of a CF is very reminiscent of the Euclidean algorithm. Indeed, the numbers that show up in a CF appear in the extended Euclidean algorithm. The **convergents** of a CF $[a_0; a_1, a_2, ..., a_n]$ are the fractions of the form $[a_0; a_1, a_2, ..., a_k]$ for $k = 0, 1, ..., n$. The kth convergent, C_k, is often denoted p_k/q_k. Here are the convergents for the CF of $612/233$.

```
cf = ContinuedFraction[612 / 233]
```

```
{2, 1, 1, 1, 2, 9, 3}
```

```
Table[Take[cf, i], {i, Length[cf]}]
```

```
{{2}, {2, 1}, {2, 1, 1}, {2, 1, 1, 1},
 {2, 1, 1, 1, 2}, {2, 1, 1, 1, 2, 9}, {2, 1, 1, 1, 2, 9, 3}}
```

```
Map[FromContinuedFraction , %]
```

$\left\{ 2, 3, \dfrac{5}{2}, \dfrac{8}{3}, \dfrac{21}{8}, \dfrac{197}{75}, \dfrac{612}{233} \right\}$

In fact, a `Convergents` function lives in the `ContinuedFractions` package.

```
Convergents[612 / 233]
```

$\left\{ 2, 3, \dfrac{5}{2}, \dfrac{8}{3}, \dfrac{21}{8}, \dfrac{197}{75}, \dfrac{612}{233} \right\}$

The `FullContinuedFraction` function gives a table showing the development of a CF. Comparing such a table with the Euclidean algorithm data shows us many connections.

FullContinuedFraction[612 / 233]
FullGCD[612, 233, ExtendedGCDValues → True]

Partial Quotients	Convergents
2	2
1	3
1	$\frac{5}{2}$
1	$\frac{8}{3}$
2	$\frac{21}{8}$
9	$\frac{197}{75}$
3	$\frac{612}{233}$

Remainders	Quotients	s	t
612		1	0
233	2	0	1
146	1	1	−2
87	1	−1	3
59	1	2	−5
28	2	−3	8
3	9	8	−21
1	3	−75	197
0		233	−612

We can now see where the continuant function gets its name from. The convergents arise from the quotients via the continuant function; this follows from the continuant discussion in Section 1.2 and the fact that the reversed s- and t-sequences are Euclidean algorithm sequences. The CNT function FullContinuant gives us the cumulative continuants.

FullContinuant[{2, 1, 1, 1, 2, 9, 3}]

{612, 233, 146, 87, 59, 28, 3, 1}

And the numerators and denominators of the convergents come from the s- and t-sequences, which occur as continuants of the reversed quotient sequence.

FullContinuant[{3, 9, 2, 1, 1, 1, 2}]

{612, 197, 21, 8, 5, 3, 2, 1}

FullContinuant[{3, 9, 2, 1, 1, 1}]

{233, 75, 8, 3, 2, 1, 1}

Proposition 7.2. (a)   The partial quotients in the CF for a/b are the quotients when the Euclidean algorithm is applied to the pair (a, b).

(b) The convergents in the CF for a/b are $C_k = -t_{k+3}/s_{k+3}$, $k \geq 2$, where the s- and t-values are those from the extended Euclidean algorithm.

(c) If the partial quotients in the CF for a/b are $\{a_k\}$, then the convergents are p_k/q_k, where these integers are defined inductively by:

$$
\begin{aligned}
p_0 &= a_0 & q_0 &= 1 \\
p_1 &= a_0 a_1 + 1 & q_1 &= a_1 \\
p_k &= a_k p_{k-1} + p_{k-2} & q_k &= a_k q_{k-1} + q_{k-2}.
\end{aligned}
$$

(d) For $k \geq 1$, $q_k \geq k$. In fact, q_k is greater than or equal to the kth Fibonacci number.

Proof. **(a)** Let $q_1 = \lfloor a/b \rfloor$ and let $r = a - q_1 b$. Then q_1 is the first Euclidean algorithm quotient for (a, b) and $a/b = q_1 + 1/(b/r)$. Because (b, r) has a shorter Euclidean algorithm sequence then (a, b), we may assume inductively that the quotients in the CF for b/r are the Euclidean algorithm quotients for (b, r). But the latter are the rest of the Euclidean algorithm quotients for (a, b) and the former are the rest of the quotients in the CF of a/b.

(b) Use the continuant function Q from Section 1.2. We claim that

$$
[a_0; a_1, a_2, ..., a_k] = \frac{Q[a_0, a_1, a_2, ..., a_k]}{Q[a_1, a_2, ..., a_k]}.
$$

This suffices because of (a) and the fact that the $|s|$- and $|t|$-sequences are Euclidean algorithm remainder sequences and so are continuant functions of the quotients; the match of the indices is as desired because s_0 is not part of the remainder sequence, while t_0 is, and the sign is right because s_k and t_k have opposite signs. We prove the claim by induction on k. The induction hypothesis is used in the middle equality below, and the inductive definition of Q is used at the end.

$$
[a_0; a_1, a_2, ..., a_i] = a_0 + \frac{1}{[a_1; a_2, ..., a_i]} = a_0 + \frac{1}{\frac{Q[a_1, a_2, ..., a_i]}{Q[a_2, ..., a_i]}}
$$

$$
= \frac{a_0 Q[a_1, a_2, ..., a_i] + Q[a_2, ..., a_i]}{Q[a_1, a_2, ..., a_i]} = \frac{Q[a_0, a_1, a_2, ..., a_i]}{Q[a_1, a_2, ..., a_i]}
$$

(c) This follows from (b) because the inductive definition in (c) agrees with the definition of the s- and t-sequences.

(d) Part (c) tells us that $q_k = Q[a_2, a_3, ..., a_k]$, which is clearly larger than $Q[1, 1, 1, ..., 1]$ (with $k-1$ 1s). But this continuant expression was shown in Section 1.2 to be the kth Fibonacci number. $\square$

This inductive definition leads to relatively prime pairs (Exercise 1.18), and so we can assume there is no cancellation in the formation of C_k as p_k / q_k.

Proposition 7.3. For a finite CF with convergents $C_k = p_k / q_k$:

(a) $p_k q_{k-1} - p_{k-1} q_k = (-1)^{k-1}$.

(b) $\frac{p_k}{q_k} - \frac{p_{k-1}}{q_{k-1}} = \frac{(-1)^{k-1}}{q_k q_{k-1}}$.

(c) $\frac{p_k}{q_k} - \frac{p_{k-2}}{q_{k-2}} = \frac{a_k(-1)^k}{q_k q_{k-2}}$ when $k \geq 2$.

(d) $C_0 < C_2 < C_4 < \cdots < C_5 < C_3 < C_1$.

Proof. Part (a) follows from Proposition 7.2(b) and Exercise 1.18(e); see also Exercise 7.1. For (b) just divide (a) by $q_k q_{k-1}$. Part (c) follows from (a) by straightforward algebra (Exercise 7.2). (d). Examining the parity of k in (c) tells us that $C_k < C_{k-2}$ when k is odd and $C_k > C_{k-2}$ when k is even. From (b) we get $C_{2k} < C_{2k-1}$, and this is sufficient to tell us that every odd-indexed convergent is larger than any even-indexed one. □

Parts (b) and (c) can be interpreted as saying that the differences between nearby convergents is small.

Exercises for Section 7.1

7.1. Prove Proposition 7.3(a) by induction on k.

7.2. Prove Proposition 7.3(c).

7.3. Prove that for any infinite sequence of positive integers $\{a_0, a_1, a_2, \ldots\}$, the convergents of the corresponding finite CFs actually converge to a single number as k approaches infinity.

7.4. Consider the finite CF for $a_0 = 1$ and $a_1 = a_2 = \cdots = a_k = 2$. Show that $p_0 / q_0 = 1$, $p_1 / q_1 = 3/2$, and

$$q_k = p_{k-1} + q_{k-1}$$
$$p_k = 2q_k - p_{k-1}.$$

7.5. With the same finite CF as in Exercise 7.4, show that

$$\frac{p_k^2}{q_k^2} = 2 - \frac{(-1)^k}{q_k^2}.$$

Hint: Use the fact that $p_k q_k = p_k (p_k - q_{k-1}) = q_k (2q_k - p_{k-1})$, and then use Proposition 7.3(a).

7.6. The solar year is $365 \frac{20929}{86400}$ days. The Gregorian calendar that we use approximates the fraction by adding 97 days every 400 years (on the leap years that occur every fourth year except for years divisible by 100 and not divisible by 400, so that 1700, 1800, and 1900 were not leap years, but 2000 is). Find a better approximation with a denominator smaller than 400.

7.2 Infinite Continued Fractions

Any real number can be decomposed into a single, infinitely long continued fraction as follows. Let's use π as an example. The equation $\pi = 3 + \frac{1}{7}$ is not exactly true; to make it true, 7 would have to be replaced by a larger number, whose value we can find by solving $3 + \frac{1}{x} = \pi$. This leads to $x = 1/(\pi - 3) \approx 7.06251$. But we want integers in our continued fraction, so we write 7.0625... as $7 + \frac{1}{x}$, where x is obtained from $1/(\pi - 3)$ in the same way that 7.0625... was obtained from π. Continue, replacing x by $\lfloor x \rfloor$, and so on. In this case, one would obtain, to six terms: $\pi = 3 + 1/(7 + 1/(15 + 1/(1 + 1/(292 + 1/(1 + ...)))))$. Using the notation introduced in the preceding section, we have $\pi = [3; 7, 15, 1, 292, 1, ...]$.

To use the built-in function we need to specify the number of terms. Here are some famous continued fractions.

ContinuedFraction[π, 20]

{3, 7, 15, 1, 292, 1, 1, 1, 2, 1, 3, 1, 14, 2, 1, 1, 2, 2, 2, 2}

Convergents[π, 10]

$$\left\{3, \frac{22}{7}, \frac{333}{106}, \frac{355}{113}, \frac{103993}{33102}, \right.$$
$$\left. \frac{104348}{33215}, \frac{208341}{66317}, \frac{312689}{99532}, \frac{833719}{265381}, \frac{1146408}{364913} \right\}$$

ContinuedFraction[E, 10]

{2, 1, 2, 1, 1, 4, 1, 1, 6, 1}

ContinuedFraction[$\sqrt{2}$, 10]

{1, 2, 2, 2, 2, 2, 2, 2, 2, 2}

ContinuedFraction[$\sqrt{3}$, 10]

{1, 1, 2, 1, 2, 1, 2, 1, 2, 1}

ContinuedFraction[GoldenRatio, 10]

{1, 1, 1, 1, 1, 1, 1, 1, 1, 1}

An ***infinite simple continued fraction*** (we will use just ***CF*** for these) is a continued fraction of the form

$$a_0 + \cfrac{1}{a_1 + \cfrac{1}{a_2 + \cdots}}$$

which we can refer to as $[a_0; a_1, a_2, ...]$. The value of this CF is taken to be the limit of the finite approximations (the convergents). Given α irrational, we can define the CF $[a_0; a_1, a_2, ...]$ for α as follows: Start with $\alpha_0 = \alpha$ and let $a_k = \lfloor \alpha_k \rfloor$ and let $\alpha_{k+1} = 1/(\alpha_k - a_k)$.

Theorem 7.4.   Existence and Uniqueness of Continued Fractions

(a) Every CF is convergent.

(b) The CF obtained from a real number by using the procedure illustrated in the paragraph preceding this theorem converges to the real number.

(c) Every infinite CF converges to an irrational number.

(d) The CF obtained from an irrational is the only CF that converges to that irrational.

Proof. **(a)** Let C_i denote the convergents of $[a_1; a_2, a_3, ...]$. Applying Proposition 7.3(d) to the finite approximations tells us that

$$C_0 < C_2 < C_4 < \cdots < C_5 < C_3 < C_1 .$$

This means that the even convergents have a limit, as do the odd convergents. Thus, it remains to show that $C_{2k+1} - C_{2k} \to 0$. But this follows from Proposition 7.3(d) and (e).

(b) Because each a_k is an integer, induction yields that each α_k is irrational. Therefore, $a_k < \alpha_k < a_k + 1$ and $\alpha_{k+1} = 1/(\alpha_k - a_k) \geq 1$, so each a_k is positive. Let $C_k = p_k/q_k$ be the kth convergent of the infinite CF $[a_1; a_2, a_3, ...]$.

$$\alpha = \alpha_0 = [a_0; \alpha_1] = [a_0; a_1, \alpha_2]$$
$$= \cdots = [a_0; a_1, a_2, ..., a_k, \alpha_{k+1}] = \frac{\alpha_{k+1} p_k + p_{k-1}}{\alpha_{k+1} q_k + q_{k-1}}$$

$$|\alpha - C_k| = \left| \alpha - \frac{p_k}{q_k} \right|$$
$$= \frac{|(-1)^k|}{(\alpha_{k+1}q_k + q_{k-1})q_k} < \frac{1}{(a_{k+1}q_k + q_{k-1})q_k} = \frac{1}{q_{k+1}q_k} < \frac{1}{k(k+1)} \to 0 .$$

(c)   Suppose that $\alpha = [a_0; a_1, a_2, ...]$. Part (a) tells us that α lies between each even- and odd-indexed convergent. This, combined with Proposition 7.3(b), gives us $0 < \alpha - (p_{2n}/q_{2n}) < C_{2n+1} - C_{2n} < 1/(q_{2n+1}q_{2n})$. It follows that $0 < \alpha q_{2n} - p_{2n} < 1/q_{2n+1}$. Now suppose that α is rational, say $\alpha = x/y$. Then $0 < xq_{2n} - yp_{2n} < y/q_{2n+1} < y/(2n+1)$, a contradiction, because the last expression converges to 0 as $n \to \infty$, while $xq_{2n} - yp_{2n}$ is an integer.

(d) See Exercise 7.7.   □

The preceding results are not very surprising as they just tell us that CFs behave in the way one would hope they would. However, the next theorem is more sophisticated, for it gives some surprising information about the relationship between α and the convergents of its CF. Note that from earlier work — proof of Proposition 7.4(b) — we know that

$$\left| \alpha - \frac{p_k}{q_k} \right| < \frac{1}{q_k q_{k+1}} \, .$$

This means that the convergents are very good approximations to the limit. For any α it is easy to find a fraction of the form p/q that it is closer to α than any other fraction with denominator q: $^{314}/_{100}$ is closer to π than any other $p/100$. But in such cases one can state only that p/q will be within $1/(2q)$ of α. The convergents are much closer than this. Look at some of the convergents for π.

con = Convergents[π, 5]

$$\left\{ 3, \ \frac{22}{7}, \ \frac{333}{106}, \ \frac{355}{113}, \ \frac{104348}{33215} \right\}$$

The third convergent is within $^1/_{12000}$ of π, consistent with the theory's prediction that it should be within $^1/_{106 \cdot 113}$. The next convergent — the famous $^{355}/_{113}$ — is even better, because it must lie within $^1/_{113 \cdot 33215}$ of π.

$$\left\{ \textbf{Round} \left[\frac{1}{\pi - \frac{333}{106}} \right], \ \textbf{106} \ast \textbf{113} \right\}$$

$\{12016, 11978\}$

$$\left\{ \textbf{Round} \left[\frac{1}{\frac{355}{113} - \pi} \right], \ \textbf{113} \ast \textbf{33215} \right\}$$

$\{3748629, 3753295\}$

The reason that $^{355}/_{113}$ is so good is that the fifth partial quotient of π is 292, which is much larger than average. The next result tells us that rational approximations that are especially good must be one of the convergents. Note that part (b) refers to q_k, not q_k^2; we can see why by looking at the convergents of π. The third convergent, $^{333}/_{106}$, is within 1/12,000th of π, but the next convergent, whose denominator is only a little larger ($^{355}/_{113}$), is much better.

Theorem 7.5. (a) If α is irrational, p_k and q_k are its CF convergents, and r and s are integers with $s > 0$ and $|s\alpha - r| < |q_k\alpha - p_k|$ then $s \geq q_{k+1}$.

(b) If p_k / q_k is a convergent for α, then p_k / q_k is closer to α than any other rational with denominator less than or equal to q_k.

(c) If $|\alpha - r/s| < 1/(2s^2)$, then r/s must be a convergent in the CF of α.

Proof. **(a)** Assume not, so $s < q_{k+1}$. Consider the following system:

$$p_k\, x + p_{k+1}\, y = r$$
$$q_k\, x + q_{k+1}\, y = s\,. \tag{1}$$

The determinant of this system is $p_{k+1}\, q_k - p_k\, q_{k+1}$, which equals $(-1)^k$. This means that there is a unique solution for x and y and the solution consists of integers (alternatively, use standard techniques to solve the equations explicitly). We claim that neither x nor y is 0. For if $x = 0$, then $s\, p_{k+1} = r\, q_{k+1}$ and so q_{k+1} divides s, a contradiction. And $y = 0$ would yield that $r = p_k\, x$ and $s = q_k\, x$, implying

$$|s\alpha - r| = |q_k\, x\alpha - p_k\, x| = |x|\,|q_k\, \alpha - p_k| \ge |q_k\, \alpha - p_k|$$

in contradiction to the assumption.

We next claim that x and y have opposite signs. For if $y < 0$, then $x > 0$ by the second equation in (1). And if $y > 0$ then, because q_{k+1} is assumed to be greater than s, the same equation implies that x is negative.

Now, convergents alternate, so either $p_k/q_k < \alpha < p_{k+1}/q_{k+1}$, or the reverse. In either case $q_k\, \alpha - p_k$ and $q_{k+1}\, \alpha - p_{k+1}$ have opposite signs. Finally:

$$\begin{aligned}
|s\alpha - r| &= |(q_k\, x + q_{k+1}\, y)\alpha - (p_k\, x + p_{k+1}\, y)| \\
&= |x(q_k\, \alpha - p_k) + y(q_{k+1}\, \alpha - p_{k+1})| \\
&= |x|\,|q_k\, \alpha - p_k| + |y|\,|q_{k+1}\, \alpha - p_{k+1}| \\
&\ge |x|\,|q_k\, \alpha - p_k| \ge |q_k\, \alpha - p_k|
\end{aligned}$$

where the claims about opposite signs are used to get the breakup of absolute values. This is a contradiction.

(b)   Suppose $|\alpha - r/s| < |\alpha - p_k/q_k|$ and $s \le q_k$. Then $|s\alpha - r| < |q_k\, \alpha - p_k|$, which contradicts (a).

(c) If not, we can find consecutive convergents, say the kth and $(k+1)$st, so that $q_k \le s < q_{k+1}$. Then, by (a), we can say that

$$|q_k\, \alpha - p_k| \le |s\alpha - r| \le s\left|\alpha - \frac{r}{s}\right| < \frac{1}{2s}$$

which implies that

$$\left|\alpha - \frac{p_k}{q_k}\right| < \frac{1}{2s\, q_k}\,.$$

We know that $|s\, p_k - r\, q_k| \ge 1$ because it is an integer and not 0 ($r/s \ne p_k/q_k$). Therefore

$$\frac{1}{sq_k} \le \frac{|sp_k - rq_k|}{sq_k} = \left|\frac{p_k}{q_k} - \frac{r}{s}\right| \le \left|\alpha - \frac{p_k}{q_k}\right| + \left|\alpha - \frac{r}{s}\right| < \frac{1}{2sq_k} + \frac{1}{2s^2} \ .$$

So $1/(2sq_k) < 1/(2s^2)$, or $q_k > s$, a contradiction. $\square$

Exercises for Section 7.2

7.7. Prove Proposition 7.4(d). *Hint*: Suppose that $\alpha = [a_0; a_1, a_2, \ldots] = [b_0; b_1, b_2, \ldots]$. First show that $a_0 = b_0$, and then use induction to show that each $a_k = b_k$.

7.8. If $\alpha > 1$ and $\alpha = [a_0; a_1, a_2, \ldots]$, then $1/\alpha = [0; a_0, a_1, a_2, \ldots]$. The convergents of $1/\alpha$ are the reciprocals of the convergents of α.

7.9. Use continued fractions to find some good rational approximations to $\log_2 3$. This number and its rational approximations are relevant to the 12-tone scale used in western music; see [DM].

7.10. Hurwitz proved that for any rational number α, there are always infinitely many convergents to α, p_k/q_k, such that

$$\left|\frac{p_k}{q_k} - \alpha\right| < \frac{1}{q_k^2 \sqrt{5}} \ .$$

For τ, the golden ratio (which is $(1 + \sqrt{5})/2$), find an infinite set of convergents that satisfy the inequality. *Hints*: First use Proposition 7.2(b) to see that the convergents for τ are ratios of Fibonacci numbers. Also Binet's formula (Proposition 1.1) will be useful.

7.11. Show that the inequality of Exercise 7.10 is the best possible in the sense that if $c > \sqrt{5}$, then there are at most finitely many convergents to the golden ratio for which

$$\left|\frac{p_k}{q_k} - \tau\right| < \frac{1}{cq_k^2} \ .$$

7.12. Show that the inequality of Exercise 7.10 is not the best possible for all irrationals. In particular, prove that there are infinitely many convergents to $\sqrt{2}$ for which

$$\left|\frac{p_k}{q_k} - \sqrt{2}\right| < \frac{1}{\sqrt{8} \, q_k^2} \ .$$

7.13. Let p_k/q_k be a convergent to $\alpha = [a_0; a_1, a_2, \ldots]$. Show that

$$p_k = \det \begin{pmatrix} a_0 & -1 & 0 & \cdots & 0 & 0 \\ 1 & a_1 & -1 & \cdots & 0 & 0 \\ 0 & 1 & a_2 & \cdots & 0 & 0 \\ \vdots & & & \ddots & & \vdots \\ 0 & 0 & 0 & \cdots & a_{k-1} & -1 \\ 0 & 0 & 0 & \cdots & 1 & a_k \end{pmatrix}$$

Find a similar expression for q_k.

7.14. Let $\alpha = [a_0; a_1, \ldots, a_n]$. Show that for $k > 0$

$$\alpha = \frac{r\alpha_k + s}{t\alpha_k + u}$$

where

$$\begin{pmatrix} r & s \\ t & u \end{pmatrix} = \begin{pmatrix} a_0 & 1 \\ 1 & 0 \end{pmatrix}\begin{pmatrix} a_1 & 1 \\ 1 & 0 \end{pmatrix}\cdots\begin{pmatrix} a_{k-1} & 1 \\ 1 & 0 \end{pmatrix}.$$

7.15. If β is a positive real number, $a_0, a_1, \ldots$ are positive integers, and n is odd, show that

$$[a_0; a_1, \ldots, a_n] > [a_0; a_1, \ldots, a_{n-1}, a_n + \beta]$$

Is this ever true if n is even?

7.16. Let α be a real number chosen uniformly at random from the interval $(0, 1)$ and let $[0; a_1, a_2, \ldots]$ be its continued fraction. Given a positive integer m, what is the probability that $a_1 > m$? What is the probability that $a_1 = m$?

7.17. Given α as in the preceding exercise, show that the probability that $a_2 \geq m$ is

$$\sum_{a_1=1}^{\infty} \frac{1}{a_1(ma_1+1)}.$$

Find the probability that $a_2 = 1$.

7.18. Let d be a positive integer that is not a perfect cube. Show that if (x_0, y_0) is a solution in positive integers to $x^3 - dy^3 = n$ and if $y_0 > 2|n|/d^{2/3}$, then x_0/y_0 is a convergent to $d^{1/3}$. *Hint*: Factor $x_0^3 - dy_0^3$.

7.3 Periodic Continued Fractions

Our goal in this section is to prove one of the prettiest theorems about continued fractions (discovered by Lagrange): a CF is periodic if and only if it represents a quadratic irrational number. A **quadratic irrational** is an irrational real number that is a root of the quadratic polynomial $Ax^2 + Bx + C$, where A, B, and C are integers. This polynomial will have two roots, and the other one is called the **conjugate** of α, denoted $\overline{\alpha}$.

Proposition 7.6. (a) Every quadratic irrational α can be written as $(a + \sqrt{b})/c$, where a, b, and c are integers and b is not a perfect square.

(b) The conjugate of $(a + \sqrt{b})/c$ is $(a - \sqrt{b})/c$.

(c) If α is a quadratic irrational, so is $(m\alpha + n)/(s\alpha + t)$, where m, n, s, and t are integers.

(d) $\overline{\alpha + \beta} = \overline{\alpha} + \overline{\beta}$.

(e) $\overline{\alpha\beta} = \overline{\alpha}\,\overline{\beta}$.

(f) $\overline{\alpha/\beta} = \overline{\alpha}/\overline{\beta}$.

(g) Every quadratic irrational α can be written as $(X + \sqrt{d})/Y$ where X, Y, and d are integers, $Y \neq 0$, d is not a square, and Y divides $d - X^2$. This representation is called a **normal form** of α (but it is not unique: $\sqrt{2}/2 = \sqrt{8}/4$). Note that X or Y might be negative.

Proof. (a) The use of the quadratic formula in the following computer solution shows that $(a + \sqrt{b})/c$ is a quadratic irrational.

```
x /. Solve[c² x² - 2 a c x + (a² - b) == 0, x]
```

$$\left\{ \frac{a - \sqrt{b}}{c}, \frac{a + \sqrt{b}}{c} \right\}$$

And it is easy to see from the quadratic formula that if α is a root of the quadratic equation $Ax^2 + Bx + C$, then α has the desired form.

(b) This follows from the quadratic formula. This allows us to program a QuadraticConjugate function to switch the sign of any square roots. This is done in the CNT package.

```
QuadraticConjugate[a + √b]
```

$$a - \sqrt{b}$$

(c) We can use QuadraticConjugate to prove this.

```
α = (a + √b̄) / c;
β = Simplify[(m α + n) / (s α + t)]
```

$$\frac{a\,m + \sqrt{b}\,m + c\,n}{a\,s + \sqrt{b}\,s + c\,t}$$

Now we multiply top and bottom by the conjugate of the bottom; the result has the proper form.

```
qc = QuadraticConjugate[Denominator[β]];
ExpandDenominator[
  Collect[Numerator[β] qc, √b̄] / (Denominator[β] qc)]
```

$$\frac{a^2\,m\,s - b\,m\,s + a\,c\,n\,s + a\,c\,m\,t + c^2\,n\,t + \sqrt{b}\,(-c\,n\,s + c\,m\,t)}{a^2\,s^2 - b\,s^2 + 2\,a\,c\,s\,t + c^2\,t^2}$$

Parts (d)–(f) can be proved by easy algebra. Here is one approach to (f).

```
α = (a + √b̄) / c;
β = (d + √ē) / f;
QuadraticConjugate[α / β]
```

$$\frac{(a - \sqrt{b})\,f}{c\,(d - \sqrt{e})}$$

```
QuadraticConjugate[α] / QuadraticConjugate[β]
```

$$\frac{(a - \sqrt{b})\,f}{c\,(d - \sqrt{e})}$$

(g) We know by (a) that $\alpha = (a + \sqrt{b})/c$. Then $\alpha = (a|c| + \sqrt{b\,c^2})/(c\,|c|)$, which satisfies the desired conditions; in particular, $d - X^2 = b\,c^2 - a^2\,c^2$, which is divisible by c^2. □

This normal form of part (g) is easily implemented; the CNT package function QuadraticNormalForm gives a triple $\{X, d, Y\}$ that represents a given quadratic irrational in a normal form.

```
{X, d, Y} = QuadraticNormalForm[α = 31/178 - 2√176/3]
```

```
{14161417272, 51720201228984299583897 6, 81313944336}
```

The built-in RootReduce function transforms quadratic irrationals to a slightly different normal form. This allows us to check the preceding result.

```
RootReduce[{α, (x + √d̄) / Y}]
```

$$\left\{ \frac{1}{534}\,(93 - 1424\,\sqrt{11}),\ \frac{1}{534}\,(93 + 1424\,\sqrt{11}) \right\}$$

Theorem 7.7. Periodicity of Quadratic Irrational CFs The CF corresponding to a quadratic irrational is eventually periodic; conversely, every CF that is eventually periodic converges to such a real.

The fact that periodic CFs represent quadratic irrationals is easy to understand. Consider the periodic CF $[1; 2, 1, 2, 1, 2, 1, 2, ...]$, which we can denote by $[\overline{1; 2}]$. If this represents α, then

$$\alpha = 1 + \cfrac{1}{2 + \frac{1}{\alpha}} .$$

But the right side quickly simplifies to a rational expression in α.

$$\mathtt{Map}\left[\mathtt{Together,}\ \alpha == 1 + \cfrac{1}{2 + \frac{1}{\alpha}}\right]$$

$$\alpha == \frac{1 + 3\,\alpha}{1 + 2\,\alpha}$$

And this means that α is a root of a quadratic equation. The general proof is no harder than simplifying the CF as we did in this example. The other direction, going from a quadratic irrational to a periodic CF, is more difficult and relies on recognizing certain patterns in the sequences of remainders α_i that show up in the construction of a CF.

Proof of Theorem 7.7. We do the easier direction first. Suppose $[a_0; a_1, a_2, ..., a_s, \overline{b_1, b_2, ..., b_n}]$ converges to α. It suffices to show that the periodic part represents a quadratic irrational, because the effect of the nonperiodic part is simply to add integers and invert, properties that leave one in the domain of quadratic irrationals by Proposition 7.6(c). So let $\beta = [\overline{b_1; b_2, ..., b_n}]$. Then $\beta = [b_1; b_2, b_3, ..., b_n, \beta]$, a finite CF, albeit one having a noninteger as its last entry. Collapsing the finite CF leads to an expression of the form $(i\beta + j)/(k\beta + l)$. We can be more precise by using Proposition 7.2(c), which tells us that the convergents give us the exact relationship:

$$\beta = \frac{p_{n+1}}{q_{n+1}} = \frac{p_n\,\beta + p_{n-1}}{q_n\,\beta + q_{n-1}} .$$

Because the p- and q-values are integers, this simplifies to a quadratic equation having β as a solution, which is what we want. We know that β (and hence α) will be irrational by Proposition 7.4(c).

It remains to show that a quadratic irrational having the normal form $\alpha = \left(X_0 + \sqrt{d}\right)/Y_0$ leads to a periodic CF. This will be done by means of an algorithm. Consider the following inductive definitions of a_k, α_k and integers X_k, Y_k:

$$\alpha_k = \frac{X_k + \sqrt{d}}{Y_k}$$
$$a_k = \lfloor \alpha_k \rfloor$$
$$X_{k+1} = a_k Y_k - X_k$$
$$Y_{k+1} = Y_{k-1} + a_k(X_k - X_{k+1}); \ \text{except}: \ Y_1 = \frac{d - X_1^2}{Y_0}.$$

Note that Y_1 is an integer because Y_0 divides $d - X_0^2$ and $X_1^2 \equiv X_0^2 \ (\text{mod } Y_0)$. One could define Y_{k+1} to be $(d - X_{k+1}^2)/Y_k$; but, for computation, it is always wise to avoid division when possible. Note that X_k or Y_k might be negative. Here are the first few values of these sequences for the case of $(2 - \sqrt{2})/4$, which has the normal form $(128 + \sqrt{8192})/256$.

$\{X, Y\}$	α	Partial Quotients	Convergents
$\{128, 256\}$	$\frac{128 + \sqrt{8192}}{256}$	0	0
$\{-128, -32\}$	$\frac{-128 + \sqrt{8192}}{-32}$	1	1
$\{96, 32\}$	$\frac{96 + \sqrt{8192}}{32}$	5	$\frac{5}{6}$
$\{64, 128\}$	$\frac{64 + \sqrt{8192}}{128}$	1	$\frac{6}{7}$
$\{64, 32\}$	$\frac{64 + \sqrt{8192}}{32}$	4	$\frac{29}{34}$
$\{64, 128\}$	$\frac{64 + \sqrt{8192}}{128}$	1	$\frac{35}{41}$
	Periodic Form:	$\langle\langle 0\ 1\ 5\ |\ 1\ 4 \rangle\rangle$	

We will show that:

1. $Y_k \neq 0$ and $Y_{k+1} = (d - X_{k+1}^2)/Y_k$.
2. Y_k divides $d - X_k^2$. [Because Y_k divides $d - X_{k+1}^2$ and $X_{k+1} \equiv -X_k \ (\text{mod } Y_k)$.]
3. $(X_k + \sqrt{d})/Y_k$ is a normal form of α_k. [By (2).]
4. $\alpha_{k+1} = 1/(\alpha_k - a_k)$.
5. The a_k are the partial quotients of the CF for α. [By (4).]
6. $0 < Y_k$ for sufficiently large k.
7. $Y_k \leq d$ for sufficiently large k. [Use (6). If $Y_k > 0$ beyond N, then for $k \geq n$ we have $Y_k \leq Y_k Y_{k+1} = d - X_{k+1}^2 \leq d$.]
8. $-\sqrt{d} < X_k < \sqrt{d}$ for sufficiently large k.
9. The pairs $\{X_k, Y_k\}$ must repeat at some point. [By 6, 7, and 8, there are only finitely many possiblilities.]
10. The α_k must repeat at some point. [By (9).]
11. The a_k must repeat at some point. [By (10).]

Before completing the proof, let us look at some data. The following table shows that $\alpha = 1 + \sqrt{3}/2$ leads to X_4 and Y_4 that agree with X_2 and

Y_2; the repetition is shown in bold. Thus, the CF for α consists of the preperiodic part $\{1, 1\}$ followed by repetitions of 6, 2; in short, $\alpha = [1; 1, \overline{6, 2}]$

FullContinuedFraction $\left[1 + \sqrt{3} \,/\, 2\right]$

$\{X, Y\}$	α	Partial Quotients	Convergents
$\{16, 16\}$	$\frac{16+\sqrt{192}}{16}$	1	1
$\{0, 12\}$	$\frac{\sqrt{192}}{12}$	1	2
{12, 4}	$\frac{12+\sqrt{192}}{4}$	**6**	$\frac{13}{7}$
$\{12, 12\}$	$\frac{12+\sqrt{192}}{12}$	**2**	$\frac{28}{15}$
{12, 4}	$\frac{12+\sqrt{192}}{4}$	6	$\frac{181}{97}$
	Periodic Form:	$\langle\langle\, 1\ 1\ |\ 6\ 2\, \rangle\rangle$	

We can get such periodic forms directly using the built-in `Continued-Fraction`.

ContinuedFraction $\left[1 + \sqrt{3} \,/\, 2\right]$

`{1, 1, {6, 2}}`

And we can check our answers by going in the other direction.

FromContinuedFraction[{1, 1, {6, 2}}]

$\frac{1}{2}\, (2 + \sqrt{3}\,)$

Perhaps the simplest periodic CF is $[\overline{1}]$. It is not hard to see that it represents the golden ratio because of $x = 1/(1 + x)$.

FromContinuedFraction[{{1}}]

$\frac{1}{2}\, (1 + \sqrt{5}\,)$

We now return to the proof of claims 1–11. Only 1, 4, 6, and 8 remain to be proved.

1. Use induction on k as follows.

$$Y_k a_k = X_k + X_{k+1}$$
$$\frac{Y_{k+1} - Y_{k-1}}{a_k} = X_k - X_{k+1} \qquad \text{(by the definition)}$$
$$Y_k(Y_{k+1} - Y_{k-1}) = X_k^2 - X_{k+1}^2 \quad \text{(multiply preceding equations)}$$
$$Y_{k-1} Y_k = d - X_k^2 \qquad \text{(the induction hypothesis)}$$
$$Y_k Y_{k+1} = d - X_{k+1}^2 \qquad \text{(add preceding two equations)}$$

Because d cannot equal the perfect square X_{k+1}^2, Y_{k+1} is not 0, which proves point 1.

4. $\alpha_{k+1} = \dfrac{X_{k+1} + \sqrt{d}}{Y_{k+1}} = \dfrac{Y_k(X_{k+1} + \sqrt{d})}{Y_k Y_{k+1}} = \dfrac{Y_k(X_{k+1} + \sqrt{d})}{d - X_{k+1}^2}$

$\qquad = \dfrac{Y_k}{\sqrt{d} - X_{k+1}} = \dfrac{Y_k}{\sqrt{d} - a_k Y_k + X_k} = \dfrac{1}{\frac{X_k + \sqrt{d}}{Y_k} - a_k} = \dfrac{1}{\alpha_k - a_k}$

6. By (4), (5), and the proof of Proposition 7.4(b), $\alpha = [a_0; a_1, a_2, \ldots, a_k, \alpha_k]$. Letting p_k / q_k be the convergents of this CF, Proposition 7.2(c) tells us that

$$\alpha = \frac{p_{k-1}\alpha_k + p_{k-2}}{q_{k-1}\alpha_k + q_{k-2}} .$$

We can take conjugates throughout, solve for $\overline{\alpha_k}$, and see what happens as k gets large. The first step is accomplished by the two substitutions in the code that follows.

```
α̅_k /. First[Solve[α ==  (p_{k-1} α_k + p_{k-2})/(q_{k-1} α_k + q_{k-2})  /. {α → ᾱ, α_k → α̅_k}, α̅_k]]
```

$$\frac{p_{k-2} - \overline{\alpha} \, q_{k-2}}{\overline{\alpha} \, q_{k-1} - p_{k-1}}$$

This quantity is the same as the following

$$-\frac{q_{k-2}\left(\overline{\alpha} - \frac{p_{k-2}}{q_{k-2}}\right)}{q_{k-1}\left(\overline{\alpha} - \frac{p_{k-1}}{q_{k-1}}\right)} .$$

```
Simplify[% -  (q_{k-2}/q_{k-1}) (ᾱ - p_{k-2}/q_{k-2})/(ᾱ - p_{k-1}/q_{k-1}) ]
```

$$\frac{2 \, (p_{-2+k} - \overline{\alpha} \, q_{-2+k})}{-p_{-1+k} + \overline{\alpha} \, q_{-1+k}}$$

But because the convergents approach α, this approaches

$$-\frac{q_{k-2}}{q_{k-1}} \frac{(\overline{\alpha} - \alpha)}{(\overline{\alpha} - \alpha)}, \quad \text{or just} \quad -\frac{q_{k-2}}{q_{k-1}} .$$

Because $q_{k-1} > q_{k-2}$, this last fraction is less than -1. Therefore, there is an N such that $\overline{\alpha_k} < 0$ whenever $k \geq N$. But $\alpha_k > 0$ by (4). So $0 < \alpha_k - \overline{\alpha_k} = 2\sqrt{d} / Y_k$, as desired.

8. If k is large enough that Y_k and Y_{k+1} are positive, then $0 < Y_k Y_{k+1} = d - X_{k+1}^2$, so $X_{k+1}^2 < d$. □

The preceding proof yields an algorithm for getting the exact periodic CF for any quadratic irrational: Form the sequences of X and Y values and check for a repeat. It must occur eventually. This is how *Mathematica's* ContinuedFraction function is able to find the periodic forms, though that built-in algorithm uses some storage tricks to speed up the comparisons with past values.

The following table shows the periodic CFs for the simple square roots. These will play a critical role in the solution to the Pell equation, discussed in detail in the next section.

$\sqrt{2}$	{1, {2}}
$\sqrt{3}$	{1, {1, 2}}
$\sqrt{4}$	2
$\sqrt{5}$	{2, {4}}
$\sqrt{6}$	{2, {2, 4}}
$\sqrt{7}$	{2, {1, 1, 1, 4}}
$\sqrt{8}$	{2, {1, 4}}
$\sqrt{9}$	3
$\sqrt{10}$	{3, {6}}
$\sqrt{11}$	{3, {3, 6}}
$\sqrt{12}$	{3, {2, 6}}
$\sqrt{13}$	{3, {1, 1, 1, 1, 6}}
$\sqrt{14}$	{3, {1, 2, 1, 6}}
$\sqrt{15}$	{3, {1, 6}}
$\sqrt{16}$	4
$\sqrt{17}$	{4, {8}}
$\sqrt{18}$	{4, {4, 8}}
$\sqrt{19}$	{4, {2, 1, 3, 1, 2, 8}}
$\sqrt{20}$	{4, {2, 8}}

You may not see any obvious patterns, but in fact there are several. For square roots near perfect squares the patterns are quite striking.

$\sqrt{r^2 - 1}$		$\sqrt{r^2 + 1}$	
$\sqrt{3}$	{1, {1, 2}}	$\sqrt{5}$	{2, {4}}
$\sqrt{8}$	{2, {1, 4}}	$\sqrt{10}$	{3, {6}}
$\sqrt{15}$	{3, {1, 6}}	$\sqrt{17}$	{4, {8}}
$\sqrt{24}$	{4, {1, 8}}	$\sqrt{26}$	{5, {10}}

$\sqrt{r^2-2}$		$\sqrt{r^2+2}$	
$\sqrt{2}$	{1, {2}}	$\sqrt{6}$	{2, {2, 4}}
$\sqrt{7}$	{2, {1, 1, 1, 4}}	$\sqrt{11}$	{3, {3, 6}}
$\sqrt{14}$	{3, {1, 2, 1, 6}}	$\sqrt{18}$	{4, {4, 8}}
$\sqrt{23}$	{4, {1, 3, 1, 8}}	$\sqrt{27}$	{5, {5, 10}}

$\sqrt{r^2-4}$		$\sqrt{r^2+4}$	
$\sqrt{0}$	{0}	$\sqrt{8}$	{2, {1, 4}}
$\sqrt{5}$	{2, {4}}	$\sqrt{13}$	{3, {1, 1, 1, 1, 6}}
$\sqrt{12}$	{3, {2, 6}}	$\sqrt{20}$	{4, {2, 8}}
$\sqrt{21}$	{4, {1, 1, 2, 1, 1, 8}}	$\sqrt{29}$	{5, {2, 1, 1, 2, 10}}
$\sqrt{32}$	{5, {1, 1, 1, 10}}	$\sqrt{40}$	{6, {3, 12}}
$\sqrt{45}$	{6, {1, 2, 2, 2, 1, 12}}	$\sqrt{53}$	{7, {3, 1, 1, 3, 14}}
$\sqrt{60}$	{7, {1, 2, 1, 14}}	$\sqrt{68}$	{8, {4, 16}}

Figure 7.1 shows the period lengths for the CFs of $\sqrt{n}$. It has been suggested by H. C. Williams [Wil1] that the period-length is bounded by approximately $\sqrt{n}\,\log(\log(4n))$; $\sqrt{n}\,\log(\log(4n))+1$, whose graph is superimposed on the data in Figure 7.1, does the job up to $n = 1000$.

```
CFPeriodLength[z_] := Length[ContinuedFraction[z][[-1]]];

ListPlot[Table[{n, CFPeriodLength[√n]}, {n, 1, 1000}]];
```

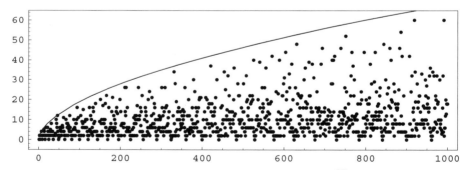

Figure 7.1. The period lengths of the continued fractions of $\sqrt{n}$, for n up to 1000. The longest period is 60, for $\sqrt{919}$ and $\sqrt{991}$, and the period-lengths are bounded by $\sqrt{n}\,\log(\log(4n))+1$ in this range.

The next theorem summarizes some properties of CFs of $\sqrt{n}$ that will be important in the next section. These results are esthetically pleasing as they explain several features of these CFs that jump out when one does computations; for example, it always takes exactly only one step to reach the periodic part. And they are important algorithmically, as we will discuss after the proof.

Theorem 7.8. (a) The quadratic irrationals that are ***purely periodic*** — that is, have the form $[\overline{a_0; a_1, ..., a_n}]$ — have a simple characterization: A quadratic irrational α has a purely periodic CF if and only if $\alpha > 1$ and $-1 < \overline{\alpha} < 0$.

(b) If α is purely periodic, then so is $-1/\overline{\alpha}$, and its periodic part is the reverse of that of α.

(c) The CF of $\sqrt{n}$ has the form $[a_0; \overline{a_1, a_2, ..., a_r, 2a_0}]$ where each $a_i \le a_0$. Moreover, the sequence $a_1, a_2, ..., a_r$ is symmetric about its center.

Proof. The proofs of (a) and (b) are somewhat long, and we refer the reader to [NZM] or [Ros]. However, part (c) can be quickly derived from (a) and (b). Let $a_0 = \lfloor \sqrt{n} \rfloor$; then, by (a), $\sqrt{n} + \lfloor \sqrt{n} \rfloor$ is purely periodic. So we have

$$\begin{aligned}
\sqrt{n} + a_0 &= [\overline{2a_0, a_1, a_2, ..., a_r}] \\
&= [2a_0; a_1, a_2, ..., a_r, \overline{2a_0, a_1, a_2, ..., a_r}]
\end{aligned} \tag{2}$$

$$\sqrt{n} = [a_0; a_1, a_2, ..., a_r, \overline{2a_0, a_1, a_2, ..., a_r}] = [a_0; \overline{a_1, a_2, ..., a_r, 2a_0}] \tag{3}$$

For the palindromic result, let $\alpha_1 = 1/(\sqrt{n} - a_0)$. Apply (b) to (2) to get $\alpha_1 = [\overline{a_r, a_{r-1}, ..., a_1, 2a_0}]$. But (3) tells us that $\sqrt{n} - a_0 = [0; \overline{a_1, a_2, ..., a_r, 2a_0}]$, whence $\alpha_1 = [\overline{a_1, a_2, ..., a_r, 2a_0}]$. Comparing the two representations of α_1 yields the palindrome.

It remains to prove that $a_i \le a_0$. Suppose that the CF of $\sqrt{d}$ is $[a_0; a_1, ..., a_r, 2a_0]$. We claim that when $0 < i < r$, $a_i \le a_0$. To prove this, we first claim that $Y_i \ne 1$ when $0 < i < r$. Using the notation of the proof of Theorem 7.7, recall that, for $\sqrt{n}$, $X_0 = 0$, $Y_0 = 1$, and $X_1 = a_0$. Now suppose that $Y_i = 1$. Then $\alpha_i = X_i + \sqrt{d}$; and then $a_i = \lfloor \alpha_i \rfloor = a_0 + X_i$. Then $X_{i+1} = a_i Y_i - X_i = a_0 + X_i - X_i = a_0 = X_1$ and $Y_{i+1} = d - X_{i+1}^2 = d - a_0^2 = Y_1$. This means that periodicity in the quotients starts with a_{i+1}, which contradicts our hypothesis that periodicity starts at $r + 2$. Now, knowing that $Y_i \ge 2$ and using the fact from Theorem 7.7's proof that $X_i < \sqrt{d}$, tells us that

$$a_i = \frac{X_{i+1} + X_i}{Y_i} \le \frac{X_{i+1} + X_i}{2} < \frac{2\sqrt{d}}{2}$$

which means that $a_i \le \lfloor \sqrt{d} \rfloor = a_0$. □

Part (c) has interesting consequences for a computation of the periodic form of the CF of $\sqrt{d}$. One could use it in a direct manner, by computing the CF until $2a_0$ shows up. But this requires knowing $\sqrt{d}$ to a certain precision, and we do not know that precision in advance. It is better to interpret part (c) as follows: Compute the Y-sequence until $Y_r = 1$ shows up for a second time. Then the periodic form of the CF of $\sqrt{d}$ is

$[a_0; \overline{a_1, a_2, ..., a_{r-1}, 2a_0}]$. The relationships of the proof of Theorem 7.7 tell us how to compute the Y-sequence, but here is an important simplification: do not use $\alpha_k = (X_k + \sqrt{d})/Y_k$ and $a_k = \lfloor \alpha_k \rfloor$ to get a_k. Rather use $a_k = \lfloor (X_k + a_0)/Y_k \rfloor$.

We leave it as an exercise to show that for CFs of $\sqrt{d}$,

$$\left\lfloor \frac{X_k + a_0}{Y_k} \right\rfloor = \left\lfloor \frac{X_k + \sqrt{d}}{Y_k} \right\rfloor$$

(see Exercise 7.21). The advantage is that this eliminates real-number arithmetic from the computation of the sequences $\{a_i\}$, $\{X_i\}$, and $\{Y_i\}$; the numerical value of $\sqrt{d}$ gets used once only, to find $a_0 = \lfloor \sqrt{d} \rfloor$. Because the CF of pure square roots arises often (Pell equation, factoring algorithms), this yields an important speed-up. *Mathematica* has a `ContinuedFraction` function that does this, but here is complete code anyway, to show how to optimize in the case of a pure square root. Note the use of `Quotient`, as that is faster than dividing in cases where the integer quotient is what is wanted. And we avoid the slow `Append`, using an `ans = {ans, q}` construction instead, flattening at the end.

```
CFSqrt[d_] :=
  Module[{k = 0, P = 0, Q = 1, a = Floor[√d], ans}, ans = {a};
    While[P = Last[ans] Q - P; Q = Quotient[d - P^2, Q];
      ans = {ans, Quotient[P + a, Q]}; k = 0 || Q ≠ 1, k++];
    {a, Rest[Flatten[ans]]}] /; ! IntegerQ[√d]
```

```
CFSqrt[919]
```

```
{30, {3, 5, 1, 2, 1, 2, 1, 1, 1, 2, 3, 1, 19, 2, 3, 1, 1, 4, 9, 1, 7,
  1, 3, 6, 2, 11, 1, 1, 1, 29, 1, 1, 1, 11, 2, 6, 3, 1, 7, 1, 9,
  4, 1, 1, 3, 2, 19, 1, 3, 2, 1, 1, 1, 2, 1, 2, 1, 5, 3, 60}}
```

And here's an example that arises in the puzzle of Archimedes that will be discussed in Section 7.5.

```
CFSqrt[4729494]
```

```
{2174, {1, 2, 1, 5, 2, 25, 3, 1, 1, 1, 1, 1, 1, 15, 1, 2, 16,
  1, 2, 1, 1, 8, 6, 1, 21, 1, 1, 3, 1, 1, 1, 2, 2, 6, 1, 1, 5,
  1, 17, 1, 1, 47, 3, 1, 1, 6, 1, 1, 3, 47, 1, 1, 17, 1, 5, 1,
  1, 6, 2, 2, 1, 1, 1, 3, 1, 1, 21, 1, 6, 8, 1, 1, 2, 1, 16,
  2, 1, 15, 1, 1, 1, 1, 1, 1, 3, 25, 2, 5, 1, 2, 1, 4348}}
```

We can now prove many of the patterns in the CFs of $\sqrt{n}$. For example, the cases where the period-length is 1 or 2 are easy to discover. The next result shows how symbolic algebra can lead to automated proofs of many of these patterns.

Proposition 7.9. (a) The CF of $\sqrt{n}$ has period-length 1 if and only if $n = r^2 + 1$. The CF then is $[r; \overline{2r}]$.

(b) Given n, a nonsquare, let $r = \lfloor \sqrt{n} \rfloor$; then $n = r^2 + k$, where $0 < k \le 2r$. The CF of $\sqrt{n}$ has period-length 2 if and only if k divides $2r$.

(c) The CF of $\sqrt{r^2 - 1}$ is $[r - 1; \overline{1, 2r - 2}]$.

(d) The CF of $\sqrt{r^2 + 2}$ is $[r; \overline{r, 2r}]$.

(e) The CF of $\sqrt{r^2 - 2}$ is $[r - 1; \overline{1, r - 2, 1, 2r - 2}]$ if $r \ge 3$.

(f) The CF of $\sqrt{r^2 + 4}$ is $\left[r; \overline{\frac{r-1}{2}, 1, 1, \frac{r-1}{2}, 2r}\right]$ if r is odd and $r \ge 3$.

(g) The CF of $\sqrt{r^2 + 4}$ is $\left[r; \overline{\frac{r}{2}, 2r}\right]$ if r is even and $r \ge 2$.

(h) The CF of $\sqrt{r^2 - 4}$ is $\left[r - 1; \overline{1, \frac{r-3}{2}, 2, \frac{r-3}{2}, 1, 2r - 2}\right]$ if r is odd and $r \ge 5$.

(i) The CF of $\sqrt{r^2 - 4}$ is $\left[r - 1; \overline{1, \frac{r-4}{2}, 1, 2r - 2}\right]$ if r is even and $r \ge 6$.

(j) The CF of $\sqrt{r^2 + 8}$ is $\left[r; \overline{\frac{r-2}{4}, 1, 1, \frac{r-2}{2}, 1, 1, \frac{r-2}{4}, 2r}\right]$ if $r \equiv 2 \pmod 4$ and $r \ge 6$.

(k) The CF of $\sqrt{r^2 + 8}$ is $\left[r; \overline{\frac{r}{4}, 2r}\right]$ if $r \equiv 0 \pmod 4$.

(l) The CF of $\sqrt{r^2 - 8}$ is $\left[r - 1; \overline{1, \frac{r-6}{4}, 2, \frac{r-2}{2}, 2, \frac{r-6}{4}, 2r - 2}\right]$ if $r \equiv 0 \pmod 4$ and $r \ge 10$.

(m) The CF of $\sqrt{r^2 - 8}$ is $\left[r - 1; \overline{1, \frac{r-8}{4}, 1, 2r - 2}\right]$ if $r \equiv 0 \pmod 4$ and $r \ge 12$.

Proof. In all the proofs we will make use of Theorem 7.8, which tells us the general form of the CF of $\sqrt{n}$. Thus, in (a), we know that the CF must look like $[r; \overline{2r}]$. So we can set up a quadratic equation by observing that if

$$x = 2r + \cfrac{1}{2r + \cfrac{1}{2r + \cfrac{1}{2r + \cdots}}}$$

then $x = 2r + 1/x$ and $[r; \overline{2r}] = r + 1/x$. We can now easily solve for x, squaring the result.

```
FullSimplify[(r + 1/x /. Solve[x == 2 r + 1/x, x])^2]
{1 + r^2, 1 + r^2}
```

This proves (a) because of uniqueness of CFs for irrationals. For the remaining parts we can automate the symbolic algebra just outlined. The following routine takes a CF in the form $\{a_1, a_2, \ldots, a_n, \{b_1, b_2, \ldots, b_n\}\}$ and returns the square of the quadratic irrational it represents. This routine can be extremely helpful in discovering patterns in CFs.

```
FromCFSquared[a_] := Module[{x, periodicsoln, radical},
  periodicsoln =
    x /. Solve[x == FromContinuedFraction[Append[a[[-1]], x]], x];
  radical = Expand[FromContinuedFraction[
    Append[Drop[a, -1], periodicsoln[[2]]]]];
  RootReduce[FullSimplify[radical^2]]]
```

Now we have a one-line machine proof of part (a).

FromCFSquared[{r, {2 r}}]

$r^2 + 1$

Here is a proof of part (b).

FromCFSquared[{r, {s, 2 r}}] // Expand

$r^2 + \dfrac{2\,r}{s}$

And here are proofs of (c)–(i). The extra conditions in (e)–(i) follow from the form of the CF in each case.

FromCFSquared[{r - 1, {1, 2 (r - 1)}}]

$r^2 - 1$

FromCFSquared[{r, {r, 2 r}}]

$r^2 + 2$

FromCFSquared[{r - 1, {1, r - 2, 1, 2 (r - 1)}}]

$r^2 - 2$

The next two cases each break into subcases depending on the parity of r.

FromCFSquared[{r, {(r - 1) / 2, 1, 1, (r - 1) / 2, 2 r}}]

$r^2 + 4$

FromCFSquared[{r - 1, {1, (r - 3) / 2, 2, (r - 3) / 2, 1, 2 (r - 1)}}]

$r^2 - 4$

FromCFSquared[{r - 1, {1, (r - 4) / 2, 1, 2 (r - 1)}}]

$r^2 - 4$

FromCFSquared[{r, {r / 2, 2 r}}]

$r^2 + 4$

The $\sqrt{r^2 \pm 8}$ cases are left as exercises.   □

There are many, many other patterns in CFs and, armed with the symbolic CF-reverser, one can find and prove them (see Exercise 7.26). The next fact about the convergents of CFs of pure square roots will be useful in subsequent sections.

Proposition 7.10. Let p_k/q_k be the kth convergent in the CF of $\sqrt{n}$, where n is not a perfect square. If Y_k is as in Theorem 7.7, then $p_k^2 - nq_k^2 = (-1)^{k-1}Y_{k+1}$.

Proof. Letting a_k denote the partial quotients and α_k denote $1/(\alpha_{k-1} - a_{k-1})$, we have the following two equations.

$$\sqrt{n} = \frac{p_k\,\alpha_{k+1} + p_{k-1}}{q_k\,\alpha_{k+1} + q_{k-1}} \quad \text{and} \quad \alpha_{k+1} = \left(X_{k+1} + \sqrt{n}\right)\big/Y_{k+1}$$

We can simplify the first equation using the second as follows.

```
0 == Together[√n‾ - (pk αk+1 + pk-1)/(qk αk+1 + qk-1) /. αk+1 → (Xk+1 + √n‾) / Yk+1]
```

```
0 == (-Xk+1 pk - √n‾ pk + n qk + √n‾ qk Xk+1 - pk-1 Yk+1 + √n‾ qk-1 Yk+1) / (Xk+1 qk + √n‾ qk + qk-1 Yk+1)
```

This means that the numerator is 0, which, by Exercise 7.19, tells us that the two coefficients — the rational part and the coefficient of $\sqrt{n}$ — must each be 0. This yields

$$n\,q_k = p_k\,X_{k+1} + p_{k-1}\,Y_{k+1}$$
$$p_k = q_k\,X_{k+1} + q_{k-1}\,Y_{k+1}$$

Multiplying the first equation by q_k and the second by p_k, and subtracting:

```
pk (qk Xk+1 + qk-1 Yk+1) - qk (pk Xk+1 + pk-1 Yk+1) // Expand // Factor
```

```
(pk qk-1 - pk-1 qk) Yk+1
```

By Proposition 7.3(a) this is $(-1)^k Y_{k+1}$. □

Exercises for Section 7.3

7.19. Suppose that r, s, t, and u are rationals and that n is not a perfect square. Prove that if $r + s\sqrt{n} = t + u\sqrt{n}$, then $r = t$ and $s = u$.

7.20. Implement an algorithm that writes any quadratic irrational in the normal form of Proposition 7.6(g).

7.21. Show that in the definition of the quotients a_k for the CF of $\sqrt{d}$, one can use $a_k = \lfloor (X_k + a_0)/Y_k \rfloor$, as opposed to the more complicated $a_k = \lfloor (X_k + \sqrt{d})/Y_k \rfloor$ of Theorem 7.7.

7.22. Implement an algorithm that returns `True` if and only if the input is a quadratic irrational. Use the ideas of Theorem 7.7.

7.23.   Use Theorem 7.9(b) to implement a `CFPeriod2Q[n]` function that returns `True` or `False` according to whether the CF of $\sqrt{n}$ has period 2 or not.

7.24. Prove that the CF of $\sqrt{r^2 + r}$ is $[r; \overline{2, 2r}]$.

7.25. Use `FromCFSquared` to prove parts (j)–(m) of Proposition 7.9.

7.26. Use `FromCFSquared` to prove that the CF of $\sqrt{r^2 + 12}$ is

$$\left\{ r, \left\{ \frac{r-3}{6}, 1, 1, \frac{r-1}{2}, \frac{2r}{3}, \frac{r-1}{2}, 1, 1, \frac{r-3}{6}, 2r \right\} \right\}$$

if $r \geq 9$ and $r \equiv 3 \pmod 6$, and that the CF of $\sqrt{r^2 - 12}$ is

$$\left\{ r - 1, \left\{ 1, \frac{r-9}{6}, 2, \frac{r-3}{2}, 1, \frac{1}{3}(r-3)2, 1, \frac{r-3}{2}, 2, \frac{r-9}{6}, 1, 2r-2 \right\} \right\}$$

if $r \geq 15$ and $r \equiv 3 \pmod 6$.

7.27. The first few recordholders among $\sqrt{n}$ for long periods are: 2, 3, 7, 13, 19, 31, 43, 46, 94, 139, 151, 166, 211, 331, 421, 526, 571, 604, 631, 751, 886, 919. Extend this list.

7.28. Show that the following construction leads to n such that the period-length of $\sqrt{n}$ is 3. Let s be even, let k be arbitrary, and let $r = k(1 + s^2) - (2s)^{-1}$ where the inverse is the least positive inverse modulo $s^2 + 1$. Let $n = r^2 + (2rs + 1)/(s^2 + 1)$. If r is positive and $s < 2r$, then the CF of $\sqrt{n}$ is $[r; \overline{s, s, 2r}]$. Prove that all period-3 examples arise in this way. Implement a function `CFPeriod3Q[n]` that returns `True` or `False` according to whether the CF period of $\sqrt{n}$ is 3 or not.

7.29. We will find integers s_n such that $[s_n; \overline{2, 2, 2, \ldots 2, 2s_n}]$, with n twos, represents $\sqrt{d}$, with d an integer. This will show that every period length occurs in the CFs of square roots of integers. Let p_i/q_i be the convergents in the CF of $\sqrt{2} - 1$: starting with p_1/q_1, they are $^1/_2$, $^2/_5$, $^5/_{12}$, $^{12}/_{29}$, $^{29}/_{70}$, $^{70}/_{169}$, ...

(a) $2p_n + p_{n-1} = q_n$

(b) $q_{n-1} = p_n$

(c) Let $s_n = q_n + 1$ and let $r_n = [s_n; \overline{2, 2, 2, \ldots 2, 2s_n}]$; then

$$r_n - s_n = \left[0; 2, 2, 2, \ldots, 2, 2s_n, \frac{1}{r_n - s_n} \right] = [0; 2, 2, 2, \ldots, 2, r_n + s_n].$$

(d) Use (c) to get

$$r_n - s_n = \frac{p_n(r_n + s_n) + p_{n-1}}{q_n(r_n + s_n) + p_n} .$$

(e) Solve (d) for r and use (a) and (b) to get $r_n^2 = s_n^2 + 2p_n + 1$; this proves that $\sqrt{s_n^2 + 2p_n + 1} = [q_n; \overline{2, 2, 2, ...2, 2(q_n + 1)}]$.

(f) Write a program to generate integers d such that the CF of $\sqrt{d}$ is as above, and verify by examining their CFs.

7.4 Pell's Equation

The Diophantine equation $x^2 - dy^2 = \pm 1$ is called **_Pell's equation_**; this equation and its relatives where the right side is n have a long and interesting history. The solutions (we consider positive integer solutions only) are very closely related to good rational approximations to $\sqrt{d}$. For example, $x = 1351$, $y = 780$ form a solution to the $d = 3$ case and $^{1351}/_{780}$ is a very good approximation to $\sqrt{3}$ (it is one of the convergents in the CF of $\sqrt{3}$). In this section we will study the relationship more closely and see why any Pell equation has a solution in positive integers, and why those solutions must occur among the convergents of a CF. We restrict the discussion to positive, nonsquare values of d, to avoid some simple cases where there are no positive solutions (Exercise 7.30).

The Pell equation has played a large role in some ancient puzzles. In 1657, Fermat challenged British mathematicians with $x^2 - 433y^2 = 1$. The smallest solution is

$$x = 104564907854286695713 \quad \text{and} \quad y = 5025068784834899736 .$$

More famous is Archimedes's immortal challenge regarding the number of cattle in the herd of the Sun god. It reduces to a Pell equation for which the smallest solution is truly gigantic. In the next section we will see how one can compute an exact form for the solution to the cattle problem. Many much simpler puzzles also reduce to the Pell equation, and some are mentioned in the exercises (Exercises 7.32, 7.34, 7.36, and 7.37) so knowledge of its solutions and how fast they grow can be useful. Finally, Matijasevic's famous solution to Hilbert's Tenth Problem was based on solutions to the Pell equation (see [Mat]; the main sequence occurring on p. 19 is related to $x^2 - 5y^2 = 4$).

Theorem 7.11. Suppose that d is a positive nonsquare integer. If $x^2 - dy^2 = \pm 1$ and x and y are positive, then x/y is a convergent of the CF for $\sqrt{d}$.

Proof. Assume first that the target is $+1$. Factoring the Pell equation as $(x + y\sqrt{d})(x - y\sqrt{d}) = 1$ yields two positive factors (because we assume x and y are positive), so $x/y - \sqrt{d} > 0$. Then

$$\frac{x}{y} - \sqrt{d} = \frac{x - y\sqrt{d}}{y} = \frac{x^2 - dy^2}{y(x + y\sqrt{d})} < \frac{1}{2y^2\sqrt{d}} < \frac{1}{2y^2}.$$

Therefore, Theorem 7.5(c) tells us that x/y is a convergent of $\sqrt{d}$. The -1 case is similar, except that we get y/x being a convergent of $1/\sqrt{d}$; by Exercise 7.8 this means that x/y is a convergent of $\sqrt{d}$. □

The preceding immediately gives us an algorithm for finding the smallest solution to the Pell equation $x^2 - dy^2 = \pm 1$: Just search through the convergents of $\sqrt{d}$. In fact, one can be perfectly precise about which convergents will lead to the solution: If the CF of $\sqrt{d}$ is periodic with period length n, then the $(n-1)$st convergent gives the first Pell solution, and subsequent convergents in steps of n yield the rest.

Theorem 7.12. Suppose that d is a positive nonsquare, n is the period length of the CF of $\sqrt{d}$, and p_k/q_k denotes the kth convergent. The smallest solution to $x^2 - dy^2 = \pm 1$ is $x = p_{n-1}$ and $y = q_{n-1}$. If n is even this is a $+1$ solution (and there are no -1 solutions). If n is odd this is a -1 solution, and there are both $+1$ and -1 solutions. Solutions beyond the first occur at the convergents p_k/q_k where $k = 2n - 1, 3n - 1, \ldots$.

Proof. Theorem 7.11 tells us that any solution is a convergent. And we know that $p_k^2 - dq_k^2 = (-1)^{k+1} Y_{k+1}$ (see Theorem 7.10). We now appeal to Theorem 7.8(c), which tells us that the preperiodic part of the CF of $\sqrt{d}$ has length 1. Knowing this, the fact that the periodic part has length n means that $(X_{n+1}, Y_{n+1}) = (X_1, Y_1)$ (using the notation of Theorem 7.7). Therefore,

$$Y_n = \frac{d - X_{n+1}^2}{Y_{n+1}} = \frac{d - X_1^2}{Y_1} = Y_0 = 1.$$

But then $p_{n-1}^2 - dq_{n-1}^2 = (-1)^n Y_n = (-1)^n$. So this gives us one solution: (p_{n-1}, q_{n-1}).

To see that no earlier convergent is a solution, suppose that the kth convergent is a solution. Then $Y_{k+1} = \pm 1$. Suppose first that it is $+1$. Then $\alpha_{k+1} = X_{k+1} + \sqrt{d}$ so the CF of α_{k+1} is $[X_{k+1} + a_0; \overline{a_1, a_2, \ldots, a_n}]$ which is purely periodic. Theorem 7.8(a) then yields $-1 < X_{k+1} - \sqrt{d} < 0$, so $\lfloor \sqrt{d} \rfloor = X_{k+1}$. This means that $\alpha_{k+2} = \alpha_1$; but because the α-sequence governs the a-sequence, this would mean that the a-sequence was periodic with period $k + 1$. But we know the period is n, so n would divide $k + 1$, which means that $k + 1 \geq n$ or $k \geq n - 1$. In fact, this argument tells us that the solutions are the convergents $C_{n-1}, C_{2n-1}, C_{3n-1}, \ldots$; a different approach to getting all the solutions will be discussed in Theorem 7.13.

Finally, we will show that Y_k never equals -1, for any k. For if it did, α_k, which we know has a purely periodic CF (every α does except α_0) would equal $-X_k - \sqrt{d}$, and Theorem 7.8(a) would yield $-X_k + \sqrt{d} < 0$ and $-X_k - \sqrt{d} > 1$. These two inequalities contradict each other. □

Here is an example. The period-length is 6 so we want the convergents C_5, C_{11}, and so on. Because indices in lists start at 1 (C_0 will be the first entry in the convergent list), we use `Range[6, 48, 6]` to get C_5, C_{11}, ..., C_{47}.

ContinuedFraction$\left[\sqrt{19}\,\right]$

{4, {2, 1, 3, 1, 2, 8}}

solns = Convergents$\left[\sqrt{19}\,, 50\right]$[[Range[6, 48, 6]]]

$$\left\{ \frac{170}{39}, \frac{57799}{13260}, \frac{19651490}{4508361}, \frac{6681448801}{1532829480}, \frac{2271672940850}{521157514839}, \right.$$
$$\left. \frac{772362118440199}{177192022215780}, \frac{262600848596726810}{60244766395850361}, \frac{89283516160768675201}{20483043382566906960} \right\}$$

The numerators and denominators solve the Pell equation with $d = 19$.

Map$\left[\text{Numerator}[\#]^2 - 19\, \text{Denominator}[\#]^2 \,\&, \text{solns}\right]$

{1, 1, 1, 1, 1, 1, 1, 1}

Theorem 7.13. The complete solution set to $x^2 - dy^2 = \pm 1$ is generated by the least positive solution (x_1, y_1) by taking the pairs (x_k, y_k) arising from $\left(x_1 + y_1 \sqrt{d}\,\right)^k = x_k + y_k \sqrt{d}$, $k = 1, 2, 3, \ldots$.

Proof. It is easy to see that the pairs (x_k, y_k) are solutions, with the help of the fact that the conjugate of a power is the power of a conjugate:

$$x_k^2 - dy_k^2 = \left(x_k + y_k \sqrt{d}\,\right)\left(x_k - y_k \sqrt{d}\,\right)$$
$$= \left(x_1 + y_1 \sqrt{d}\,\right)^k \left(x_1 - y_1 \sqrt{d}\,\right)^k = (x_1^2 - y_1^2 d)^k = \pm 1.$$

It remains to show that every positive solution arises in this way. Suppose (X, Y) is a solution not equal to any (x_k, y_k). Let n be an integer such that

$$\left(x_1 + y_1 \sqrt{d}\,\right)^n < X + Y\sqrt{d} < \left(x_1 + y_1 \sqrt{d}\,\right)^{n+1}.$$

Divide all by the leftmost to get

$$1 < \left|\left(X + Y\sqrt{d}\,\right)\left(x_1 - y_1 \sqrt{d}\,\right)^n\right| < x_1 + y_1 \sqrt{d}.$$

Write $\left(X + Y\sqrt{d}\,\right)\left(x_1 - y_1 \sqrt{d}\,\right)^n$ as $a + b\sqrt{d}$.

$$a^2 - db^2 = \left(a - b\sqrt{d}\right)\left(a + b\sqrt{d}\right)$$
$$= \left(X + Y\sqrt{d}\right)\left(x_1 - y_1\sqrt{d}\right)^n \left(X - Y\sqrt{d}\right)\left(x_1 + y_1\sqrt{d}\right)^n$$
$$= \left(x_1^2 - y_1^2 d\right)^n \left(X^2 - dY^2\right)$$
$$= \pm 1$$

So (a, b) is a solution to the equation and $1 < |a + b\sqrt{d}| < x_1 + y_1\sqrt{d}$. This is almost a contradiction to the minimality of (x_1, y_1), but we do not yet know that a and b have the same sign. But $|a + b\sqrt{d}| > 1$, so its reciprocal is less than 1 in absolute value. But its reciprocal is either $a - b\sqrt{d}$ or $-a + b\sqrt{d}$. It follows in all cases that a and b have the same sign. For example, suppose $-a - b\sqrt{d} > 1$ and its reciprocal, in absolute value, is $a - b\sqrt{d}$, which is less than 1. Adding and subtracting yields that both $-2b\sqrt{d}$ and $-2a$ are positive. The other cases are identical. So either (a, b) or $(-a, -b)$ is a subminimal solution, contradiction. $\square$

The pairs of the preceding theorem can be generated very quickly using matrix powers. To each irrational $m + n\sqrt{d}$ associate the 2×2 matrix $\begin{pmatrix} m & n \\ dn & m \end{pmatrix}$. Then we can easily see that $\text{matrix}(\alpha) \cdot \text{matrix}(\beta) = \text{matrix}(\alpha\beta)$.

```
matr[m_ + n_ √d_] := {{m, n}, {d n, m}}
α = a + b √d; β = m + n √d;
prod = Collect[Expand[α β], √d]
```

$$a m + b d n + \sqrt{d}\ (b m + a n)$$

```
Map[MatrixForm , {matr[prod], matr[α].matr[β]}]
```

$$\left\{ \begin{pmatrix} a m + b d n & b m + a n \\ d\ (b m + a n) & a m + b d n \end{pmatrix}, \begin{pmatrix} a m + b d n & b m + a n \\ b d m + a d n & a m + b d n \end{pmatrix} \right\}$$

Here is a specific example. The reader can run some timing tests to see that the matrix-power approach is much faster then expanding the radical.

```
Expand[(111 + 888 √7)^12]
```

$$3252111235440571787021146876924275512 9409 +$$
$$6310924416267739109586884270928718516320\ \sqrt{7}$$

```
MatrixPower[matr[111 + 888 √7], 12][[1]]
```

$$\{3252111235440571787021146876924275512 9409,$$
$$6310924416267739109586884270928718516320\}$$

Now the Pell equation $x^2 - dy^2 = \pm 1$ can be completely solved by getting the smallest solution from the CF, forming the matrix, and using powers of the matrix to get all solutions. There are two CNT package functions for solving Pell equations. The first, `PellSolve`, returns the generating matrix. Then powers of this matrix can be used to generate all solutions to $x^2 - dy^2 = \pm 1$.

```
d = 13;
mat = PellSolve[d]
```

{{18, 5}, {65, 18}}

```
solns = Table[MatrixPower[mat, i][[1]], {i, 5}]
{xvals, yvals} = Transpose[solns];
xvals² - d yvals²
```

{{18, 5}, {649, 180}, {23382, 6485},
 {842401, 233640}, {30349818, 8417525}}

{-1, 1, -1, 1, -1}

`PellSolutions` can be used to get a set of solutions.

```
PellSolutions[13, {1, 5}]
```

{{18, 5}, {649, 180}, {23382, 6485},
 {842401, 233640}, {30349818, 8417525}}

```
PellSolutions[13, {1, 5}, RightSide → -1]
```

{{18, 5}, {23382, 6485}, {30349818, 8417525},
 {39394040382, 10925940965}, {51133434066018, 14181862955045}}

Here is the answer to Fermat's challenge of 1657. An omitted second argument is assumed to be 1, so the first solution only is given.

```
PellSolutions[433, RightSide → 1]
```

{104564907854286695713, 5025068784834899736}

Exercises for Section 7.4

7.30. Show that the Pell equation $x^2 - dy^2 = \pm 1$ has no solutions in positive integers if d is either negative or a positive perfect square.

7.31. Our discussion of solutions to the Pell equation $x^2 - dy^2 = \pm 1$ often talks about one solution being smaller than another. But these solutions are pairs, not just single numbers. Explain why the concept of order makes sense for Pell solutions.

7.32. A triangular number is one such as 6, which equals $1 + 2 + 3$, or $10 = 1 + 2 + 3 + 4$. Their general form is $n(n+1)/2$. 36 is both triangular and square. Find some other numbers that are simultaneously square and triangular. *Hint*: If $n(n+1)/2 = m^2$ then a simple substitution can transform this equation to the Pell equation $x^2 - 2y^2 = 1$. But finding the substitution is a little tricky.

7.33. (a) Consider the Pell equation $x^2 - 2y^2 = \pm 7$. Let $\alpha = 3 + \sqrt{2}$ and $\beta = 1 + \sqrt{2}$. Prove that $\alpha \beta^i$, when expanded, yields solutions to the equation $x^2 - 2y^2 = \pm 7$. Show that the same is true for $\overline{\alpha} \beta^i$.

(b) Show that the solutions found in part (a) are all the solutions. *Hint*: Proceed as in Theorem 7.13. Given a solution (a, b), let i be such that $(1 + \sqrt{2})^i < a + b\sqrt{2} < (1 + \sqrt{2})^{i+1}$. Divide by the leftmost term, let $c + d\sqrt{2}$ be the radical in the middle, and show how to get a bound on the possible values of c and d. Then find all possibilities and deduce that, in each case, $a + b\sqrt{2} = (c + d\sqrt{2})(1 + \sqrt{2})^i$.

7.34. (I. Vardi) A rack of 15 billiard balls has the pleasant property that they fit into a triangle and when the cue ball is added they fit into a square. Find some other numbers with this property by reducing the problem to a Pell equation with right side equal to -7. Then use Exercise 7.33.

7.35. A certain Belgian street has houses numbered consecutively along one side of the street starting with the number 1. A friend who lives on this street invites you to his house. He won't tell you the house number, but says that the sum of all the house numbers that come before his house is equal to the sum of all the house numbers that come after his house. If you know that there are more than 50 and less than 500 houses on this street, what is the house number?

7.36. An urn contains n balls, some black, some white. Two balls are drawn at random without replacement, and both of them turn out to be black. Find values of n for which the chance of this happening is exactly $1/2$.

7.37. When a regular hexagon of side-length 1 is inscribed in a circle, the circle's radius is 1. Find other integers m such that when a hexagon with sides (in order) $1, 1, 1, m, m, m$ is inscribed in a circle, the circle has an integer radius. *Hint*: Reduce the question to a Pell equation.

7.5 Archimedes and the Sun God's Cattle

Approximately 2000 years ago, Archimedes asked:

> If thou are diligent and wise, O stranger, compute the number of cattle of the Sun, who once upon a time grazed on the fields of the Thrinacian isle of Sicily, divided into four herds of different colors, one milk white, another glossy black, the third yellow, and the last dappled. In each herd were bulls, mighty in number according to these proportions: Understand, stranger, that the white bulls were equal to half and a third of the black together with the whole of the yellow, while the black were equal to the fourth part of the dappled and a fifth, together with, once more, the whole of the yellow. Observe further that the remaining bulls, the dappled, were equal to a sixth part of the white and a seventh, together with all the yellow. These were the proportions of the cows: The white were precisely equal to the third

part and a fourth of the whole herd of the black; while the black were equal to the fourth part once more of the dappled and with it a fifth part, when all, including the bulls, went to pasture together. Now the dappled in four parts were equal in number to a fifth part and a sixth of the yellow herd. Finally, the yellow were in number equal to a sixth part and seventh of the white herd.

If thou canst accurately tell, O stranger, the number of Cattle of the Sun, giving separately the number of well-fed bulls and again the number of females according to each color, thou wouldst not be called unskilled or ignorant of numbers, but not yet shalt thou be numbered among the wise.

But come, understand also all these conditions regarding the cows of the Sun. When the white bulls mingled their number with the black, they stood firm, *equal in depth and breadth,* and the plains of Thrinacia, stretching far in all ways, were filled with their multitude. Again, when the yellow and the dappled bulls were gathered into one herd they stood in such a manner that their number, beginning from one, grew slowly greater till it completed a triangular figure, there being no bulls of other colors in their midst nor one of them lacking.

If thou art able, O stranger, to find out all these things and gather them together in your mind, giving all the relations, thou shalt depart crowned with glory and knowing that thou hast been adjudged perfect in this species of wisdom.

This problem has an interesting history, but only recently has symbolic computation been brought to bear. In this section we describe Ilan Vardi's approach to the problem [Var]. Quoting Vardi: "The simple nature of the question and the difficulty of its solution make this a perfect example of a challenge problem and show once more that Archimedes is one of the greatest mathematicians of all time." See [Var] for more information on attempts over the ages to solve the puzzle and for connections with algebraic number theory.

■ Wurm's Version: Using Rectangular Bulls

J. F. Wurm in 1830 cleverly assumed that a bull had a rectangular shape, so that the underlined condition says only that a certain number isn't prime. The problem then translates into the eight equations that follow, where X, Y, Z, and W denote number of white, black, yellow, and dappled bulls, respectively, and x, y, z, w similarly denote the cows.

eqns = {

$$W == \left(\frac{1}{2} + \frac{1}{3}\right) X + Y, \quad X == \left(\frac{1}{4} + \frac{1}{5}\right) Z + Y, \quad Z == \left(\frac{1}{6} + \frac{1}{7}\right) W + Y,$$

$$w == \left(\frac{1}{3} + \frac{1}{4}\right) (X + x), \quad x == \left(\frac{1}{4} + \frac{1}{5}\right) (Z + z),$$

$$z == \left(\frac{1}{5} + \frac{1}{6}\right) (Y + y), \quad y == \left(\frac{1}{6} + \frac{1}{7}\right) (W + w),$$

$$Y + Z == \frac{1}{2} q (q + 1) \};$$

The first seven equations form a linear system in eight unknowns, and so can be solved by standard linear algebra techniques.

```
rawSoln = {W, X, Y, Z, w, x, y, z} /.
  First[Solve[Drop[eqns, -1], {X, Y, Z, x, y, z, w}]]
```

$$\left\{W, \frac{267\,W}{371}, \frac{297\,W}{742}, \frac{790\,W}{1113}, \right.$$
$$\left. \frac{171580\,W}{246821}, \frac{815541\,W}{1727747}, \frac{1813071\,W}{3455494}, \frac{83710\,W}{246821}\right\}$$

We do not want fractional cows, so W must be a multiple of all the denominators.

```
Apply[LCM, Denominator[rawSoln]]
```

```
10366482
```

So now we can express the solution in terms of W_1, where $W = 10366482\,W_1$.

```
{W₀, X₀, Y₀, Z₀, w₀, x₀, y₀, z₀} = rawSoln /. W → 10366482 W₁
```

```
{10366482 W₁, 7460514 W₁, 4149387 W₁, 7358060 W₁,
  7206360 W₁, 4893246 W₁, 5439213 W₁, 3515820 W₁}
```

This means that the total size of the herd is $25194541\,W/5183241$. We next try to find q, the number of rows in the large triangle.

```
q /. Solve[Y₀ + Z₀ == 1/2 q (q + 1), q]〚2〛
```

$$\frac{1}{2}\left(-1 + \sqrt{1 + 92059576\,W_1}\right)$$

This tells us that we seek W_1 so that $92059576\,W_1 + 1 = s^2$; or $s^2 \equiv 1 \pmod{92059576}$. We can use the modular square root algorithm from Section 6.3 to find all the possibilities for s.

```
SqrtModAll[1, 92059576]
```

```
{1, 3287843, 4303069, 7590911, ≪26≫, 88771733, 92059575}
```

We cannot take s to be 1, because that leads to $W_1 = 0$ and no cattle. So let $s = 3287843$.

```
s = 3287843;
(s² - 1) / 92059576
```

```
117423
```

This gives us the value of W_1, which in turn gives us everything else.

```
{W₀, X₀, Y₀, Z₀, w₀, x₀, y₀, z₀} /. W₁ → 117423
```

```
{1217263415886, 876035935422, 487233469701, 864005479380,
  846192410280, 574579625058, 638688708099, 412838131860}
```

```
Apply[Plus, %]
```

```
5916837175686
```

Because $W_0 + X_0$ is even (and so not prime), this is a valid solution — the smallest one — to Wurm's version of the problem; the Sun's herd consisted of almost six trillion head of cattle. Here's a check, where `Thread` is used to turn `list1 → list2` into the list of coordinatewise replacements.

```
eqns /. q → 1/2 (-1 + √(1 + 92059576 * 117423)) /.
   Thread[{W, X, Y, Z, w, x, y, z} →
      {W₀, X₀, Y₀, Z₀, w₀, x₀, y₀, z₀}] /. W₁ → 117423
```

```
{True, True, True, True, True, True, True, True}
```

■ The Real Cattle Problem

A proper reading of the problem says that $W + X$ is a square integer. We know that

$$W + X = 10366482\,W_1 + 7460514\,W_1 = 17826996\,W_1$$

from Wurm's work. Because $17826996 = 4 \cdot 3 \cdot 11 \cdot 29 \cdot 4657$, W_1 must have the form $3 \cdot 11 \cdot 29 \cdot 4657\,m^2$, or $4456749\,m^2$. And $Y_0 + Z_0$ must be triangular.

```
Y₀ + Z₀ /. W₁ → 4456749 m²
```

```
51285802909803 m²
```

So $51285802909803\,m^2$ must be $q(q+1)/2$. It is easy to find q.

```
q /. Last[Solve[51285802909803 m² == q (q + 1) / 2, q] ]
```

$$\frac{1}{2}\left(-1 + \sqrt{1 + 410286423278424\,m^2}\right)$$

So we must solve $1 + 410286423278424\,m^2 = k^2$. One could try directly for the CF of $\sqrt{410286423278424}$, but its period is very large. The problem can be much simplified if we factor 410286423278424.

```
FactorForm[FactorInteger[410286423278424]]
```

```
2³ 3 7 11 29 353 4657²
```

This means that we can replace m in the Pell equation by $m_1 = 2 \cdot 4657\,m$. Then the equation becomes $1 + 2 \cdot 3 \cdot 7 \cdot 11 \cdot 29 \cdot 353\,m_1^2 = k^2$, or $k^2 - 4729494\,m_1^2 = 1$. We can generate solutions to this, but we are seeking a very special solution, one that has m_1 divisible by $2 \cdot 4657$. Even with this reduction, the CF has a large period, so the first Pell solution will be large.

```
d = 4729494;
ContinuedFraction[√d] // Timing
```

```
{0.0666667 Second,
  {2174, {1, 2, 1, 5, 2, 25, 3, 1, 1, 1, 1, 1, 1, 15, 1, 2, 16,
      1, 2, 1, 1, 8, 6, 1, 21, 1, 1, 3, 1, 1, 1, 2, 2, 6, 1, 1, 5,
      1, 17, 1, 1, 47, 3, 1, 1, 6, 1, 1, 3, 47, 1, 1, 17, 1, 5, 1,
      1, 6, 2, 2, 1, 1, 1, 3, 1, 1, 21, 1, 6, 8, 1, 1, 2, 1, 16,
      2, 1, 15, 1, 1, 1, 1, 1, 1, 3, 25, 2, 5, 1, 2, 1, 4348}}}
```

Here is the matrix whose powers generate the solutions.

```
mat = PellSolve[d]
```

```
{{10993198673282973497986623282143354390108 8049,
    50549485234315033074477819735540408986340},
  {23907348711878154303554440157231995105844 1111960,
    10993198673282973497986623282143354390108 8049}}
```

Now, we seek a solution $\{k, m_1\}$ where m_1 is divisible by 2 and by 4567. Because $4729494 \equiv 2 \pmod 4$, k must be odd and m_1 must be even, so only 4657 need be considered. The m_1 just obtained is not divisible by 4657. Here is the radical form of the smallest solution.

```
ε = mat[[1]] . {1, √d}
```

$$10993198673282973497986623282143354390108 8049 +$$
$$50549485234315033074477819735540408986340 \sqrt{4729494}$$

Theorem 7.13 tells us that all solutions are given by powers of ε. We can use the matrix power approach to find the first solution that has a radical coefficient that is divisible by 4657. Note that such a solution *must* exist, because it is equivalent to a solution of the larger Pell equation, and we know that all Pell equations do have solutions! We can search for one by looking at them all in order. Of course, a reduction modulo 4657 is necessary at each step so the computation will avoid gigantic integers. And it helps to perform a modular reduction on the multiplying matrix right at the start too.

```
i = 1;
ans = mat1 = Mod[mat, 4657];
While[ans[[1, 2]] ≠ 0, i++; ans = Mod[mat1.ans, 4657]];
{ans, i}
```

```
{{{4656, 0}, {0, 4656}}, 2329}
```

The preceding computation takes under a second, but in fact we could have been even more efficient. Let $p = 4657$. Because ε is a Pell solution, $1/\varepsilon^k = \overline{\varepsilon^k}$; therefore finding a power of ε with radical coefficient divisible by p is the same as finding k such that $\varepsilon^k - \varepsilon^{-k} \equiv 0 \pmod p$. But this is the same as asking that $\varepsilon^{2k} \equiv 1 \pmod p$. It can be shown (see [Var]) that the smallest such k must divide $(p+1)/2$. Because those divisors of are $\{1, 17, 137, 2329\}$, one can simply check that 17 and 137 do not work, and it is then certain that 2329 will work. This is precisely what the exhaustive search yielded.

In any case, the 2329th solution is what we want. This is given by ε^{2329}, with m_1 being the radical coefficient in the expression. So all that remains is to use the information we have to go back and build the actual solution. Recall that $1/\varepsilon$ is the conjugate of ε. So

$$\frac{(\varepsilon^{2329} - \frac{1}{\varepsilon^{2329}})}{2\sqrt{4729494}} = \frac{\varepsilon^{2329} - (\overline{\varepsilon})^{2329}}{2\sqrt{4729492}} = m_1 \,.$$

This allows us to get m and then W_1 and W_0.

$$m^2 = \left(\frac{m_1}{2\cdot 4567}\right)^2 = \frac{(\varepsilon^{2329} - \frac{1}{\varepsilon^{2329}})^2}{(2\cdot 4657\cdot 2\sqrt{4729494})^2} = \frac{1}{4\cdot 410286423278424}(\varepsilon^{4658} + \frac{1}{\varepsilon^{4658}} - 2)$$

$$W_1 = 4456749\,m^2$$

$$W_0 = 10366482\,W_1 = 10366482\cdot 4456749\,m^2$$

$$= \frac{10366482\cdot 4456749}{4\cdot 410286423278424}(\varepsilon^{4658} + \frac{1}{\varepsilon^{4658}} - 2)$$

$$= \frac{159\varepsilon^{4658}}{5648} + \frac{159}{5648}(\frac{1}{\varepsilon^{4658}} - 2) \approx \frac{159\varepsilon^{4658}}{5648} - \frac{159}{2824} = \frac{159\varepsilon^{4658}}{5648} - 0.056\ldots$$

where the number ε^{-4658}, which is extremely small, is ignored. We can now eliminate the 0.056 by using a ceiling function. In short, we can express W_0 *exactly* using ceiling, square root, and exponentiation.

$$W_0 = \lceil \tfrac{159}{5648}\big(10993198673282973497986623282143354390108 8049 +$$

$$5054948523431503307447781973554040898 6340\sqrt{4729494}\,\big)^{4658}\rceil$$

This is a perfectly exact solution for the smallest number of dappled bulls in the Archimedes challenge. The same argument using the ceiling function applies to the other variables, so that the complete solution vector is

$$\lceil(\tfrac{159}{5648}, \tfrac{801}{39536}, \tfrac{891}{79072}, \tfrac{395}{19768}, \tfrac{128685}{6575684}, \tfrac{2446623}{184119152}, \tfrac{5439213}{368238304}, \tfrac{125565}{13151368})\varepsilon^{4658}\rceil$$

Therefore the total number of cattle is $\lceil\frac{25194541\cdot 159}{5183241\cdot 5648}\varepsilon^{4658}\rceil$, which we can approximate as follows.

```
approx = N[ε, 100]

2.1986397346565946995973246564286708780217609799999999999999
  99999999999999999999999999999954517332502 × 10⁴⁴
```

$$\frac{25194541}{184119152} \text{approx}^{4658}$$

7.76027140648681826953023283321388666423232240592337610315061922690321593061406953194348955323833 × 10^{206544}

That's a lot of cattle!

A natural question is whether Archimedes could have solved the problem. He did know about ideas related to the Pell equation, because his famous book *The Measurement of the Circle* contains the rational approximation $^{1351}/_{780}$ for $\sqrt{3}$. This is the twelfth convergent for the CF of $\sqrt{3}$ and the sixth solution to the Pell equation.

Convergents $[\sqrt{3}, 12]$

$$\left\{1, 2, \frac{5}{3}, \frac{7}{4}, \frac{19}{11}, \frac{26}{15}, \frac{71}{41}, \frac{97}{56}, \frac{265}{153}, \frac{362}{209}, \frac{989}{571}, \frac{1351}{780}\right\}$$

PellSolutions $[3, \{1, 6\}]$

$$\{\{2, 1\}, \{7, 4\}, \{26, 15\}, \{97, 56\}, \{362, 209\}, \{1351, 780\}\}$$

But this is a far cry from understanding the theory well enough to know that $x^2 - 4729494y^2 = 1$ has a solution. As Vardi observes, Archimedes could not have solved this problem, but he might well have gathered enough evidence to be convinced that all Pell equations have solutions.

Exercises for Section 7.5

7.38. Find the rightmost ten digits in the number representing the total number of cattle in the Sun's herd.

7.39. What is the approximate size of the second-smallest solution to Archimedes's problem?

7.6 Factoring via Continued Fractions

In 1926 Maurice Kraitchik proposed the following scheme to factor n. Search for "random" pairs of integers that satisfy $x^2 \equiv y^2 \pmod{n}$. Kraitchik suggested taking values of x close to $\sqrt{n}$ and trying to factor $x^2 - n$ just using a relatively small set of primes. The set of primes over which we try to factor $x^2 - n$ is called the *factor base*; the values $x^2 - n$ are called *residues*. We focus only on those values of $x^2 - n$ that factor completely using the primes in our factor base. If for any of these factorizations, all of the powers are even, then we have found integers x and y for which $x^2 \equiv y^2 \pmod{n}$. If p and q are distinct primes that divide n, then pq divides n which divides $x^2 - y^2 = (x - y)(x + y)$, and there is a 50% chance that exactly one of these two primes divides $\gcd(x - y, n)$. If this happens, then $\gcd(x - y, n)$ is a proper, nontrivial divisor of n.

To illustrate the idea, let $n = 30{,}929$ and consider values of x in the range $166 \le x \le 185$. Seven values of $x^2 - n$ factor completely using primes less than 50:

$$167^2 - 30929 = -1 \cdot 2^5 \cdot 5 \cdot 19$$
$$169^2 - 30929 = -1 \cdot 2^6 \cdot 37$$
$$173^2 - 30929 = -1 \cdot 2^3 \cdot 5^3$$
$$175^2 - 30929 = -1 \cdot 2^4 \cdot 19$$
$$176^2 - 30929 = 47$$
$$177^2 - 30929 = 2^4 \cdot 5^2$$
$$183^2 - 30929 = 2^9 \cdot 5 \,.$$

In this case, we immediately find an (x, y) pair:

$$177^2 \equiv \left(2^2 \cdot 5\right)^2 \ (\mathrm{mod}\ 30929) \,.$$

This is great, for it tells us that $30{,}929 = (177 - 20)(177 + 20)$

We were lucky. What usually happens is that each factorization of $x^2 - n$ will contain at least one odd power of a prime. But one can then compare the factorizations that are over a small set of primes, searching for a product of the factorizations in which all primes appear to even powers. This example is simple enough that such a combination can be found by inspection, but we will see shortly how to automate the search. Here is a combination of three factors that works.

$$167^2 \cdot 175^2 \cdot 183^2 \equiv (-1)^2 \cdot 2^{18} \cdot 5^2 \cdot 19^2 \ (\mathrm{mod}\ 30929)$$

So we set $x = 167 \cdot 175 \cdot 183$ and $y = 2^9 \cdot 5 \cdot 19$, the square root of the square we found. We then use gcd to discover a factor.

GCD[167 175 183 – 2⁹ 5 19, 30929]

157

To summarize, this method tries to find integers x_i so that the product of the integers $x_i^2 - n$ is a perfect square, say

$$\prod_{i \in S} (x_i^2 - n) = Y^2 \,.$$

Then X is set to $\prod x_i$. It follows that $X^2 \equiv Y^2 \,(\mathrm{mod}\, n)$, and we can look at $\gcd(n, X - Y)$ in the hope of splitting off a nontrivial factor; if the X and Y

are independent then there is at least a 50% chance that this splitting will occur.

Note that we can restrict the factor base to primes for which n is a quadratic residue: for if p divides $x^2 - n$, then $n \equiv x^2 \pmod{p}$.

The difficulty with Kraitchik's approach is that $|x^2 - n|$ grows fairly quickly. It is approximately equal to $2|x - \sqrt{n}|\sqrt{n}$. This reduces the probability that $x^2 - n$ factors completely using only primes from the factor base.

In 1931 D. H. Lehmer and R. E. Powers published an algorithm that uses the continued fraction expansion of $\sqrt{n}$ to get around the problem of the size of the residue. Let p_k/q_k be the kth convergent to $\sqrt{n}$. We know that

$$p_k^2 - n q_k^2 = (-1)^{k+1} Y_{k+1}$$

where Y_k is as in the proof of Theorem 7.7, and so

$$p_k^2 \equiv (-1)^{k+1} Y_{k+1} \pmod{n}.$$

We have already seen in the proof of Theorem 7.7 that $|Y_k| < n$ for general quadratic irrationals. In fact, as we shall prove, much more is true in the case of a pure square root:

$$0 < Y_k < 2\sqrt{n}.$$

The algorithm for finding the CF for $\sqrt{n}$ generates, as a byproduct, lots of perfect squares — the p_k^2 — whose residues (mod n) — $\pm Y_k$ — are close to 0. We will generate the sequence of Y-values and keep track of the ones we are able to factor using the factor base, as well as the corresponding p-values. Then we will try, using linear algebra, to combine the factorizations so that each prime goes evenly into the final product. If we can do this, then the product of the Y-values used, call it Z, will be a square, and we can let $Y = \sqrt{Z}$ and X be the product of the corresponding p-values. We will than have that $X^2 \equiv Y^2 \pmod{n}$. The linear algebra part would be tedious to program in its full detail; fortunately the built-in `NullSpace` command of *Mathematica* will do the job for us.

First we take care of the generation of the factor base, making sure at the same time that no small prime divides n.

```
fb[n_Integer, len_Integer] := Module[{smallQ = False, fac, j, u},
    If[EvenQ[n], Return[2]];
    u = Select[Prime[Range[2 len]], # == 2 || (j = JacobiSymbol[n, #];
        If[j == 0, smallQ = True; fac = GCD[#, n]]; j == 1) &];
    If[smallQ, fac, u]]
```

```
fb[7919, 20]
```

```
{2, 5, 7, 31, 37, 71, 73, 79, 89,
 97, 103, 107, 109, 113, 131, 149, 163}
```

```
fb[7919 * 11, 20]
```

```
11
```

The next function returns the parities of the exponents of the primes in y, under the assumption that all the primes that occur in y lie in the helper list. `MapIndexed` is like `Map` but it is stronger in that it allows us to refer to both `#1`, the element of the list being scanned, and `#2`, the position of the scanned element within the list. We use this, via `Cases`, to change the appropriate entry of `ans` to the mod-2 value of the power of the prime.

```
exponentsBinary[y_, helpers_] := Module[{prs, ans = helpers, p =
    DeleteCases[Map[{#, IntegerExponent[y, #]} &, helpers],
      {_, 0}]},
  prs = First /@ p;
  MapIndexed[(ans[[#2[[1]]]] = If[! MemberQ[prs, #1], 0,
      Cases[p, {#, e_} :> Mod[e, 2]][[1]]]) &, helpers]]
```

```
exponentsBinary[31 37^3 71^6, fb[7919, 20]]
```

```
{0, 0, 0, 1, 1, 0, 0, 0, 0, 0, 0, 0, 0, 0, 0, 0, 0}
```

The powers 1, 3, 6 show up as 1, 1, 0 in positions 4, 5, 6, which correspond to the factor-base primes 31, 37, and 73, respectively.

And now the main routine. We compute the p- and Y-values together, making use of various relationships discussed earlier in this chapter. Whenever a Y-value can be factored by the factor base, which we check by raising the product of the primes in the factor base modulo Y, we call on `exponentsBinary` to get the parity vector. We want to be able to add some of these to get the zero vector mod-2, for that will mean that the corresponding product of the Ys will be a square. We can then let Y be the square root of the product of the Y-values and let X be the product of the corresponding p-values. Sometimes it will happen that $X \equiv \pm Y \pmod{n}$, which is bad for then neither of $X \pm Y$ will split n; we just move on to the next combination in such a case.

In order to be sure that some combinations of the parity vectors sum to 0 (mod 2) we need to have more such vectors than there are entries in a single vector; this is standard linear algebra. And this is why the target is set to 30 more than the length of the factor base.

Other points about the code worth noting:

- Any factoring algorithm should begin by checking for small divisors. We here check for divisibility by the first 10,000 primes by using a gcd test with the product of those primes.

- `AppendTo` is avoided when accumulating the p- and y-values because it is slow. Instead we use code such as `pvals = {pvals, p}` to attach a new value, flattening when we are done. In the case of rows we must `Partition` the flattened result to get the right row-sizes (the size is one more than the length of the factor base because of the powers of -1 added). We use `count` to tell when the target of 30 more than the size of the factor base is reached.

- We reduce modulo n whenever possible.

- Because the key congruence, $p_k^2 \equiv (-1)^{k+1} Y_{k+1} \pmod{n}$, can involve a minus sign, we must be sure that the product of the y-values we take is positive; otherwise the p-product would not have the correct properties. We ensure this by adding the relevant power of -1 — called k in the code: it is either 0 or 1 — to the binary vectors. Then the `Null-Space` command will take this into account and the resulting combinations will have a 0th power of -1, and so be positive.

- `NullSpace` has a `Modulus` option, which we set to 2 because the parity is all that matters in the search for combinations that lead to squares. Here is an example where `mat` consists of some odd and some even integers.

- `Scan` is used to work through the list of combinations, breaking out of the search via the `Scan` when a factor has been found.

- Experiments have shown that $10 \cdot 1.1^{\log_{10} n}$ is a good guide to the length of the factor base, and we use that.

- We must transpose the matrix of mod-2 powers because `NullSpace` gives vectors that can be used to combine the columns of a matrix to get the zero vector. Here's an example where the matrix is not reduced modulo 2.

```
mat = {{3, 3, 1}, {3, 2, 1}, {3, 13, 12}, {3, 0, 2}, {2, 1, 32}};
nsp = NullSpace[Transpose[mat], Modulus → 2]
Map[Flatten[Position[#, 1]] &, nsp]
```

```
{{1, 1, 0, 0, 1}, {1, 1, 1, 1, 0}}
```

```
{{1, 2, 5}, {1, 2, 3, 4}}
```

And the coefficients yield vectors that are entirely even.

```
{mat[[1]] + mat[[2]] + mat[[5]], mat[[1]] + mat[[2]] + mat[[3]] + mat[[4]]}
```

```
{{8, 6, 34}, {12, 18, 16}}
```

- The method can fail on prime powers (Exercise 7.42), so we insert a special check at the beginning to handle the case that n is a prime power. We do this efficiently by checking only if n is a square, cube, or fifth power. Because of our check of primes through 10^5, any power that slips by these two tests would be greater than 10^{35} and, realistically, the CF factoring algorithm should be restricted to numbers smaller than this.

Here is the CF factoring code, which requires `exponentsBinary` and `fb` presented earlier.

```
testnumber = Apply[Times, Prime[Range[PrimePi[10000]]]];

CFFactor[n_Integer, opts___] := Module[
  {count = 0, k = 1, i = 1, p0 = 1, p1, X, Y, Z, ans = $Failed,
   g, a0 = Floor[√n], a, j = 1, pvals = {}, r, yvals = {},
   rows = {}, factorbase = fb[n, Floor[10 1.1^Log[10, n]]]},
  If[1 < (g = GCD[n, testnumber]) < n, Return[{g, n / g}]];
  While[! IntegerQ[r = n^(1/Prime[j])] && j ≤ 3, j++];
  If[IntegerQ[r], Return[{r, n / r}]];
  If[PrimeQ[n], Return[{n}]];
  X = p1 = a0; Y = n - a0^2;
  If[IntegerQ[factorbase],
   Return[Sort[{factorbase, n / factorbase}]]];
  prod = Apply[Times, factorbase];
  While[count < Length[factorbase] + 30, k = 1 - k;
   a = Quotient[X + a0, Y]; {p1, p0} = {Mod[a p1 + p0, n], p1};
   {X, Y} = {a Y - X, Mod[(-1)^k p1^2, n]};
   If[PowerMod[prod, 2^Floor[Log[2., Y]], Y] == 0,
    count++; pvals = {pvals, p1}; yvals = {yvals, Y};
    rows = {rows, k, exponentsBinary[Y, factorbase]}]];
  {pvals, yvals, rows} = Flatten /@ {pvals, yvals, rows};
  rows = Partition[rows, Length[factorbase] + 1];
  coeffs = Flatten[Position[#, 1]] & /@
   NullSpace[Transpose[rows], Modulus -> 2];
  Scan[({XX, YY} = Mod[{Times @@ (pvals[[#]]),
      √Times @@ (yvals[[#]]) }, n];
    If[XX ≠ YY && XX ≠ -YY && 1 < (g = GCD[n, XX - YY]) < n,
     Return[ans = Sort[{g, n / g}]]]) &, coeffs];
  ans]

CFFactor[Prime[100000] Prime[100001]]

{1299709, 1299721}
```

The version in the CNT package has several options that give more information.

```
Get["CNT`"]

CFFactor[Prime[10000000] Prime[10000001], ShowData -> True]

Beginning the search for 83 factorable Y-values...
It took 2092 Y-values to find the 83 good values needed.
13 of the 35 square combinations had X = ±Y;
0 of the remaining 22 failed to split n.

{179424673, 179424691}
```

The output means that there were 35 pairs of X and Y values that could be used, but 13 of them would fail to work as hoped. However, all of the 22 other values both worked to split the input. In fact, this 100% success rate for the good pairs *always* occurs whenever n is a product of two primes (Exercise 7.40).

Here is an example of an 11-digit integer where the method fails.

CFFactor[n = 12091920623, ShowData → True]

Beginning the search for 56 factorable Y-values...

It took 377 Y-values to find the 56 good values needed.

All 33 of the square combinations had X = ±Y.

$Failed

But increasing the number of squares sought to 100 more than the factor base's size gets the job done. It yields 101 squares and only 96 of them are bad.

CFFactor[n, ShowData → True, TargetMargin → 100]

Beginning the search for 126 factorable Y-values...

It took 855 Y-values to find the 126 good values needed.

96 of the 101 square combinations had X = ±Y;
0 of the remaining 5 failed to split n.

{104849, 115327}

Here is an example that shows the method is capable of handling moderately large numbers. There turn out to be 31 combinations of the Y-values that were squares, and 15 of them were bad. All of the remaining 16 succeeded in factoring n.

n = NextPrime[10^{10}] NextPrime[10^{11}];
CFFactor[n, ShowData → True]

Beginning the search for 107 factorable Y-values...

It took 9323 Y-values to find the 107 good values needed.

15 of the 31 square combinations had X = ±Y;
0 of the remaining 16 failed to split n.

{10000000019, 100000000003}

The next example shows success on a 34-digit number, though it takes several hours. The `TraceIntervalFactor` option causes percentage-done statements to be printed. Note that over 400,000 terms of the CF had to be examined in order to get the desired set of 276 Y-values.

```
CFFactor[16281705918903387455182528191 93643,
  ShowData → True, TraceIntervalFactor → 0.10] // Timing
```

Beginning the search for 276 factorable Y-values...

10% done

20% done

30% done

41% done

51% done

61% done

71% done

81% done

91% done

It took 458193 Y-values to find the 276 good values needed.

18 of the 41 square combinations had $X = \pm Y$;

0 of the remaining 23 failed to split n.

{4895.57 Second, {12050933302352249, 135107427038247107}}

Factoring is a messy problem. There are many algorithms, and each of them is somewhat complicated to program; indeed CFFactor is significantly more complicated than the other algorithms in this book. But, precisely because there is no simple method, it is very satisfying to discover the factors of a number and this explains why many people are intrigued by the problem. And, of course, much of the motivation for this work comes from the prevalence of codes that assume that factoring large numbers is essentially impossible.

The continued fraction factoring method is knows as *CFRAC*; until the early 1980s it was the method of choice for factoring large integers. But it hits a limit at around 50 digits. For integers of this size, the Y-values have 25 digits. If we use all primes under one million in our factor base, that means we want to factor Y over a set of 40,000 primes. Our chances in any particular case are only about 1 in 1,000. So the number of Y-values we must compute is running close to 50,000,000. Trial division by 40,000 integers 50,000,000 times is too time-consuming.

We conclude by proving the bound on the Y-values that arise in the CF of a pure square root.

Theorem 7.14. In the derivation of the CF for $\sqrt{n}$, n not a square, we always have $0 < X_k < \sqrt{n}$ when $k \geq 1$, and $0 < Y_k < 2\sqrt{n}$ for all k.

Proof. Assume the theorem holds for X_k and Y_{k-1}. We shall prove it holds for the next X and Y values. Note that $X_1 = \lfloor \sqrt{n} \rfloor$ and $Y_0 = 1$, so we can start the induction at $k = 1$.

We know that

$$1 < \alpha_k = \frac{X_k + \sqrt{n}}{Y_k}$$

so $Y_k > 0$ and $Y_k < X_k + \sqrt{n} < 2\sqrt{n}$. We also have

$$1 < \alpha_{k+1} = \frac{X_{k+1}^2 - n}{Y_{k+1}\left(X_{k+1} - \sqrt{n}\right)} = \frac{-Y_k Y_{k+1}}{Y_{k+1}\left(X_{k+1} - \sqrt{n}\right)} = \frac{Y_k}{\sqrt{n} - X_{k+1}} \tag{4}$$

so $X_{k+1} < \sqrt{n}$. To prove that $X_{k+1} > 0$, we note from inequality (4) that if $X_{k+1} \le 0$, then $Y_k > \sqrt{n}$. But this would mean that $X_{k+1} = a_k Y_k - X_k \ge Y_k - X_k > 0$. It follows that X_{k+1} must be strictly positive. $\square$

Exercises for Section 7.6

7.40. Suppose that n is a product of two primes and X and Y are such that $X^2 \equiv Y^2 \pmod{n}$ but $X \not\equiv \pm Y \pmod{n}$. Show that $\gcd(X - Y, n)$ will be one of the two primes.

7.41. Implement a general factoring algorithm that combines a brute force method below 10^6, the Pollard rho method between 10^6 and 10^{15}, and the continued fraction method between 10^{15} and 10^{25}. And in all cases improve the check for small divisors. Compare this method's performance with that of any single method. (Code for Pollard rho is in Chapter 5).

7.42. Show that if n is a prime power, $n = p^e$, $e > 1$, then the CF factoring method will not factor n. *Hint*: If e is even, then the square root of p^e is not irrational, so we can assume that e is odd. Show that if the method finds a pair of integers, (X, Y), for which

$$X^2 \equiv Y^2 \pmod{p^e} \quad \text{and} \quad 1 < \gcd(X - Y, p^e) < p^e$$

then p must be an element of the factor base. But the elements in the factor base are quadratic residues modulo p^e.

7.43. Use CFFactor (or the more general routine of Exercise 7.41) to get a routine that provides the complete prime factorization of an integer.

CHAPTER 8

Prime Testing with Lucas Sequences

8.0 Introduction

Lucas sequences are defined by two-term linear recurrences, the most famous example of which is the Fibonacci sequence. Édouard Lucas's generalization of the Fibonacci sequence plays an important role in computational number theory. For a comprehensive treatment of the work of Lucas, and much other historical material on prime-testing, see [Wil2]. We will see in this chapter how it can be used to:

1. yield a general-purpose prime-testing technique that is one of the best known;
2. yield a prime certification procedure that is similar to Pratt certification, except that one needs a factorization of $p + 1$ (as opposed to $p - 1$ in the Pratt case);
3. get a very efficient algorithm for finding Mersenne primes (primes of the form $2^m - 1$) called the Lucas–Lehmer algorithm.

Indeed, regarding (1) we will see how some very fast primality testing procedures based on Lucas's ideas have no known counterexamples; *Mathematica*'s `PrimeQ` uses such tests.

Lucas sequences come in pairs and are related to continued fractions. For integers P and Q we define the sequences $\{U_i\}$ and $\{V_i\}$, $i = 0, 1, 2, 3, \ldots,$ by

$$U_0 = 0, \ U_1 = 1, \text{ and } U_{i+1} = P U_i - Q U_{i-1} \text{ when } i \geq 1$$

$$V_0 = 2, \ V_1 = P, \text{ and } V_{i+1} = P V_i - Q V_{i-1} \text{ when } i \geq 1$$

We define the **discriminant** to be the largest square-free integer that divides $d = P^2 - 4Q$. The discriminant of the Fibonacci sequence (the case $P = 1$, $Q = -1$) is 5. We shall restrict our attention to Lucas sequences for which the discriminant is not 1 (i.e., $P^2 - 4Q$ is not a perfect square). Throughout this chapter, d denotes $P^2 - 4Q$ and α is $(P + \sqrt{d})/2$. Note that

α is a root of $x^2 - Px + Q$. We shall also use the notion of quadratic conjugate, defined and studied in Section 7.3: $\overline{\alpha} = (P - \sqrt{d})/2$. Note that $\alpha\overline{\alpha} = Q$. As a consequence, if p is prime and Q is relatively prime to p, then $Q^{-1}\overline{\alpha}$ is a mod-p inverse of α, where Q^{-1} is the mod-p inverse of Q.

The first critical point (Exercise 8.1) is that

$$\frac{V_i + U_i\sqrt{d}}{2} = \alpha^i \tag{1}$$

from which it follows that

$$V_i^2 - d\,U_i^2 \;=\; 4\alpha^i\,\overline{\alpha}^i = 4(\alpha\overline{\alpha})^i = 4Q^i \tag{2}$$

Note also that if $Q = \pm 1$, then the terms in the Lucas sequences that are even correspond (by halving) to solutions to Pell's equation $x^2 - d\,y^2 = \pm 1$, and therefore also arise from the continued fraction expansion of $\sqrt{d}$.

8.1 Divisibility Properties of Lucas Sequences

A critical first task is to get an efficient way of computing the mod-m value of members of a Lucas sequence. Recall from Section 1.4 how matrix powers can be used to quickly get the Fibonacci numbers. The same ideas work for any recurrence relation. Here is how one would do it for a degree-3 recurrence, where $a_n = c_1 a_{n-1} + c_2 a_{n-2} + c_3 a_{n-3}$ and a_0, a_1, and a_2 have specific values; the ideas are perfectly general.

$$\begin{pmatrix} 0 & 1 & 0 \\ 0 & 0 & 1 \\ c_1 & c_2 & c_3 \end{pmatrix} \begin{pmatrix} a_{n-3} \\ a_{n-2} \\ a_{n-1} \end{pmatrix} = \begin{pmatrix} a_{n-2} \\ a_{n-1} \\ a_n \end{pmatrix}$$

It follows that the nth power of the 3×3 matrix, when applied to the vector of three initial values, will give a vector with a_n in the third position. Because we can use `MatrixPowerMod` to get modular powers of the matrix quickly, this is an efficient, recursion-free approach to getting values of a_n for large n. The following snippet of code rotates the rows of the identity matrix (except the last row) and then places the coefficients into the third row.

```
coeffs = {c₁, c₂, c₃};
ReplacePart[
  Map[RotateRight, IdentityMatrix[Length[coeffs]]],
  coeffs, 3] // MatrixForm
```

$$\begin{pmatrix} 0 & 1 & 0 \\ 0 & 0 & 1 \\ c_1 & c_2 & c_3 \end{pmatrix}$$

And so the following short program gets the nth element of the sequence modulo m. This code will work for any linear recurrence, not just those of degree 3. The function is included in the CNT package.

```
LinearRecurrence[coeffs_, init_, n_Integer, m_: ∞] :=
  Mod[First[MatrixPowerMod[ReplacePart[
        RotateRight /@ IdentityMatrix[Length[coeffs]],
        coeffs, 1], n, m]].init, m] /; n ≥ Length[coeffs]

LinearRecurrence[coeffs_, init_, n_Integer, m_: ∞] :=
  Mod[init[[n + 1]], m] /; n < Length[coeffs]

LinearRecurrence[coeffs_, init_, n_List, m_: ∞] :=
  LinearRecurrence[coeffs, init, #, m] & /@ n
```

Here are the Fibonacci numbers, for which the basic recurrence is $a_n = a_{n-1} + a_{n-2}$.

```
LinearRecurrence[{1, 1}, {0, 1}, Range[0, 10]]
```

$\{0, 1, 1, 2, 3, 5, 8, 13, 21, 34, 55\}$

These ideas can be used to define LucasU[{P, Q}, m] as a function so that LucasU[{P, Q}, m][n] gives the mod-m value of U_n. LucasU and LucasV functions are included in the CNT package, by a somewhat faster algorithm that we will discuss later in this section.

If $Q = \pm 1$, then the Lucas sequences satisfy

$$V_i^2 - d\,U_i^2 = 4 \quad \text{or} \quad V_i^2 - d\,U_i^2 = (-1)^i 4$$

If V_i and U_i are both even, then they are the doubles of solutions to Pell's equation

$$\left(\frac{V_i}{2}\right)^2 - d\left(\frac{U_i}{2}\right)^2 = \pm 1$$

and therefore $V_i/2$ and $U_i/2$ must appear as convergents in the continued fraction of $\sqrt{d}$.

The next result shows how a version of Fermat's Little Theorem is valid for powers of α. Note that when working in a quadratic extension, $\beta \equiv \gamma \pmod{p}$ means there is a δ in the extension such that $\beta = \gamma + p\delta$. Throughout this chapter we shall use $\varepsilon(n)$ to denote the Jacobi symbol $\left(\frac{d}{n}\right)$ [defined in Section 6.2; if n is prime then this is just the Legendre symbol that encodes the quadratic character of $d \pmod{n}$].

Theorem 8.1. Fermat's Little Theorem in Quadratic Field Fix P and Q and let p be an odd prime. Then

$$\alpha^p \equiv \begin{cases} \alpha \pmod{p} & \text{if } \varepsilon(p) = 1 \\ \overline{\alpha} \pmod{p} & \text{if } \varepsilon(p) = -1 \\ P/2 \pmod{p} & \text{if } \varepsilon(p) = 0. \end{cases} \tag{3}$$

Proof. $\alpha^p = (P + \sqrt{d})^p / 2^p \equiv (P^p + d^{(p-1)/2}\sqrt{d}) / 2 \equiv (P + \varepsilon(p)\sqrt{d}) / 2 \pmod{p}$, where the well-known property of binomial coefficients — p divides $\binom{p}{k}$ when $1 < k < p$ — is used to cancel all the middle terms in the expansion, and Euler's criterion is used to eliminate the power of d. $\square$

An alternative conclusion that follows easily and makes the connection with Fermat's Little Theorem clearer is:

$$\alpha^{p-\varepsilon(p)} \equiv \begin{cases} 1 \pmod{p} & \text{if } \varepsilon(p) = 1 \\ Q \pmod{p} & \text{if } \varepsilon(p) = -1 \\ P/2 \pmod{p} & \text{if } \varepsilon(p) = 0. \end{cases} \tag{4}$$

Observe that this means that if p is prime, then $\alpha^{p-\varepsilon(p)}$ is congruent to an integer modulo p. But this does not capture the full strength, as the exact value of the integer is specified.

Here are two examples: the first shows that $\alpha^{11} \equiv \alpha \pmod{11}$, while the second shows that $\alpha^7 \equiv \overline{\alpha} \pmod 7$. These use `QuadraticPowerMod` and `QuadraticMod`, which are `CNT` functions to do modular arithmetic on quadratic irrationals. In such cases one does not wish to apply the reduction to the radical, and that is why special functions are needed.

```
P = 3; Q = 1; d = P² - 4 Q;  α = 1/2 (P + √d )
```

$$\frac{1}{2}\,(3 + \sqrt{5})$$

```
p = 11; JacobiSymbol[d, p]
```

1

```
{QuadraticPowerMod[α, p, p],
 QuadraticMod[α, p], QuadraticPowerMod[α, p - 1, p]}
```

$\{7 + 6\sqrt{5},\ 7 + 6\sqrt{5},\ 1\}$

Now we switch to $p = 7$.

```
p = 7; JacobiSymbol[d, p]
```

-1

```
{QuadraticPowerMod[α, p, p],
 QuadraticMod[QuadraticConjugate[α], p],
 QuadraticPowerMod[α, p + 1, p]}
```

$\{5 + 3\sqrt{5},\ 5 + 3\sqrt{5},\ 1\}$

Corollary 8.2. Fix P and Q, let p be an odd prime, and let $\varepsilon = \left(\frac{d}{p}\right)$. Then

$$U_{p-\varepsilon} \equiv 0 \pmod{p}$$

$$V_{p-\varepsilon} \equiv \begin{cases} 2 \pmod{p} & \text{if } \varepsilon = 1 \\ 2Q \pmod{p} & \text{if } \varepsilon = -1 \\ P \pmod{p} & \text{if } \varepsilon = 0 . \end{cases} \tag{5}$$

Proof. These follow almost immediately from equations (1) and (4). Those equations imply

$$V_{p-\varepsilon} + U_{p-\varepsilon} \sqrt{d} \equiv R \pmod{p}$$

where R is the right side of (5). But now a subtle point arises! The preceding congruence is in the context of $\mathbb{Z}\left[\sqrt{d}\right]$ and states that $V_{p-\varepsilon} + U_{p-\varepsilon} \sqrt{d} = R + p\left(a + b\sqrt{d}\right)$. But we can apply the result of Exercise 7.4 to conclude that $V_{p-\varepsilon} \equiv R$ and $U_{p-\varepsilon} \equiv 0 \pmod{p}$, as desired. $\square$

Note the special case of Fibonacci numbers, which are the U-sequence when $P = 1$ and $Q = -1$. If p is prime then p divides $F_{p-\left(\frac{5}{p}\right)}$.

This gives us an easy pseudoprime test on a candidate n. Given n, and a pair of parameters P and Q, we check whether n divides $U_{n-\varepsilon}$. If it does not, then n is proved composite. If it does, then either n is prime or n is a *{P, Q}-Lucas pseudoprime*. We will pursue this in Section 8.2.

We now continue to examine divisibility properties of the Lucas U-sequences, for fixed P and Q. To illustrate, let $P = 3$ and $Q = -1$ (so that the discriminant d is 13). We consider the sequence of (U_i, V_i) and look for those U_i that are divisible by p, as p varies through the primes from 3 through 19. Some data is shown in Table 8.1, where the 0s are in bold.

It appears that 3 divides U_i if and only if 2 divides i; 5 divides U_i if and only if 3 divides i; 7 divides U_i if and only if 8 divides i; 11 divides U_i if and only if 4 divides i; 13 divides U_i if and only if 13 divides i; 17 divides U_i if and only if 8 divides i; and 19 divides U_i if and only if 20 divides i.

The next several theorems will explain what is happening. It turns out, for any odd number m that is relatively prime to Q, there is a positive integer, called the ***rank modulo m*** and denoted $\boldsymbol{\omega(m)}$ such that m divides U_i if and only if $\omega(m)$ divides i. This is analogous to the order of b (modulo m). In that case we consider the powers $b^i \pmod{m}$ and look at the ones that are 1; the least such i is called $\mathrm{ord}_m(b)$; here P and Q play the role of b. More precisely, the powers of α underlie what we are examining when we look at the rank: When $U_i \equiv 0 \pmod{p}$, then α^i, when reduced modulo p, is an integer. We reiterate: The rank function depends on the parameters P and Q. We will learn shortly how to compute the rank. Table 8.2 shows some ω values, with the prime and composite cases distinguished. For example, the observation that 19 divides U_i if and only if 20 divides i becomes $\omega(19) = 20$.

i	mod 3	mod 5	mod 7	mod 11	mod 13	mod 17	
0	**(0,2)**	(0,2)	(0,2)	(0,2)	(0,2)	**(0,2)**	(0,2)
1	(1,0)	(1,3)	(1,3)	(1,3)	(1,3)	(1,3)	(1,3)
2	**(0,2)**	(3,1)	(3,4)	(3,0)	(3,11)	(3,11)	(3,11)
3	(1,0)	**(0,1)**	(3,1)	(10,3)	(10,10)	(10,2)	(10,36)
4	**(0,2)**	(3,4)	(5,0)	**(0,9)**	(7,2)	(16,0)	(33,119)
5	(1,0)	(4,3)	(4,1)	(10,8)	(5,3)	(7,2)	(109,393)
6	**(0,2)**	**(0,3)**	(3,3)	(8,0)	(9,11)	(3,6)	(360,1298)
7	(1,0)	(4,2)	(6,3)	(1,8)	(6,10)	(16,3)	(1189,4287)
8	**(0,2)**	(2,4)	**(0,5)**	(0,2)	(1,2)	**(0,15)**	(3927,14159)
9	(1,0)	**(0,4)**	(6,4)	(1,3)	(9,3)	(16,14)	(12970,46764)
10	**(0,2)**	(2,1)	(4,3)	(3,0)	(2,11)	(14,6)	(42837,154451)
11	(1,0)	(1,2)	(4,6)	(10,3)	(2,10)	(7,15)	(141481,510117)
12	**(0,2)**	(0,2)	(2,0)	**(0,9)**	(8,2)	(1,0)	(467280,1684802)
13	(1,0)	(1,3)	(3,6)	(10,8)	**(0,3)**	(10,15)	(1543321,5564523)
14	**(0,2)**	(3,1)	(4,4)	(8,0)	(8,11)	(14,11)	(5097243,18378371)
15	(1,0)	**(0,1)**	(1,4)	(1,8)	(11,10)	(1,14)	(16835050,60699636)
16	**(0,2)**	(3,4)	**(0,2)**	(0,2)	(2,2)	**(0,2)**	(55602393,200477279)
17	(1,0)	(4,3)	(1,3)	(1,3)	(4,3)	(1,3)	(183642229,662131473)
18	**(0,2)**	**(0,3)**	(3,4)	(3,0)	(1,11)	(3,11)	(606529080,2186871698)
19	(1,0)	(4,2)	(3,1)	(10,3)	(7,10)	(10,2)	(2003229469,7222746567)
20	**(0,2)**	(2,4)	(5,0)	**(0,9)**	(9,2)	(16,0)	(6616217487,23855111399)

Table 8.1. The Lucas sequences (U_i, V_i) for $P = 3$ and $Q = -1$ are shown in the last column, and reduced modulo 3, 5, 7, 11, 13, 17, and 19.

The main points to be gleaned from Table 8.2 are:

- If p is an odd prime , then $\omega(p)$ divides $p - \left(\frac{d}{p}\right)$.

- If p is composite then, apparently, $\omega(p)$ does not divide $p - \left(\frac{d}{p}\right)$.

We will build on these observations to develop some powerful prime tests. But first we continue to develop the proof that $\omega(n)$ always exists.

Theorem 8.4. Let $\{U_i\}$ be the Lucas sequence determined by parameters P and Q and let m be odd and relatively prime to Q. If any U_i is divisible by m, then there is an integer e such that m divides U_i if and only if e divides i.

Proof. From equation (1) we have that

$$V_{ke} + U_{ke} \sqrt{d} = 2\alpha^{ke} = (2\alpha^e)^k 2^{1-k} = \left(V_e + U_e \sqrt{d}\right)^k 2^{1-k} \tag{6}$$

If U_e is divisible by m, then the same is true for the coefficient of $\sqrt{d}$ when the kth power is expanded, and so U_{ke} is divisible by m too.

We now assume that e is the smallest positive index for which m divides U_e. To prove that if m divides U_i then e divides i, we use $\alpha^{-1} = Q^{-1}\bar{\alpha}$.

n	$\omega(n)$	$n - (\frac{13}{n})$		n	$\omega(n)$	$n - (\frac{13}{n})$
3	2	2		9	6	8
5	3	6		15	6	16
7	8	8		21	8	22
11	4	12		25	15	24
13	13	13		27	18	26
17	8	16		33	4	34
19	20	20		35	24	34
23	22	22		39	26	39
29	7	28		45	6	46
31	32	32		49	56	48
37	19	38		51	8	50
41	7	42		55	12	54
43	42	42		57	20	58
47	48	48		63	24	64
53	26	52		65	39	65
59	12	60		69	22	68
61	30	60		75	30	74
67	68	68		77	8	76
71	72	72		81	54	80

Table 8.2. Some Lucas ranks $\omega(n)$ for odd integers, where $P = 3$, $Q = -1$, $d = 13$. In the prime cases $\omega(n)$ divides $n - (\frac{13}{n})$.

Assume that m divides U_r. We choose k so that ke is the greatest multiple of e less than or equal to r. Then

$$V_{r-ke} + U_{r-ke}\sqrt{d} = 2\alpha^{r-ke} = 2\alpha^r Q^{-ke}(2\overline{\alpha}^e)^k 2^{-k}$$
$$= 2^{-k} Q^{-ke}\left(V_r + U_r\sqrt{d}\right)\left(V_e - U_e\sqrt{d}\right)^k.$$

Because m divides both U_r and U_e, it must divide U_{r-ke}. The subscript $r - ke$ is strictly less than e, and so it must be 0. □

We now begin the proof that $\omega(m)$ always exists when m is relatively prime to $2Q$. In the case of odd primes we know from equation (5) that $U_{p-\varepsilon(p)}$ is divisible by p, and therefore that $\omega(p)$ exists in this case.

The Lucas numbers satisfy some useful recurrences that involve much larger steps than the basic recurrence that defines them. These recurrences can be useful for computation of the U- and V-values. Equation (6) with $k = 2$ implies that

$$V_{2e} + U_{2e}\sqrt{d} = \frac{\left(V_e + U_e\sqrt{d}\right)^2}{2}.$$

This gives us, for any e, the following "doubling rules":

$$V_{2e} = \frac{V_e^2 + D U_e^2}{2} = \frac{V_e^2 + V_e^2 - 4Q^e}{2} = V_e^2 - 2Q^e \tag{7}$$

$$U_{2e} = U_e V_e \tag{8}$$

There are many other relations among Lucas numbers, and the following ones will all prove useful later in this chapter. Proofs are left to Exercises 8.4 and 8.5.

$$\text{Binary } U\text{-formulas:} \quad \begin{aligned} U_{2k} &= 2 U_k U_{k+1} - P U_k^2 \\ U_{2k+1} &= U_{k+1}^2 - Q U_k^2 \\ U_{2k+2} &= P U_{k+1}^2 - 2 Q U_k U_{k+1} \end{aligned} \tag{9}$$

$$\text{Binary } V\text{-formulas:} \quad \begin{aligned} V_{2k} &= V_k^2 - 2 Q^k \\ V_{2k+1} &= V_k V_{k+1} - P Q^k \end{aligned} \tag{10}$$

$$V\text{-from-}U \text{ formulas:} \quad V_k = 2 U_{k+1} - P U_k \tag{11}$$

We pause momentarily to show how these equations lead to a very fast way of computing U_n and V_n (mod m).

Algorithm 8.1 Computing Lucas Sequences

To get U_n (mod m) start by writing n in base 2. Then keep track of a pair of values that start as $(U_0, U_1) = (0, 1)$ and will finish as (U_n, U_{n+1}) (mod m). Scan the base-2 digits of n from left to right. At each step, if we have two consecutive values U_k, U_{k+1} in hand, we can use equations (9) to get U_{2k}, U_{2k+1}, U_{2k+2} (mod m), keeping the first two if the current base-2 digit of n is 0, and the last two if it is 1. The leftmost base-2 digit must be 1 so the first step gives (U_1, U_2). When we're finished we will have (U_n, U_{n+1}) and can use (11) to get V_n. Here is an implementation that uses Fold to scan the bits of n. LucasBoth returns the pair of U-V values.

```
LucasUPair[{P_, Q_}, mod_: ∞][n_] := Fold[
  Mod[If[#2 == 0, {#1[[1]] (2 #1[[2]] - P #1[[1]]), #1[[2]]^2 - Q #1[[1]]^2},
    {#1[[2]]^2 - Q #1[[1]]^2, #1[[2]] (P #1[[2]] - 2 Q #1[[1]])}], mod] &,
  {0, 1}, IntegerDigits[n, 2]]
```

```
LucasUFast[{P_, Q_}, mod_: ∞][n_] :=
  First[LucasUPair[{P, Q}, mod][n]]

LucasVFast[{P_, Q_}, mod_: ∞][n_] :=
  Mod[LucasUPair[{P, Q}, mod][n].{-P, 2}, m]

LucasBoth[{P_, Q_}, mod_: ∞][n_] :=
  Mod[{{1, 0}, {-P, 2}}.LucasUPair[{P, Q}, mod][n], mod]
```

Here is a matrix version followed by timing comparisons that illustrate the speedup. The bit method uses about 80% of the time of the matrix method in the case of a single V, and 40% in the case of both.

```
LucasUMat[{P_, Q_}, m_: ∞][n_] :=
  LinearRecurrence[{-Q, P}, {0, 1}, n, m]
LucasVMat[{P_, Q_}, m_: ∞][n_] :=
  LinearRecurrence[{-Q, P}, {2, P}, n, m]
```

$$P = 3; Q = 1; p = 10^{60} + 57;$$

$$\left\{ \mathtt{First}\left[\mathtt{Timing}\left[\mathtt{Do}\left[\mathtt{LucasVFast}[\{P, Q\}, p]\left[\frac{p+1}{2}\right], \{10\}\right]\right]\right] \middle/ \right.$$

$$\mathtt{First}\left[\mathtt{Timing}\left[\mathtt{Do}\left[\mathtt{LucasVMat}[\{P, Q\}, p]\left[\frac{p+1}{2}\right], \{10\}\right]\right]\right],$$

$$\mathtt{First}\left[\mathtt{Timing}\left[\mathtt{Do}\left[\mathtt{LucasBoth}[\{P, Q\}, p]\left[\frac{p+1}{2}\right], \{10\}\right]\right]\right] \middle/$$

$$\mathtt{First}\left[\mathtt{Timing}\left[\mathtt{Do}\left[\{\mathtt{LucasUMat}[\{P, Q\}, p]\left[\frac{p+1}{2}\right],\right.\right.\right.$$

$$\left.\left.\left.\left.\mathtt{LucasVMat}[\{P, Q\}, p]\left[\frac{p+1}{2}\right]\}, \{10\}\right]\right]\right]\right\}$$

$$\{0.809524, 0.409449\}$$

Even more speedup is possible in situations where we know that, say, P is fixed at 1, for then special code could be written with the P's deleted.

We now return to our study of the Lucas rank function. Let p be an odd prime and use ε to denote $\varepsilon(p)$. We have seen that p divides $U_{p-\varepsilon}$. But, by (8), this means that p divides $U_{(p-\varepsilon)/2} V_{(p-\varepsilon)/2}$. Assume now that p does not divide Q. Then, because $V_i^2 - d U_i^2 = 4 Q^i$, the prime p cannot divide both $U_{(p-\varepsilon)/2}$ and $V_{(p-\varepsilon)/2}$. Which factor it divides is determined by whether or not Q is a quadratic residue modulo p. Thus the quadratic character of both Q and d are critical to the divisibility properties in question.

Theorem 8.5. Let V_i be the Lucas sequence determined by parameters P and Q and let p be an odd prime that does not divide Q. Then p divides $V_{(p-\varepsilon)/2}$ if and only if Q is not a quadratic residue modulo p, and p divides $U_{(p-\varepsilon)/2}$ if and only if Q is a quadratic residue modulo p.

Proof. Equation (7) and Corollary 8.2 imply that

$$V_{\frac{p-\varepsilon}{2}}^2 = V_{p-\varepsilon} + 2\,Q^{(p-\varepsilon)/2} = 2\,Q^{(1-\varepsilon(p))/2} + 2\,Q^{(p-\varepsilon)/2}$$

$$\equiv 2\,Q^{(p-\varepsilon)/2}(Q^{(1-p)/2} + 1) \equiv 2\,Q^{(p-\varepsilon)/2}\left(\left(\frac{Q^{-1}}{p}\right) + 1\right) \pmod{p}$$

Because Q^{-1} is a quadratic residue if and only if Q is, p divides $V_{(p-\varepsilon)/2}$ if and only if $\left(\frac{Q}{p}\right) = -1$. We know that p divides $U_{p-\varepsilon}$, which means that p divides $U_{(p-\varepsilon)/2}\,V_{(p-\varepsilon)/2}$. If $\left(\frac{Q}{p}\right) = +1$, then p does not divide the V-term because of the case just handled, so p divides the U-term, as claimed. □

We next consider the rank of a Lucas sequence for prime powers.

Theorem 8.6. Let U_i be the Lucas sequence determined by parameters P and Q and let p be an odd prime that does not divide Q. Then $\omega(p^t)$ exists and divides $p^{t-1}[p - \varepsilon(p)]$.

Proof. The $t = 1$ case has been done. For the general case, use induction on t. Let $e = p^{t-2}\left(p - \left(\frac{d}{p}\right)\right)$, which turns the induction hypothesis into: p^{t-1} divides U_e. Use equation (6) with $k = p$ to get

$$2^{p-1}\left(V_{pe} + U_{pe}\sqrt{d}\right) = \left(V_e + U_e\sqrt{d}\right)^p$$

Expanding the power we see that the coefficient of $\sqrt{d}$ on the right side is

$$p\,V_e^{p-1}\,U_e + \binom{p}{3}V_e^{p-3}\,U_e^3\,d + \binom{p}{5}V_e^{p-5}\,U_e^5\,d^2 + \cdots + U_e^p\,d^{(p-1)/2}$$

By the induction hypothesis, and because each binomial coefficient is divisible by p, each term in this sum is divisible by p^t. Because p does not divide 2^{p-1}, this means that p^2 divides U_{pe}. □

We now get two easy corollaries, the second of which will lead us to a pseudoprime test and the Lucas–Lehmer test for primality of Mersenne numbers.

Corollary 8.7. Let U_i be the Lucas sequence determined by parameters P and Q and let $n = p_1^{e_1} \cdots p_r^{e_r}$ be odd and relatively prime to Q. Then $\omega(n)$ for the sequence exists and is a divisor of the least common multiple of $p_i^{e_i-1}(p_i - \varepsilon(p_i))$, $i = 1, \ldots, r$. In particular, if $r \geq 2$ and d is relatively prime to n, then $\omega(n) < n - 1$.

Proof. By Theorem 8.6, for each i and any s, $p_i^{e_i}$ divides $U_{s(p_i^{e_i-1}[p_i-\varepsilon(p_i)])}$. Therefore n divides U_k, where k is the least common multiple in the corollary; and so $\omega(n)$ divides k. The last assertion follows from the fact that $r > 2$ implies that $k < n - 1$. This is because n is odd, and so the factors $p_i - \varepsilon(p_i)$ have at least one 2 in common. □

Corollary 8.8. Fix a Lucas sequence determined by parameters P and Q, and let n be an odd integer relatively prime to Q and d. If $\omega(n)$ for the sequence is $n \pm 1$, then n is prime.

Proof. Suppose that n is divisible by two or more distinct primes. Then by Corollary 8.7, $\omega(n) < n - 1 \leq n \pm 1$, contradicting the assumption. If $n = p^t$ and $t > 1$ then Theorem 8.6 states that $\omega(n)$ is one of $p^t \pm p^{t-1}$, and neither equals $p^t \pm 1$. $\square$

If we have an integer n that we suspect is prime, then this gives us a means of proving primality by generating Lucas sequences until we find one for which $\omega(n)$ is $n - \varepsilon(n)$. This is very similar to proving primality using primitive roots where we tried to find an element whose order modulo n was $n - 1$. We will pursue this in Section 8.2.

There is one more refinement of our primality test that we need. It is exactly analogous to the observation that underlies Theorem 4.4.

Theorem 8.9. If n is odd and relatively prime to Qd, n divides $V_{(n+1)/2}$, and, for each odd prime p that divides $n + 1$, n does not divide $V_{(n+1)/(2p)}$, then $\omega(n) = n + 1$ and n is prime.

Proof. If n divides $V_{(n+1)/2}$, then by equation (8) n divides U_{n+1}, and so $\omega(n)$ divides $n + 1$. Because n is relatively prime to Qd and $V_{(n+1)/2}^2 - d\,U_{(n+1)/2}^2 = 4Q^i$, n is relatively prime to $U_{(n+1)/2}$. This means that $\omega(n)$ does not divide $(n + 1)/2$. It also means, because of equation (6), that n is relatively prime to $U_{(n+1)/(2p)}$. Because n does not divide $V_{(n+1)/(2p)}$, it cannot divide $U_{(n+1)/(2p)} \cdot V_{(n+1)/(2p)}$, which equals $U_{(n+1)/p}$. It follows that $\omega(n) = n + 1$. By Corollary 8.8, n is prime. $\square$

The next section will show how all these results can be used in both prime-testing and prime certification. We conclude this section by illustrating the analogs between Lucas sequences and powers of the ordinary integers. Recall from Chapter 4 that, by Euler's theorem, $\mathrm{ord}_n(b)$ divides $\phi(n)$; and the Carmichael function $\lambda(n)$ is the least universal exponent, so that, in fact, $\mathrm{ord}_n(b)$ divides $\lambda(n)$. The preceding results allow us to define two new functions, $\Phi_{P,Q}(n)$ and $\Lambda_{P,Q}(n)$ that have analogous properties with respect to $\omega(n)$, the main one being: $\omega(n)$ divides $\Lambda(n)$. Suppose that n is odd and $n = p_1^{e_1} p_2^{e_2} \cdots p_r^{e_r}$; then

$$\Phi(n) = \prod_{j=1}^{r} p_i^{e_i - 1}\left[p_i - \left(\frac{d}{p_i}\right)\right] \quad \text{and} \quad \Lambda(n) = \mathrm{LCM}\left[\left\{p_i^{e_i - 1}\left[p_i - \left(\frac{d}{p_i}\right)\right], \{i, 1, r\}\right\}\right].$$

Then $\omega(n)$ is the first divisor r of $\Lambda(n)$ for which $U_r \equiv 0 \pmod{n}$. Here are implementations of all these functions.

```
LucasΦ[{P_, Q_}, {p_, e_}] :=
  p^(e-1) (p - JacobiSymbol[P^2 - 4 Q, p]) /; GCD[Q, p] == 1
```

```
Lucas𝛷[{P_, Q_}, n_?OddQ] :=
 Apply[Times, Lucas𝛷[{P, Q}, #] & /@ FactorInteger[n]] /;
  GCD[Q, n] == 1

Lucas𝛬[{P_, Q_}, n_?OddQ] :=
 Apply[LCM, Lucas𝛷[{P, Q}, #] & /@ FactorInteger[n]] /;
  GCD[Q, n] == 1

LucasRank[{_, _}, 1] = 1;

LucasRank[{P_, Q_}, n_?OddQ] :=
 Select[Divisors[Lucas𝛬[{P, Q}, n]],
    LucasU[{P, Q}, n][#] == 0 &, 1][[1]] /; GCD[n, Q] == 1
```

<u>**LucasRank**</u>[{3, -1}, 19]

20

<u>**LucasRank**</u>[{3, -1}, 119]

8

<u>**LucasRank**</u>[{3, -1}, 4284179]

2146320

Exercises for Section 8.1

8.1. Given integers P and Q so that $d = P^2 - 4Q$ is not a square, define u_i and v_i by

$$\frac{v_i + u_i \sqrt{d}}{2} = \left(\frac{P + \sqrt{d}}{2} \right)^i$$

Prove that

$$\left(\frac{P + \sqrt{d}}{2} \right)^{i+1} = P \left(\frac{P + \sqrt{d}}{2} \right)^i - Q \left(\frac{P + \sqrt{d}}{2} \right)^{i-1}$$

and therefore

$u_0 = 0$, $u_1 = 1$, and $u_{i+1} = P u_i - Q u_{i-1}$ when $i \geq 1$

$v_0 = 2$, $v_1 = P$, and $v_{i+1} = P v_i - Q v_{i-1}$ when $i \geq 1$

8.2. Show that if α and $\overline{\alpha}$ are the two roots of $x^2 - Px + Q$, then the Lucas sequences determined by the parameters P and Q satisfy

$$V_i = a^i + b^i \quad \text{and} \quad U_i = \frac{a^i - b^i}{a - b}.$$

8.3. Derive equation (4) from Theorem 8.1

8.4. Prove that $U_{k+1} = (V_k + PU_k)/2$ and $V_{k+1} = (PV_k + dU_k)/2$. This proves equation (11).

8.5. Prove the binary U- and V-formulas: equations (9) and (10). *Hint:* First use $\alpha^{2k+1} = \alpha^k \alpha^{k+1}$ to prove that

$$V_{2k+1} = \frac{V_k V_{k+1} + dU_k U_{k+1}}{2} \quad \text{and} \quad U_{2k+1} = \frac{V_k U_{k+1} + U_k V_{k+1}}{2}.$$

8.6. Explain in more detail why the bit-scanning method presented in Algorithm 8.1 is a correct implementation of the equations (9) for computing U-values.

8.7. The three parameter-pairs $P = 1$, $Q = -1$; $P = -3$, $Q = 1$; and $P = 4$, $Q = -1$ all have the same discriminant (the d-values are 5, 5, and 20, respectively, but recall that the discriminant is the square-free part of d). Investigate the corresponding Lucas sequences and discover the pattern that relates them.

8.2 Prime Tests Using Lucas Sequences

▦ Lucas Certification

Theorem 8.9 gives a proof procedure for primality that is completely analogous to the proof using primitive roots of Section 4.2. Suppose that we have an integer n that we think is prime. If $n - 1$ can be factored, then we already know from our earlier work how to search for a primality proof using primitive roots. But Lucas sequences yield proofs in the case that $n + 1$ can be factored. We will attempt to prove primality by finding a Lucas sequence for which $\omega(n) = n + 1$. Note that for this to happen it must be that $\left(\frac{P^2 - 4Q}{n}\right) = -1$. We can set P to be 1 and then let Q vary until the Jacobi symbol condition is met. Now, to prove that the rank is $n + 1$ we will use Theorem 8.9, which requires the primes in $n + 1$. We perform the factorization and store the prime divisors. Then we see if our Q, chosen so that $\left(\frac{1 - 4Q}{n}\right) = -1$, yields $n + 1$ for $\omega(n)$; this is checked by using the stored primes to check whether Theorem 8.9 applies to the Lucas sequence determined by 1 and Q. If not, we simply increment Q and try again. Here is an implementation. See Exercise 8.8 for why Q is started at 2 (as opposed to 0 or 1).

A proof that a certifying Q exists for every prime p is given in Appendix B.

```
PrimeQWithLucasProof[n_ /; PrimeQ[n] && n > 2] :=
 Module[{Q = 2, nn = (n + 1) / 2,
   primes = Rest[First /@ FactorInteger[n + 1]]},
  While[JacobiSymbol[1 - 4 Q, n] ≠ -1 ||
    LucasV[{1, Q}, n][nn] ≠ 0 || Apply[Or,
     Map[LucasV[{1, Q}, n][nn / #] == 0 &, primes]], Q++];
  StringForm["n = `` is prime because \nrank_n
    for the Lucas sequence defined from
    P = 1 and Q = `` equals n +1 .", n, Q]]
```

PrimeQWithLucasProof[17]

n = 17 is prime because
rank_n for the Lucas sequence
  defined from P = 1 and Q = 3 equals n +1 .

PrimeQWithLucasProof[47]

n = 47 is prime because
rank_n for the Lucas sequence
  defined from P = 1 and Q = 13 equals n +1 .

PrimeQWithLucasProof[10^{16} + 61]

n = 10000000000000061 is prime because
rank_n for the Lucas sequence
  defined from P = 1 and Q = 10 equals n +1 .

Note that the preceding are not really proofs of primality because we have not certified the primality of the prime factors of $p + 1$. But that can be addressed recursively exactly as was done for Pratt certificates in Section 4.2 (Exercise 8.10). Here is an example of such a Lucas certificate. The witnesses, to the right of the arrows, are the Q values, it being understood that $P = 1$ in this implementation. The complexity is quite similar to that of Pratt certificates.

LucasCertificate[7919, Tree → True]

```
7919 → 7
   3 → 2
   5 → 2
      3 → 2
  11 → 7
      3 → 2
```

LucasCertificate$[10^{16} + 61$, Tree $\rightarrow$ True$]$

```
10000000000000061  →  10
                3  →  2
               13  →  2
                7  →  3
   42735042735043  →  3
                    11  →  7
                     3  →  2
                    67  →  5
                        17  →  3
                         3  →  2
                 119803993  →  17
                        59901997  →  5
                                13  →  2
                                 7  →  3
                           2303923  →  3
                                    7  →  3
                                  107  →  5
                                    3  →  2
                                  769  →  21
                                      5  →  2
                                      3  →  2
                                      7  →  3
                                     11  →  7
                                      3  →  2
```

Because the Lucas method requires that $p + 1$ be factored, it is especially appropriate for numbers of some special forms, such as $2^n - 1$ (see Exercise 8.9). Here is how it works on a large Mersenne prime. There are no odd primes in 2^n, so the tree has but a single row.

LucasCertificate$[2^{127} - 1$, Tree $\rightarrow$ True$]$

```
170141183460469231731687303715884105727  →  3
```

For numbers such as Euclid numbers $(p_1 p_2 \cdots p_r + 1)$ one would use the primitive root method. For numbers with no special form and of moderate size (30–50 digits), the timing comparison will depend entirely on the factoring difficulties. For smaller numbers (16–25 digits), it seems as if the primitive root method is generally faster.

▪ The Lucas–Lehmer Algorithm Explained

The introduction to Chapter 1 mentions the Lucas–Lehmer algorithm for determining the primality of Mersenne numbers, $2^n - 1$. The algorithm is a beauty because it is exceedingly simple and can be used efficiently on very large numbers. Indeed, the largest known prime has almost always been a Mersenne prime, and the Lucas–Lehmer test has been used to prove the primality of these number. The current world record is $2^{6,972,593} - 1$, a number with over two million digits that was proved prime in 1999 by Nayan Hajratwala, a participant in the Great Internet Mersenne Prime Search (GIMPS; see www.mersenne.org/prime.htm).

Exponent	Year	Digits
2	—	1
3	—	1
5	—	2
7	—	3
13	1461	4
17	1588	6
19	1588	6
31	1750	10
61	1883	19
89	1911	27
107	1913	33
127	1876	39
521	1952	157
607	1952	183
1279	1952	386
2203	1952	664
2281	1952	687
3217	1957	969
4253	1961	1281
4423	1951	1332
9689	1963	2917
9941	1963	2993
11213	1963	3376
19937	1971	6002
21701	1978	6533
23209	1979	6987
44497	1979	13395
86243	1982	25962
110503	1988	33265
132049	1983	39751
216091	1985	65050
756839	1992	227832
859433	1993	258716
1257787	1996	378632
1398269	1996	420920
2976221	1997	895932
3021377	1998	909526
6972593	1999	2098960

Table 8.3. The history of the discovery of the Mersenne primes. Sometimes the discoveries have not been in chronological order. Thus the following were not record-setters at their time of discovery: 61, 89, 107, 521, 607, 1279, 2203, 2281, 3217, 4253, 110503.

From Theorem 8.9, we see that $n = 2^p - 1$ is proven to be prime if we can find a Lucas sequence for which n divides $V_{2^{p-1}}$, for that will yield $\omega(n) = n + 1$. For this to happen it will have to be the case that $\left(\frac{d}{n}\right)$ and $\left(\frac{Q}{n}\right)$ are not quadratic residues modulo n. The situation is reminiscent of Pépin's test for Fermat primes: There is a Lucas sequence that is *guaranteed to work* when n is prime; it is given by $P = 2$ and $Q = -2$, which gives $d = 12$. We know that $\left(\frac{12}{n}\right) = \left(\frac{4}{n}\right)\left(\frac{3}{n}\right) = \left(\frac{3}{n}\right)$. And by quadratic reciprocity, 3 is not a quadratic residue modulo $2^p - 1$ if and only if $2^p - 1$ is a quadratic residue modulo 3. We can assume that p is an odd prime ($2^2 - 1$ is prime and so we do not need a test for this case), and therefore $2^p - 1 \equiv 2 - 1 = 1 \pmod 3$. This shows that 12 is not a quadratic residue for n.

We also know (by Theorem 6.3 and Corollary 6.7) that $Q = -2$ is a quadratic residue if and only if $2^p - 1 \equiv 1$ or $3 \pmod 8$. Because p is at least 3, $2^p - 1 \equiv 7 \pmod 8$, and therefore $Q = -2$ is not a quadratic residue. We have therefore proven that if $n = 2^p - 1$ is prime then n divides $V_{(n+1)/2}$. But by Theorem 8.9, if $n = 2^p - 1$ divides $V_{(n+1)/2}$, then n is prime. Thus we have an infallible test for primality of $2^p - 1$, when p is an odd prime. We not only have a test for primality, we have an algorithm that infallibly determines whether $n = 2^p - 1$ is prime or composite.

Some algebraic manipulation turns the preceding ideas into a very fast algorithm. Recall from equation (7) that

$$V_{2e} = V_e^2 - 2(-2)^e .$$

With our choice of parameters, $V_2 = 2 \cdot 2 + 2 \cdot 2 = 8$. We can therefore define a new sequence:

$$S_t = \frac{V_{2^t}}{2^{(2^{t-1})}} .$$

It follows that

$$S_1 = \frac{V_2}{2} = 4 \quad \text{and} \quad S_t = \frac{V_{2^t}}{2^{(2^{t-1})}} = \frac{V_{2^{t-1}}^2 - 2(-2)^{2^{t-1}}}{\left(2^{(2^{t-2})}\right)^2} = S_{t-1}^2 - 2 .$$

Because $n = 2^p - 1$ is odd, n divides $V_{2^{p-1}}$ if and only if n divides S_{p-1}. And this can be checked in $p - 2$ steps by forming the sequence $S_1, S_2, \ldots, S_p$, reducing modulo n at each step.

Here is an implementation of the Lucas–Lehmer test (it is in the CNT package). We do not have to bother checking that the exponent is prime, because the test works properly in the case of $2^m - 1$ even when m is composite, provided m is odd and at least 3 (Exercise 8.11).

```
MersennePrimeQ[3] = True;
MersennePrimeQ[1] = False;
MersennePrimeQ[n_Integer] :=
  IntegerQ[m = Log[2, n + 1]] && Nest[Mod[#^2 - 2, n] &, 4, m - 2] == 0
```

Here is a list of the primes p among the first 100 that yield Mersenne primes $2^p - 1$.

```
Select[Prime[Range[100]], MersennePrimeQ[2^# - 1] &]
```

```
{2, 3, 5, 7, 13, 17, 19, 31, 61, 89, 107, 127, 521}
```

We could do a timing comparison to see whether `MersennePrimeQ` outperforms `PrimeQ`. But that would miss the point. Remember that for large values `PrimeQ`, when it returns `True`, is not guaranteed to be accurate. The Lucas–Lehmer test is an absolutely correct test of primality. So even though it is true that `MersennePrimeQ` is a little faster that `PrimeQ`, that is not a fair comparison. Actually, for the negative cases `MersennePrimeQ` is only slightly faster than `PrimeQ`. That makes sense from a complexity point of view: a single `PowerMod` to check whether $2^n - 1$ is, say, a 3-pseudoprime would require $n - 1$ squarings and $n - 1$ multiplications by 3. But the Lucas–Lehmer test requires only $n - 2$ squarings and $n - 2$ subtractions of 2. (Recall from Proposition 2.7 and Exercise 2.37 that Mersenne numbers are often 2-pseudoprimes, and that is why 3 should be used if one is using a pseudoprime test on them.)

■ Lucas Pseudoprimes

There are several ways to use Lucas sequences to get pseudoprime tests. Recall from Corollary 8.2 that if p is an odd prime then $U_{p-\varepsilon(p)} \equiv 0 \pmod{p}$. This leads to an odd integer n being called a *{P, Q}-Lucas pseudoprime* if n is not prime and $U_{n-\varepsilon(n)} \equiv 0 \pmod{n}$. As always, d denotes $P^2 - 4Q$ and $\varepsilon(n) = \left(\frac{d}{n}\right)$. Note that this condition implies that $\alpha^{n-\varepsilon(n)}$ is congruent to an integer modulo n. The special case of Fibonacci pseudoprimes, which corresponds to $P = 1$, $Q = -1$, was mentioned in Exercise 1.39: n is a *Fibonacci pseudoprime* if n is composite and n divides $F_{n-\left(\frac{5}{n}\right)}$. These tests are by no means infallible, but they form a foundation for some very satisfactory methods that seem to approach infallibility. Here is an implementation. The first function serves as a prime test; the second detects pseudoprimes.

```
LucasPseudoprimeTest[{P_, Q_}, n_] := Module[{d = P^2 - 4 Q},
   Which[1 < (g = GCD[n, 2 Q d]) < n, False, g == n, PrimeQ[n],
     g == 1, 0 == LucasU[{P, Q}, n][n - JacobiSymbol[d, n]]]]
```

```
LucasPseudoprimeQ[{P_, Q_}, n_] := ! PrimeQ[n] &&
   LucasPseudoprimeTest[{P, Q}, n]
```

```
LucasPseudoprimeQ[{P_, Q_}, n_List] :=
  Map[LucasPseudoprimeTest[{P, Q}, #] &, n]
```

We could use these functions in a `Select[Range[], …]` construction, but let us introduce the `CNT` package function `PseudoprimeSearch`. It takes a bound and various options control the type of pseudoprime searched for, Here are the Fibonacci pseudoprimes.

>**PseudoprimeSearch**[10000, Type → Fibonacci]
>
>{323, 377, 1891, 3827, 4181, 5777, 6601, 6721, 8149}

And here is another example.

>**PseudoprimeSearch**[10000, Type → Lucas, LucasParameters → {3, -1}]
>
>{119, 649, 1189, 1763, 3599, 4187, 5559, 6681}

Unlike the case for ordinary pseudoprimes, it is not known that in all cases there are infinitely many Lucas pseudoprimes. However, one step in this direction is due to Emma Lehmer, who proved in 1964 that F_{2p} is a Fibonacci pseudoprime whenever p is prime, except $p = 3$ or 5.

>**LucasPseudoprimeQ**[{1, -1}, Fibonacci[2 Prime[Range[20]]]]
>
>{True, False, False, True, True, True, True, True, True, True,
> True, True, True, True, True, True, True, True, True, True}

Recalling the Fermat's Little Theorem generalization, it makes sense to invoke the stronger property that looks at both U- and V-values. Indeed, it is no more difficult to compute both values than one of them, as explained in Algorithm 8.1. Thus, we will call n a *{P, Q}-quadratic pseudoprime* if the following holds, where a power of Q is used to combine the two cases.

$$U_{n-\varepsilon(n)} \equiv 0 \pmod{n} \quad \text{and} \quad V_{n-\varepsilon(n)} \equiv 2Q^{\frac{1-\varepsilon(n)}{2}} \pmod{n}$$

Note that this is equivalent to the condition $\alpha^{n-\varepsilon(n)} \equiv Q^{(1-\varepsilon(n))/2} \pmod{n}$. The word *quadratic* is used because this notion is directly related to powers of a quadratic irrational. It would be natural to call this a Lucas pseudoprime, but the term Lucas psp has the history just presented. The implementation that follows uses `Lucas`, which computes the {U, V} pair.

```
QuadraticPseudoprimeTest[{P_, Q_}, n_] := Module[{d = P^2 - 4 Q},
  Which[1 < (g = GCD[n, 2 Q d]) < n, False,
   g == n, PrimeQ[n], g == 1, ε = JacobiSymbol[d, n];
   Lucas[{P, Q}, n][n - ε] == {0 , Mod[2 Q^((1-ε)/2), n]}]];

QuadraticPseudoprimeQ[{P_, Q_}, n_Integer] := ! PrimeQ[n] &&
  QuadraticPseudoprimeTest[{P, Q}, n]

QuadraticPseudoprimeQ[par_, n_List] :=
  Map[QuadraticPseudoprimeQ[par, #] &, n]
```

In the Fibonacci case this extension reduces the number of bad examples less than 10,000 from nine to three.

PseudoprimeSearch[10000, Type → QuadraticFibonacci]

{4181, 5777, 6721}

But even for this test, the number of pseudoprimes can be numerous.

PseudoprimeSearch[10000,
  Type → Quadratic, LucasParameters → {14, 1}]

{65, 209, 629, 679, 901, 989, 1241,
 1769, 1961, 1991, 2509, 2701, 2911, 3007, 3439,
 3869, 5249, 5719, 5777, 6061, 6767, 6989, 9869}

Using ± 1 for Q turns out to be very bad, as we will explain in a moment; switching to 2 yields a much better test. Do you recognize the number in the following output? 2047 is the first 2-strong pseudoprime. This is not a coincidence (see Exercise 8.15).

PseudoprimeSearch[10000,
  Type → Quadratic, LucasParameters → {14, 2}]

{2047}

In addition to Lucas psps and quadratic psps, we can also examine what we will call ***LucasV pseudoprimes***, by using only the V-equation of the pair given in (5). These pseudoprimes, which are somewhat rarer than Lucas psps, are pursued in Exercise 8.16.

Now we address two important points in the theory of quadratic pseudoprimes: one should avoid using $Q = \pm 1$, and one should avoid choices that lead to $\left(\frac{d}{n}\right) = +1$. First, the Q-problem: The case $P = -1$, $Q = 1$ is an extreme example, because $\alpha^3 = 1$, and this implies that *every* composite not divisible by 3 is a quadratic psp.

More generally, the Q-problem has to do with the order of α (mod p). This order exists because $\alpha^{p-\varepsilon(p)} \equiv 2$ or $2Q$ (mod p), and so $\alpha^{[p-\varepsilon(p)][p-1]} \equiv (\alpha^{p-\varepsilon(p)})^{p-1} \equiv 1$ (mod p). Now, suppose that $Q = 1$ and $d = P^2 - 4$. If p is a prime such that $\left(\frac{d}{p}\right) = -1$, then we know that $\alpha^p \equiv \overline{\alpha}$ (mod p) and so $\alpha^{p+1} \equiv \alpha\overline{\alpha} \equiv 1$ (mod p). This means that integers of the form $n = mp$ where p does not divide m, $m \equiv 1$ (mod $p+1$), and $\varepsilon(n) = -1$ have an above-average chance of being a quadratic psp. This is because for such m, $\alpha^n \equiv (\alpha^p)^m \equiv \overline{\alpha}^m = \overline{\alpha^m} \equiv \overline{\alpha}$ (mod p), and so we would need only that $\alpha^{pm} \equiv \overline{\alpha}$ (mod m). If, say, m is prime, then $\alpha^{pm} \equiv (\alpha^m)^p \equiv \alpha^p$ (mod m) because $\left(\frac{d}{m}\right) = +1$. Thus we need only find such a prime, m, so that $\alpha^p \equiv \overline{\alpha}$ (mod m), which is equivalent to $\alpha^{p+1} \equiv 1$ (mod m). In fact, this happens fairly often, as a simple search for quadratic psps will show.

Table[{P, **PseudoprimeSearch**[2000, Type → Quadratic,
    LucasParameters → {P, 1}]}, {P, 3, 10}] // ColumnForm

{3, {323, 377, 1891}}
{4, {209, 901, 989}}
{5, {527, 551, 1807, 1919}}

```
{6, {35, 169, 385, 779, 899, 961, 1121, 1189}}
{7, {323, 329, 377, 451, 1081, 1819, 1891}}
{8, {527, 559, 781, 1763, 1921}}
{9, {9, 703, 711, 989, 1343}}
{10, {49, 323, 385, 1067, 1079, 1595, 1763}}
```

The 323 example fits the pattern exactly, where $P = 3$, $Q = 1$, $d = 5$, $p = 17$, $m = 19$: $\alpha^{18} \equiv 1 \pmod{19}$

```
n = 323; P = 3; Q = 1; d = P² - 4 Q; p = 17; m = 19; α = (P + √d̄) / 2;
JacobiSymbol[d, n]
```

    −1

```
QuadraticPowerMod[α, p + 1, m]
```

    1

Similar heuristics apply to $Q = -1$. Table 8.4 shows the number of quadratic pseudoprimes below 150000 in several cases where $|Q|$ is and is not 1. The performance difference is clear. Note the large numbers for $(2, 2)$ and $(3, 3)$. That is because α^4 and α^6, respectively, are integers in those cases, thus increasing the chances for a modular power of α to be an integer. Note also that several cases are perfect through 150,000. But perfection does not last long: the first pseudoprimes for $(3, -2)$, $(9, 2)$, $(9, -2)$, $(13, -2)$ are, respectively, 220729, 514447, 642001, and 658801.

$Q = -1$ P	Number of psps	$Q = -2$ P	Number of psps	$Q = 1$ P	Number of psps	$Q = 2$ P	Number of psps	$Q = 3$ P	Number of psps
2	50	2	5	2	□	2	43	2	2
3	28	3	——	3	61	3	□	3	47
4	63	4	2	4	69	4	7	4	□
5	28	5	1	5	53	5	3	5	3
6	47	6	2	6	105	6	1	6	2
7	41	7	2	7	103	7	2	7	1
8	26	8	4	8	88	8	3	8	2
9	40	9	——	9	86	9	——	9	3
10	36	10	10	10	95	10	8	10	3
11	59	11	2	11	61	11	6	11	4
12	29	12	3	12	75	12	2	12	1
13	50	13	——	13	93	13	3	13	2
14	111	14	2	14	127	14	5	14	2
15	52	15	6	15	85	15	3	15	3

Table 8.4. The number of quadratic pseudoprimes for various choices of P and Q. When Q is ± 1 there are a lot; for other values of Q, pseudoprimes are much rarer. A small square indicates that $P^2 - 4Q$ is a square, in which case the theory does not apply. Dashes mean that there are no pseudoprimes. The counts are based on integers below 150,000.

We next consider what might be called the square root problem: Suppose that P and Q lead to d such that $\left(\frac{d}{n}\right) = +1$. This does not mean that d is a square modulo n, but it does increase the likelihood that that is so; for if d is a square modulo n, then it is a square modulo each prime divisor of n, and so $\left(\frac{d}{n}\right) = +1$. Before explaining why squareness is bad, we need a lemma that connects ordinary pseudoprimes and quadratic pseudoprimes.

Lemma 8.10. *If an odd integer n is a b-psp and a c-psp, then n is a $\{P, Q\}$-quadratic psp whenever $P \equiv b + c$ and $Q \equiv bc$ (mod n).*

Proof. Define α from P and Q as usual. Then $\{\alpha, \overline{\alpha}\} \equiv \{b, c\}$ (mod n) because the quadratic polynomials $x^2 - Px + Q$ and $x^2 - (b+c)x + bc$ have coefficients that are congruent modulo n. Using Exercise 8.2, $V_n \equiv b^n + c^n \equiv b + c \equiv P$ (mod n), and $U_n \equiv (b^n - c^n)/(b - c) \equiv (b - c)/(b - c) \equiv 1$ (mod n); because $\left(\frac{P^2 - 4Q}{n}\right) = \left(\frac{(b-c)^2}{n}\right) = 1$, this shows that n is a $\{P, Q\}$-quadratic psp.   □

Now, if P and Q are such that $d \equiv r^2 \pmod{n}$, then the pair of simultaneous equations $P = b + c$ and $Q = bc$ can be easily solved *modulo n* for b and c to get $b = (P - r)(n + 1)/2$ and $c = (P + r)(n + 1)/2$. If, for example, n is a 2-pseudoprime, then it might well be a b-psp and a c-psp (assuming $\gcd(n, bc) = 1$), because it might be a Carmichael number, or one of b and c might be ± 1. This would mean, by the lemma, that n will be a quadratic $\{P, Q\}$-psp. This is bad because it means that the quadratic test will not be independent of the ordinary Fermat pseudoprime test. We strongly desire such independence because, if we have a quadratic psp, it is very fast to apply, say, a 2-psp test in the hope that that will prove compositeness. In short, our pseudoprime tests are more powerful if they are independent. Here's an example, where n is 2047, the first strong 2-psp. If we allowed $\left(\frac{d}{n}\right) = +1$, we might choose $P = 1$ and $Q = -2$, which yield $d = 9$. Of course, this is the square of 3.

```
n = 2047;
P = 1; Q = -2; d = P^2 - 4 Q
r = SqrtModAll[d, n][[1]]

9

3
```

Now, b and c turn out to be -1 and 2, respectively, and so the lemma tells us that n is a quadratic $\{P, Q\}$-psp.

```
{b, c} = Mod[(n + 1) / 2 {(P - r), (P + r)}, n]
GCD[n, b c]

{2046, 2}

1
```

QuadraticPseudoprimeQ[{P, Q}, n]

`True`

This desire to make the quadratic test as independent as possible from the Fermat pseudoprime test is why we always try to use P and Q so that $\varepsilon(n) = -1$. A curiosity is that, if we actively seek d such that $\left(\frac{d}{n}\right) = +1$, we seem to get fewer pseudoprimes than in the -1 case. But quality, not quantity is the issue: These pseudoprimes are very likely to be 2-psps or quadratic psps (see the entries under method A^+ in Table 8.5).

The preceding discussion about good and bad choices of P and Q brings us now to a most intriguing point about quadratic pseudoprimes. Given n, we can try to choose P and Q to maximize the chance that the quadratic test will tell the truth about n. Such a customization procedure is entirely absent from the classical pseudoprime contexts. There one either chooses a few b-values randomly (the Miller–Rabin test), or just takes a couple of small values of b, such as 2 and 3 and runs b-strong pseudoprime tests. But in the quadratic situation, we can carefully choose P and Q to maximize the chance of success.

Here is a method suggested by J. Selfridge [PSW], called **method A**: Given an odd, nonsquare integer n, let d be the first number among $5, -7, 9, -11, \ldots$ such that $\left(\frac{d}{n}\right) = -1$; then let $P = 1$ and $Q = (1-d)/4$. If, during this search, one finds $\left(\frac{d}{n}\right) = 0$, then one has solved the problem because either d proves n composite or $d = n$ and n is so small that it can be checked by elementary means. Here is how one can handle this in code. `MethodA[n]` either returns a pair of parameters that can be used, or, in the $\left(\frac{d}{n}\right) = 0$ case, it simply returns `True` or `False`. The calling routine, typically `LucasPseudoprimeQ` or `QuadraticPseudoprimeQ` must be taught to recognize that a `True` or `False` settles the problem.

```
MethodA[n_ /; OddQ[n] && !IntegerQ[√n]] := Module[{d = 5, j},
  While[j = JacobiSymbol[d, n] == 1, d = -d - 2 Sign[d]];
  If[j == -1, {1, (1 - d)/4}, If[GCD[d, n] < n, False, PrimeQ[d]]]]

LucasPseudoprimeQ[a_, _] := a /; MemberQ[{True, False}, a]
QuadraticPseudoprimeQ[a_, _] := a /; MemberQ[{True, False}, a]
```

One problem with method A is that, very often, d will turn out to be 5, and that leads to $Q = -1$, which we have seen should be avoided. Because of this Baillie and Wagstaff [BW] introduced **method A^***: same as A, but use $P = 5$ and $Q = 5$ in the cases where $d = 5$. And we can also introduce **method A^{**}**: same as A but start the search at $d = -7$ instead of $d = 5$. One can consider these methods in the context of Lucas pseudoprimes, but of course it makes more sense to consider the full quadratic version of the test. These methods are in the CNT package as `MethodA` and `MethodAStar`; method A^{**} is available in the package as `MethodA[n, -7]` (that is, any starting point can be used).

These methods are not so good in the Lucas situation, as there turn out to be 219 bad examples below one million. The CNT package has access to various lists of pseudoprimes, including a master list through 10^9 of any integer that satisfies either the U- or V-equation using any of methods A, A^*, or A^{**}. That large list, called LucaPseudoprimesAllTypes, can be filtered to get complete lists satisfying various other conditions. For example, here is how to find the 219 Lucas psps under one million using method A.

```
Lucaspsps = Select[LucasPseudoprimesAllTypes,
  # < 10^6 && LucasPseudoprimeTest[MethodA[#], #] &]
```

{323, 377, 1159, 1829, 3827, 5459, 5777, 9071, 9179, 10877,
11419, 11663, 13919, 14839, 16109, 16211, 18407, 18971,
19043, 22499, 23407, 24569, 25199, 25877, 26069, 27323,
32759, 34943, 35207, 39059, 39203, 39689, 40309, 44099,
46979, 47879, 50183, 51983, 53663, 56279, 58519, 60377,
63881, 69509, 72389, 73919, 75077, 77219, 79547, 79799,
82983, 84419, 86063, 90287, 94667, 97019, 97439, 100127,
101919, 103739, 104663, 113573, 113849, 115439, 115639,
120581, 121103, 121393, 130139, 142883, 150079, 155819,
157079, 158399, 158717, 161027, 162133, 162719, 164699,
167969, 176399, 176471, 178949, 182513, 184781, 189419,
192509, 195227, 197801, 200147, 201871, 203699, 216659,
218129, 223901, 224369, 226529, 230159, 230691, 231703,
238999, 242079, 243629, 250277, 253259, 256409, 265481,
268349, 271991, 275099, 277399, 283373, 284171, 288919,
294527, 306287, 308699, 309959, 313499, 317249, 324899,
324911, 327359, 345913, 353219, 364229, 366799, 368351,
380393, 381923, 383921, 385307, 391169, 391859, 396899,
427349, 429263, 430127, 436409, 436589, 441599, 454607,
455519, 475799, 480689, 487199, 500207, 507527, 510479,
519689, 531439, 543449, 548627, 554651, 566579, 569087,
572669, 575279, 575819, 600767, 606761, 611399, 622169,
626639, 635627, 636199, 636707, 636999, 653939, 656011,
667589, 676367, 676499, 678869, 685583, 686279, 697883,
701999, 708749, 721279, 732887, 736163, 749699, 753059,
753377, 761039, 765687, 770783, 775207, 794611, 807971,
812239, 819839, 823439, 824879, 828827, 833111, 835999,
839159, 850859, 851927, 871859, 875879, 885119, 887879,
895439, 895679, 907061, 948433, 950821, 960859, 961999,
967591, 972311, 980099, 980699, 983903, 994517, 997919}

The quadratic test is much better. The quadratic method-A test admits only 38 psps below 10^7.

```
Select[LucasPseudoprimesAllTypes,
  # < 10^7 && QuadraticPseudoprimeTest[MethodA[#], #] &]
```

{5777, 10877, 75077, 100127, 113573, 161027, 162133,
231703, 430127, 635627, 851927, 1033997, 1106327, 1256293,
1388903, 1697183, 2263127, 2435423, 2662277, 3175883,

```
    3399527, 3452147, 3774377, 3900797, 4109363, 4226777,
    4403027, 4828277, 4870847, 5208377, 5942627, 6003923,
    7353917, 8518127, 9401893, 9713027, 9793313, 9922337}
```

The LucasV test using method A^* is impressive, as 913 is the only pseudoprime below $5 \cdot 10^8$.

```
Select[LucasPseudoprimesAllTypes,
  LucasVPseudoprimeTest[MethodAStar[#], #] &]

 {913}
```

But the results are even more impressive when method A^{**} is used: There are no counterexamples known.

Table 8.5 summarizes the situation with respect to various combinations of these tests, and includes data for some variations discussed in the exercises.

It turns out that each of the 219 Lucas method-A pseudoprimes is unmasked by a simple 2-psp test.

```
Or @@ PseudoprimeQ[2, Lucaspsps]

 False
```

This leads to a famous unsolved problem.

The PSW Challenge. Find a counterexample to: n is prime if and only if n passes the 2-psp test and n passes the Lucas test with $P = 1$ and $Q = (1 - d)/4$ where d is the first element of 5, -7, 9, -11, 13, ... such that $\left(\frac{d}{n}\right) = -1$.

But the preceding points fade somewhat when we look at the superb performance of methods A^* and A^{**}. Baillie and Wagstaff [BW] checked the integers below 10^8 and found *no pseudoprimes at all* with respect to the quadratic method-A^* test! Let us use the term ***pseudoperfect*** to refer to a pseudoprime test for which no counterexample to its perfection is known. So the quadratic method A^* test is pseudoperfect. The method A^{**} approach has the implementation advantage that P is fixed at 1. Like method A^*, this approach leads to a quadratic test that is pseudoperfect. Indeed, even the LucasV version of this is pseudoperfect. Thus we have the following challenges.

The Lucas Pseudoprime Challenges.

1. Find a counterexample to: n is prime if and only if n is not a quadratic pseudoprime for $P = 1$ and $Q = (1 - d)/4$, where d is the first element of 5, -7, 9, -11, 13, ... such that $\left(\frac{d}{n}\right) = -1$, but with $(P, Q) = (-1, 1)$ replaced by (5, 5).

2. Find a counterexample to: n is prime if and only if n is not a {1, Q}-quadratic pseudoprime where $Q = (1 - d)/4$, and d is the first element of -7, 9, -11, 13, ... such that $\left(\frac{d}{n}\right) = -1$.

	Method	First two pseudoprimes	Number of pseudoprimes
1	A, Lucas	323, 377	$219\,;\,659\,;\,1911\,;\,3999$
2	A*, Lucas	323, 377	$219\,;\,659\,;\,1911\,;\,3999$
3	A** (= A start at −7) Lucas	1127, 1159	$203\,;\,649\,;\,1976\,;\,4146$
4	A, LucasV	1127, 5777	$18\,;\,55\,;\,145\,;\,280$
5	A*, LucasV	913	1
6	**A**, LucasV**	——	0
7	A, Quadratic	5777, 10877	$11\,;\,38\,;\,105\,;\,214$
8	**A*, Quadratic**	——	0
9	A, Strong quadratic	5777, 10877	$11\,;\,38\,;\,105\,;\,214$
10	**A*, Strong Quadratic**	——	0
11	A$^+$ (Exercise 9.17), Lucas	1891, 2047	134
12	A$^+$, LucasV	1003, 2047	103
13	A$^+$, Quadratic	2047, 4181	67
14	2-pseudoprime	341, 561	$245\,;\,750\,;\,2057\,;\,5597$
15	2-strong pseudoprime	2047, 3277	$46\,;\,162\,;\,488\,;\,1282$
16	2-strong psp & 3-strong psp	1373653, 1530787	$7\,;\,21\,;\,58\,;\,52593$
17	Perrin, basic	271441, 904631	2
18	Perrin, symmetric	27664033, 46672291	55
19	Fibonacci	323, 377	155
20	FibonacciV	1127, 3751	271
21	Quadratic Fibonacci	4181, 5777	$16\,;\,56$
22	Strong Quadratic Fibonacci	4181, 5777	$14\,;\,41$
23	Carmichael numbers	561, 1105	$43\,;\,105\,;\,255\,;\,646$

Table 8.5. A summary of the efficacy of several customized tests. The strong versions are discussed in the next section. A^+ is used for the variation of method A that actively seeks d so that $(\frac{d}{n}) = +1$.

3. Find a counterexample to: n is prime if and only if n is not a $\{1,\,Q\}$-LucasV pseudoprime where $Q = (1 - d)/4$, and d is the first element of $-7,\,9,\,-11,\,13,\,\ldots$ such that $\left(\frac{d}{n}\right) = -1$.

And here is a final challenge. It applies only to numbers ending in a 3 or 7, because, for the cases 1 and 9, n would be such that $(\frac{5}{n}) = +1$, and that has an impact on the independence of the tests. But no counterexample is known for the 3 and 7 cases, so one might use a 2-psp test followed by a Fibonacci psp test on such numbers; this would be a very fast test that is applicable half the time.

Search bound	Comments	
10^6; 10^7; 10^8; $5 \cdot 10^8$	Fixes $P = 1$; none are 2-psps	1
10^6; 10^7; 10^8; $5 \cdot 10^8$	P is 1 or 5; none are 2-psps	2
10^6; 10^7; 10^8; $5 \cdot 10^8$	Fixes $P = 1$; none are 2-psps	3
10^6; 10^7; 10^8	Fixes $P = 1$	4
$5 \, 10^8$	P is 1 or 5; neither a 2-psp nor a 5-psp	5
$5 \cdot 10^8$	The fastest pseudoperfect test? Fixes $P = 1$	6
10^6; 10^7; 10^8; $5 \cdot 10^8$	Intersection of 1 and 4	7
$5 \cdot 10^8$	Intersection of 2 and 5	8
10^6; 10^7; 10^8; $5 \cdot 10^8$	Curiosity: same data as in 7	9
$5 \cdot 10^8$		10
10^6	38 are 2-psps; 9 are strong 2-psps	11
10^6	29 are 2-psps; 8 are strong 2-psps	12
10^6	Intersection of 11 and 12	13
10^6; 10^7; 10^8; 10^9	There are infinitely many	14
10^6; 10^7; 10^8; 10^9	There are infinitely many	15
10^7; 10^8; 10^9; 10^{16}	There are infinitely many	16
10^6	The 3, 0, 2, 3, 2, 5, ... sequence; Ex. 8.18	17
$50 \cdot 10^9$	Both directions, plus signature	18
10^6	There are infinitely many	19
10^6	Surprise: V yields a worse test than U	20
10^5; 10^6		21
10^5; 10^6		22
10^6; 10^7; 10^8; 10^9	There are asymptotically $x^{2/7}$ under x	23

The Fibonacci Pseudoprime Challenge. Find n such that $n \equiv 3$ or 7 (mod 10), and n is both a 2-psp and a Fibonacci psp.

▣ Strong Quadratic Pseudoprimes

Just as the notion of a strong pseudoprime provided a major improvement over the simple pseudoprime test involving a single exponentiation, so too can we improve our quadratic criteria. One traditional notion of strong Lucas pseudoprime fixes $Q = 1$ and looks at the V-sequence only. But we will consider both the U- and V-sequences, which is equivalent to looking at properties of powers of α.

Recall that the quadratic pseudoprime test checks whether $\alpha^{n-\varepsilon(n)} \equiv 2Q$ (mod n). We turn this into a strong test with the simple observation that if n is prime and α^e squares to an integer modulo n, then α^e is congruent (mod n) to either an integer or an integer times $\sqrt{d}$. This is just a restatement of the doubling rule, which says that if $U_{2e} \equiv 0$ (mod n) then one of U_e or V_e is 0 (mod n). So we can follow the strong pseudoprime scenario: Write $n - \varepsilon(n)$ as $2^e m$ where m is odd, look at the U and V values for indices m, $2m$, $4m$, ..., $n+1$. If $U_m \equiv 0$ (mod n), we learn nothing new and can proceed directly to a check that $V_{n-\varepsilon} \equiv 2Q$ (mod n). But if the first U-value is not 0, we examine the U-sequence until we get a 0; then we check that the preceding V-value is 0 (mod n); and again we then proceed to the end to make sure $2Q$ shows up.

Here is some code to compute the UV-sequences. Equations (8) and (10) are used to move up at each step. For V, this requires knowing certain powers of Q, and they are kept along and updated (by simply squaring) as the third entry of the list and discarded at the end.

```
UVSequence1[{P_, Q_}, n_?OddQ] :=
 Module[{m = n - JacobiSymbol[P^2 - 4 Q, n]},
  e = IntegerExponent[m, 2]; m /= 2^e;
  Transpose[(Drop[#, -1] & /@
    NestList[Mod[{#[[1]] #[[2]], #[[2]]^2 - 2 #[[3]], #[[3]]^2}, n] &,
     Append[Lucas[{P, Q}, n][[m]], PowerMod[Q, m, n]], e])]]]
```

The version in the CNT package can produce nice gridboxes. The three examples that follow are for primes; note how the 0s slide from the U-sequence to the V-sequence at one point, except for the case where the U-sequence is only 0.

UVSequence[{3, 13}, 539039, UseGridBox → True]

Powers	m	$2m$	$2^2 m$	$2^3 m$	$2^4 m$	$2^5 m = 539040$
U	280023	144566	383211	153412	236877	0
V	55311	283184	516152	474252	0	26

UVSequence[{3, 13}, 533261, UseGridBox → True]

Powers	m	$2^1 m = 533262$
U	0	0
V	445985	26

UVSequence[{3, 13}, 914161, UseGridBox → True]

Powers	m	$2m$	$2^2 m$	$2^3 m$	$2^4 m = 914160$
U	411591	875637	187752	0	0
V	596925	243841	0	2	2

Recall from earlier work that 658801 was the first quadratic psp for the parameters $(13, -2)$. The strong test unmasks it.

UVSequence[{13, -2}, 658801, **UseGridBox** → True]

Powers		m	$2m$	$2^2 m$	$2^3 m$	$2^4 m = 658800$
U		244970	0	0	0	0
V		320593	378589	113280	116261	2

Now here is how to implement the strong quadratic test using the UV-sequence.

```
StrongQuadraticPseudoprimeTest[{P_, Q_}, n_?OddQ] :=
 Module[{d = P^2 - 4 Q, g, u, v}, g = GCD[Q d, n];
  Which[1 < g < n, False, g == n, PrimeQ[n],
   True, {u, v} = UVSequence[{P, Q}, n];
    v[[-1]] == Mod[2 Q^(1-JacobiSymbol[d,n])/2, n] &&
    (u[[1]] == 0 || MemberQ[v, 0])]]

StrongQuadraticPseudoprimeQ[{P_, Q_}, n_?OddQ] :=
 !PrimeQ[n] && StrongQuadraticPseudoprimeTest[{P, Q}, n]
```

It works nicely on the cases of fixed P, Q. Here are 39 large numbers that are quadratic pseudoprimes for the Fibonacci case ($P = 1$, $Q = -1$).

```
fibpsps = Select[StrongPseudoprimes[{2, 3}, 10^11],
 QuadraticPseudoprimeTest[{1, -1}, #] &]
```

```
{101649241, 117987841, 579606301, 927106561, 1157839381,
 2217879901, 2626783921, 2693739751, 3215031751, 4060942381,
 4710862501, 5755495201, 8214723001, 9974580661, 15114550951,
 15579919981, 17475044329, 21276028621, 25366866661,
 27716349961, 29118033181, 30481338889, 42550716781,
 43536545821, 44453586241, 44732778751, 48354810571,
 52139147581, 53700690781, 74190097801, 75285070351,
 75350936251, 79696887661, 82318050361, 83828294551,
 91609762861, 96133341781, 98071561561, 98515393021}
```

We unmask almost half of them when we move to the strong test.

```
Length[fibpsps]
Length[
 Select[fibpsps, StrongQuadraticPseudoprimeTest[{1, -1}, #] &]]
```

39

22

It is hard to judge the benefits of the strong test for the situation where the P and Q are chosen in a customized fashion for each n. This is because we know of no quadratic pseudoprimes whatsoever for this test! Still it does make sense to use the strong test because no extra work is involved.

We can now describe how *Mathematica*'s `PrimeQ` works. The description below is not completely accurate in the sense that it uses a slightly different strong Lucas test than the one presented here.

1. If n has a small divisor, return `False`. Check this via a gcd with the product of primes under 100.
2. If n is proved to be composite by the 2-strong pseudoprime test, return `False`.
3. If n is proved to be composite by the 3-strong pseudoprime test, return `False`.
4. If n is proved to be composite by the strong Lucas pseudoprime test, return `False`.
5. Return `True`.

We do not know if this test always tells the truth. Note that it is not a probabilistic test: There is no randomness involved. The question is simply whether there is a composite that is so well disguised that it slips by the three (really, four) tests. There are heuristic reasons for believing that such examples exist. But we know, thanks to Bleichenbacher's data set of 2- and 3-strong pseudoprimes to 10^{16}, that this test is valid to that bound.

🔳 Primality Testing's Holy Grail

It is generally believed that no simple combination of a fixed number of pseudoprime tests will characterize primality. On the other hand, probably $2(\log n)^2$ strong-pseudoprime tests do the job; this follows from a conjecture known as the extended Riemann Hypothesis (Section 4.2) and has as a consequence that primality testing can be done in polynomial time. Thus it is tempting to hope that some simple test might indeed characterize the primes, and there are several for which the first counterexample is quite large. We list here some open questions in this direction.

Find counterexamples to the assertion that any of the following conditions characterize primality:

1. The LucasV pseudoprime test with P and Q chosen by method A^{**}.
2. The quadratic pseudoprime test using method A^*.
3. The strong quadratic pseudoprime test using method A^*.
4. The quadratic pseudoprime test using method A^{**}.
5. The strong quadratic pseudoprime test using method A^{**}.
6. The 2- and 3-strong pseudoprime tests and also the Lucas pseudoprime test using method A (or any of the other quadratic tests just mentioned).

While the last assertion seems the most complicated, the existence of a database of all the 2- and 3-spsps below 10^{16} allows us to say that there is no counterexample below that bound. No such comparable results exist (yet) for the other criteria. This explains why *Mathematica* uses (a variation of) Method 6 in its `PrimeQ`: It is guaranteed correct below 10^{16}. Also, when considering how to define a general-purpose prime-testing function, attention must be paid to speed. For almost all composite inputs, a 2-psp or a 2-strong psp test will suffice; so it makes sense to use such a simple test first. And it is always wise, right at the beginning, to check the input for small factors, perhaps by doing a GCD with some precomputed large product of small primes.

Exercises for Section 8.2

8.8. Why, in the Lucas certification procedure described in the text, would it be foolish to begin the Q-search at either 0 or 1?

8.9. Determine, with a proof via Lucas certificates, which numbers of the form $p_1 p_2 \cdots p_n - 1$ are prime, where $n \le 100$ and p_i is the ith prime.

8.10. Write a program that uses the Lucas prime-proving method to generate a true certificate of primality, in the sense that all primes in the factorization of $p + 1$ are certified by the same method, and so on recursively down to 2.

8.11. Show that the Lucas–Lehmer algorithm will return `False` if the input has the form $2^k - 1$ where k is not prime.

8.12. Explain why, in the Lucas–Lehmer algorithm, one could replace the choice $P = 2$ and $Q = -2$ by $P = 2$ and $Q = 3$.

8.13. Prove that if n is a $\{P, Q\}$ quadratic psp then n is also a quadratic pseudoprime for $\{-P, Q\}$.

8.14. Show that if n is a quadratic pseudoprime for $\{P, Q\}$, then n is a Q-pseudoprime.

8.15. Use Exercise 2.40 to explain why 2-pseudoprimes are likely to be b-pseudoprimes for a greater than average number of bs. Investigate this by counting the number of such bs when $n = 341$. For how many bases less than 2047 is 2047 a b-psp? Show how this phenomenon combines with the perfect square problem discussed in the text to explain why the 2-strong pseudoprime 2047 is a $\{14, 2\}$-quadratic pseudoprime.

8.16. Do some computations to confirm some of the values in Table 8.5. In particular, verify that 913 is the only small LucasV pseudoprime under method A^*.

8.17. To see first-hand the difference between the $\left(\frac{d}{n}\right) = +1$ and $\left(\frac{d}{n}\right) = -1$ situations, modify the functions that define the method A and method A^* parameter choices so that they search for a d with $\left(\frac{d}{n}\right) = +1$; call this ***method A^+***. Count the number of Lucas pseudoprimes for method A^+, confirming some of the entries in Table 8.5. Check how many of the examples you find are 2-pseudoprimes. This should confirm the discussion in the text that these two tests are not independent when $\left(\frac{d}{n}\right) = +1$.

8.18. (Perrin's Test) The following test for primality — essentially a cubic test — is not quite as good as the quadratic tests of this chapter, but it is noteworthy for its simplicity. **Perrin's sequence**, $\{a_n\}$, is defined by $a_n = a_{n-2} + a_{n-3}$, where $a_0 = 3$, $a_1 = 0$, and $a_2 = 2$.

(a) Implement a routine `Perrin[n,m]` that quickly returns the mod-m value of a_n.

(b) Show that each of the three roots of $x^3 - x - 1$, the characteristic polynomial of the recurrence, is such that its powers (complex numbers, perhaps) satisfy the Perrin recurrence (though not the initial conditions).

(c) Show that if α, β, and γ are the three roots of $x^3 - x - 1$ and c_i are constants, then $b_n = c_1 \alpha^n + c_2 \beta^n + c_3 \gamma^n$ satisfies the Perrin recurrence.

(d) Show that taking $c_1 = c_2 = c_3 = 1$ in (c) yields a solution to both the recurrence and the initial conditions.

(e) Observe experimentally that if p is prime, than $a_p \equiv 0 \pmod{p}$; this is called **Perrin's test** for primality. Prove it by establishing $(\alpha + \beta + \gamma)^p \equiv \alpha^p + \beta^p + \gamma^p \pmod{p}$. A subtle point: Just as in the Lucas case, we are working in $\mathbb{Z}[\alpha, \beta, \gamma]$, and some algebraic knowledge is required to conclude that p divides n in $\mathbb{Z}[\alpha, \beta, \gamma]$ means that p divides n in $\mathbb{Z}$. You may assume this fact in this problem. For the record, it can be proven by first proving that every element in $\mathbb{Z}[\alpha, \beta, \gamma]$ has the form $n_0 + n_1 \alpha + n_2 \beta + n_3 \gamma + n_4 \alpha \beta + n_5 \alpha \gamma + n_6 \beta \gamma +$ terms involving squares (terms involving cubes reduce because the polynomial allows one to reduce them to quadratic powers).

(f) Call n a **Perrin pseudoprime** if n is not prime and n passes Perrin's test. Write a program that searches for and finds the first Perrin pseudoprime. Find the second.

(g) Extend the recurrence backwards, defining a_{-n} from a_{-n+1}, a_{-n+2}, and a_{-n+3} so that the main recurrence holds. What pattern do you observe in a_{-p} when p is prime? What is the first Perrin psp if this negative condition is taken into account? (These are called **symmetric Perrin pseudoprimes**.)

(h) Generalize parts (a)–(g) to sequences defined by $a_n = r a_{n-1} - s a_{n-2} + a_{n-3}$, where $a_0 = 3$, $a_1 = r$, and $a_2 = r^2 - 2s$.

For more information on Perrin pseudoprimes see [AS] and [KSW].

CHAPTER 9

Prime Imaginaries and Imaginary Primes

9.0 Introduction

The square root of -1 is an imaginary number. We have already seen that this number has a concrete realization modulo certain primes; for example, 2 is a square root of -1 (mod 5) because $2^2 = 4 \equiv -1$ (mod 5). And in the domain of numbers using the traditional imaginary number i, one can define primes of the form $a + bi$. For example, $1 + 2i$ is a prime, but $1 + 3i$ is not. Thus, we have prime imaginaries, such as 2 (mod 5), and imaginary primes, such as $1 + 2i$. Remarkably, there are several important connections between the two concepts, and each will help us gain a deeper understanding of the other.

In this chapter we investigate in some detail the problem of writing a prime, or an arbitrary integer, as a sum of two squares. We will solve this problem with a very fast algorithm, and then apply our knowledge to the question of primes in the complex numbers, and also to a problem of chemistry involving electrical energy in the lattice structure of salt. A further application of the imaginary realm is to an extension of the notion of quadratic reciprocity to fourth powers.

9.1 Sums of Two Squares

■ Primes

Some primes are a sum of two squares: $2 = 1 + 1$, $5 = 1 + 4$, $13 = 4 + 9$, $17 = 1 + 16$. Others, such as 3 and 11, are not. A landmark result of number theory is the theorem of Fermat that tells us exactly which primes are the sums of two squares.

Theorem 9.1. A prime p is a sum of two squares if and only if $p \not\equiv 3 \pmod 4$.

One direction is quite easy: If $p \equiv 3 \pmod 4$, then p cannot have the form $a^2 + b^2$ because a square is congruent to 0 or 1 (mod 4) and so a sum of two squares is $0 + 0$, $0 + 1$, or $1 + 1$; 3 is therefore not a possibility. This applies to any integer, not just primes. Fermat's proof of the other direction, showing that all other primes can be so represented, was by a method called "infinite descent". He assumed there was a counterexample p, and showed that there would then be a smaller counterexample. Such a process would eventually lead to 5, but 5 *is* representable, as $1 + 4$. Fermat's method is pretty, and Fermat himself applied the idea to other problems of number theory, such as proving that the equation $x^4 + y^4 = z^4$ has no nontrivial solutions.

But the infinite descent method is not very useful for a computer algorithm, so we will not present the details here. We will instead present an algorithm due to H. J. S. Smith [CELV] (1855) that not only proves that p has the desired representation, but also finds, and quickly, the integers a and b that work. Some of the ideas go back to Hermite and Serret a few years earlier; they were working in the language of continued fractions, but Smith's formulation, using the Euclidean algorithm, is simpler. In fact, the solution to $p = a^2 + b^2$ is unique (up to order and sign); see Exercise 9.4.

Smith's algorithm requires that we can quickly get a square root of -1 (mod p). This was discussed in Theorem 6.4, where it was shown that the primes that admit such square roots are precisely the primes congruent to 1 (mod 4). Chapter 6 presented some general methods for modular square roots, but the case of $\sqrt{-1}$ can be handled quickly with a simple formula: If $p \equiv 1 \pmod 4$ and c is a nonresidue for p, then Euler's criterion says that $c^{(p-1)/2} \equiv -1 \pmod p$. Because $p = 4k + 1$, $(p-1)/2$ is even and so $c^{(p-1)/4}$ is the desired square root of -1. The code that follows implements this, using a trial-and-error method to find a suitable value of c. Note that `SqrtNeg-One` returns the smaller of the two values of $\sqrt{-1}$.

```
Nonresidue[p_?OddQ] := Module[{n = 0},
  While[JacobiSymbol[Prime[++n], p] == 1]; Prime[n]]

SqrtNegOne[p_] :=
 Module[{a = PowerMod[Nonresidue[p], (p - 1) / 4, p]},
   First[Sort[{a, p - a}]]] /; Mod[p, 4] == 1

SqrtNegOne[p_] := {} /; Mod[p, 4] == 3

SqrtNegOne[2] = {1};

SqrtNegOne[181]

19

Mod[19^2, 181]

180
```

Algorithm 9.1. Smith's Algorithm

Suppose that p is a prime and $p \equiv 1 \pmod{4}$. To find the (unique) solution to $p = a^2 + b^2$, let $x^2 \equiv -1 \pmod{p}$ and apply the Euclidean algorithm to the pair (p, x). The first two remainders that are less than $\sqrt{p}$ are the a and b that solve the equation.

This is a remarkable algorithm because it is both easy to implement and very fast. Note that once we have a we can immediately get b as $\sqrt{p - a^2}$. Here is some code that finds a, which turns out to be the middle remainder in the remainder sequence. The main routine has the two arguments p and x, but an auxiliary case with three arguments carries out the recursion: the extra argument keeps track of p so the stopping criterion can be checked. Thus, the code is almost identical to the recursive technique for the basic Euclidean algorithm. Note that the condition `/; x² ≥ p` is not appended to the second case, because the first line will be checked first: *Mathematica* tries to use particular cases before general cases. This saves a few multiplications.

```
CentralRemainder[p_, m_, x_] := x /; x² < p
CentralRemainder[p_, m_, x_] :=
 CentralRemainder[p, x, Mod[m, x]]
CentralRemainder[p_, x_] := CentralRemainder[p, p, x]

Sum2Squares[p_] :=
 Module[{a = CentralRemainder[p, SqrtNegOne[p]]},
  {√p - a² , a}] /; Mod[p, 4] == 1
```

To see how fast this is, we can try it on a very large prime congruent to 1 (mod 4). We use the package function `RandomPrime` with a second argument of {1, 4} to find a large random prime congruent to 1 (mod 4).

```
p = RandomPrime[10, {1, 4}]
```

```
7961617333
```

Now it takes only an eyeblink to write p as $a^2 + b^2$.

```
{a, b} = Sum2Squares[p]
```

```
{60922, 65193}
```

It is easy to check.

```
a² + b²
```

```
7961617333
```

And this algorithm is very fast on 100-digit numbers.

```
p = RandomPrime[100, {1, 4}];
{a, b} = Sum2Squares[p]
```

```
{11264074630750654943822399203687021058843986754800,
  954368018908949861698306661347532412798842924143733}
```

```
p == a² + b²
```

```
True
```

The proof that this algorithm works is a little intricate, but entirely elementary. This proof was found by an intensive investigation of the data in the extended Euclidean algorithm. Note that Theorem 9.1 is a consequence of the theorem that follows.

Theorem 9.2. Smith's Algorithm. If p is a prime and $p \equiv 1 \pmod 4$, let $x \in \mathbb{Z}_p$ be a solution to $x^2 \equiv -1 \pmod p$ and let a be the first remainder that is smaller than $\sqrt{p}$ when the Euclidean algorithm is applied to the pair p and x. Then $\sqrt{p - a^2}$ is an integer; that is, a is one member of a pair (a, b) that satisfies $p = a^2 + b^2$.

Proof. Recall the sequence $\{t_i\}$ that is produced by the extended gcd algorithm of Section 1.2; its main property is that $t_i x \equiv r_i \pmod p$, where $\{p, x, r_2, \ldots, r_n\}$ is the sequence of remainders, with $r_n = 1$. Moreover, as was shown in Section 1.2 (Exercise 1.18), the t-sequence alternates in sign, $t_{n+1} = \pm p$, and the absolute values of the t_i themselves (with the exception of the initial 0 and 1 if $x > p/2$) form a Euclidean algorithm remainder sequence in reverse, with quotients the reverse of the main sequence of quotients $\{q_i\}$. It may be helpful to focus on an example, so let $p = 73$; then x can be either 27 or 46 and we'll use 27. We shall prove that several of the patterns evident in the table that follows hold whenever x is a square root of -1 modulo p.

FullGCD[73, 27, ExtendedGCDValues → True]

Remainders	Quotients	s	t
73		1	0
27	2	0	1
19	1	1	-2
8	2	-1	3
3	2	3	-8
2	1	-7	19
1	2	10	-27
0		-27	73

The proof leans heavily on the continuant functions Q, which give each remainder as a function of the quotients that follow it (see Section 1.2). Let r_m be the first remainder that is less than $\sqrt{p}$. Then

$$p = Q(q_1, \ldots, q_n),$$
$$r_{m-1} = Q(q_m, \ldots, q_n), \qquad (1)$$
$$|t_m| = Q(q_{m-1}, \ldots, q_1) = Q(q_1, \ldots, q_{m-1}).$$

The first two equations hold because each remainder r_j begins a Euclidean algorithm remainder sequence; the third equality holds because the absolute values of the t_i are a remainder sequence with the same quotients and (Exercise 1.18(c)) Q is invariant under reversal of its arguments. (The exceptional case in which t_1 is not part of a remainder sequence occurs when $x > p/2$, and when this happens $m \geq 2$.) Now, the sum-of-products characterization of Q (Proposition 1.4) implies that

$$Q(q_1, \ldots, q_n) \geq Q(q_1, \ldots, q_{m-1}) Q(q_m, \ldots, q_n)$$

because each product occurring as a summand in the expanded right-hand side also occurs in the left-hand side. This last inequality combines with (1) to yield $p \geq r_{m-1}|t_m|$, and because $r_{m-1} > \sqrt{p}$, t_m must be less than $\sqrt{p}$.

To conclude the proof, simply observe that $t_m x \equiv r_m \pmod{p}$, so p divides $r_m^2 - t_m^2 x^2$. But $t_m^2 x^2 \equiv -t_m^2$. Thus, $r_m^2 + t_m^2$ is divisible by p. Each of r_m and t_m is less than $\sqrt{p}$, so $r_m^2 + t_m^2 < 2p$. The only positive integer less than $2p$ that is divisible by p is p itself, whence $r_m^2 + t_m^2 = p$. □

The General Problem

With very little more work, Smith's algorithm can be used to obtain a complete list of the representations of n as a sum of two squares. Two key observations allow this. First, note that the proof of Theorem 9.2 never used the primality of p; the result works in exactly the same way if p is replaced by an arbitrary integer n for which $\sqrt{-1}$ modulo n exists. And a second key idea is that one should focus first on the primitive representations; a sum $a^2 + b^2$ is called **primitive** if $\gcd(a, b) = 1$. Remember that the values of a and b can be negative.

Theorem 9.3. An integer n admits a primitive representation as $a^2 + b^2$ if and only if the square root of -1 exists modulo n.

Proof. All congruences in this proof are modulo n. We will show that there is a one-one correspondence between primitive representations $n = a^2 + b^2$ with positive a and b and pairs $\{x, n - x\}$ where $x^2 \equiv -1$ and $x < n - x$. Suppose that we are given such a pair $\{x, n - x\}$. Then we can apply Smith's algorithm to the pair n and x to get a primitive representation $n = a^2 + b^2$, with $a > b$. This is indeed primitive because a and b coincide with r_m and r_{m+1} (see Exercise 9.1(g)) and the gcd of these two remainders coincides with $\gcd(n, x)$, which is 1 (because $x^2 \equiv -1$). Moreover, the proof of Smith's algorithm yields $bx \equiv a$, which means $x \equiv ab^{-1}$ [b^{-1} exists because gcd(b, n) = 1, else a and b would have common factor] and it follows that different values of x yield different pairs a, b.

It remains to show that every primitive representation arises in this way. So suppose that $a^2 + b^2$ is a primitive representation of n with positive a, b. Then $a^2 \equiv -b^2$ and so $a\,b^{-1}$ is a mod-n square root of -1. Assume for a moment that (the reduced residue of) $a\,b^{-1}$ is the smaller of the two square roots of -1, and call this value y. Applying Smith's algorithm to n and y yields a primitive solution $\{c, d\}$ with $c > d > 0$ and such that $c\,d^{-1} \equiv y$. It follows that $a\,b^{-1} \equiv c\,d^{-1}$, so $ad \equiv bc$. But each of a, b, c, and d is less than $\sqrt{n}$, so ad and bc are less than n; therefore $ad = bc$. And gcd(a, b) = 1, so $a = c$ and $b = d$. If the assumption is false, then the other square root of -1 is the smaller. But this other square root is $-a\,b^{-1}$, which is the same as the inverse of $a\,b^{-1}$, or $b\,a^{-1}$. Now we can use the arguments from the first case to get $bd = ac$, or $b = c$ and $a = d$, which also means that the pair $\{a, b\}$ is the same as $\{c, d\}$. So in either case the pair $\{a, b\}$ arises from the Euclidean algorithm method applied to n and a mod-n square root of -1.   □

To implement a primitive representations algorithm we first make `CentralRemainder` a listable function so that we can feed a bunch of square roots of -1 to it, and then we define `SmallSqrtsNegOne` to give the smaller of each pair of such square roots.

```
Attributes[CentralRemainder] = Listable;
```

```
SmallSqrtsNegOne[n_] := Select[SqrtModAll[-1, n], # < n / 2 &]
```

```
PrimitiveReps[n_] :=
    avals = CentralRemainder[n, SmallSqrtsNegOne[n]];
    Sort /@ Transpose[{avals, √(n - avals²)}] /; n > 0
PrimitiveReps[1] = {{0, 1}};
PrimitiveReps[0] = {};
```

```
PrimitiveReps[6565]
```

```
{{47, 66}, {33, 74}, {18, 79}, {2, 81}}
```

One last step is needed to get the complete set of representations of n as a sum of two squares. If $n = a^2 + b^2$ where gcd(a, b) = $d > 1$, then $n = d^2\,(a/d)^2 + d^2\,(b/d)^2$, where the pair a/d, b/d is a primitive representation of n/d^2. So we can proceed by finding the possible values of d and then using `PrimitiveReps` on n/d^2. We find the possible values of d by looking at all divisors of n. Of course, if n is too large to be factored, then this approach will be stopped in its tracks.

```
SqrtSquareDivisors[n_] := Select[√Divisors[n], IntegerQ]
SqrtSquareDivisors[100]
```

```
{1, 2, 5, 10}
```

```
Attributes[PrimitiveReps] = Listable;

GeneralSum2Squares[0] = {{0, 0}};
GeneralSum2Squares[n_] := (d = SqrtSquareDivisors[n];
   Sort[Apply[Join, PrimitiveReps[n / d²] d]])
```

```
GeneralSum2Squares[112132]
```

```
{{24, 334}, {136, 306}, {206, 264}}
```

Some sum-of-squares functions are included in the standard Number-TheoryFunctions package.

```
SumOfSquaresRepresentations[2, 3380]
```

```
{{4, 58}, {26, 52}, {38, 44}}
```

The package also includes a function that gets representations of d squares provided d and n are not very large. The algorithm for this uses recursion on d (see Exercise 9.7).

```
SumOfSquaresRepresentations[3, 3380]
```

```
{{0, 4, 58}, {0, 26, 52}, {0, 38, 44}, {4, 40, 42}, {8, 20, 54},
 {10, 12, 56}, {10, 24, 52}, {20, 26, 48}, {22, 36, 40}}
```

■ How Many Ways

Let $r_d(n)$ be the number of ways of writing n as a sum of d squares. We take order and sign into account so, for example, $r_2(5) = 8$ because of the representations $\{1, 2\}$, $\{1, -2\}$, $\{-1, 2\}$, $\{-1, -2\}$, $\{2, 1\}$, $\{2, -1\}$, $\{-2, 1\}$, $\{-2, -1\}$. This is a natural way to view the problem because we are, in essence, counting the number of integer lattice points on the circle of radius $\sqrt{5}$ (see Figure 9.1).

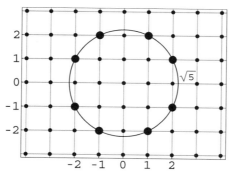

Figure 9.1. The eight representations of 5 as a sum of two squares correspond to the eight lattice points on the circle centered at (0, 0) and having radius $\sqrt{5}$.

Indeed, this geometric interpretation of the problem helps resolve a fact that might otherwise seem mysterious. On average, the value of $r_2(n)$ is π.

Here is some evidence. The r_2 function is included as `SumOfSquaresR` in *Mathematica*'s standard `NumberTheoryFunctions` package. We should be precise about what ***average*** means: It is the asymptotic average

$$\lim_{n \to \infty} \left(\frac{1}{n} \sum_{i=0}^{n} r_2(i) \right).$$

```
Needs["NumberTheory`NumberTheoryFunctions`"]
```

```
n = 5000;    Σⁿᵢ₌₀ N[SumOfSquaresR[2, i]]
             ─────────────────────────────
                          n
```

```
3.141
```

Exercise 9.11 discusses why the limiting value of this expression is π.

Now let us turn our attention to how one can quickly compute $r_2(n)$. The derivation of a useful formula breaks into several steps.

Lemma 9.4. $r_2(n) = r_2(2n)$.

Proof. We can exhibit a direct one-one and onto correspondence f between the representations of n and those of $2n$.

```
f[{a_, b_}] := {a - b, a + b}
```

It is easy to check that f is one-one; for example, this follows from the fact that the determinant of $\begin{pmatrix} 1 & -1 \\ 1 & 1 \end{pmatrix}$ is nonzero. To see that f gives representations of $2n$ just perform the necessary algebra.

```
(a - b)² + (a + b)² // Simplify
2 (a² + b²)
```

To prove that f is onto we need a function g that goes from representations of $2n$ to representations of n and is such that $f[g(a, b)] = (a, b)$. Here is such a function.

```
g[{x_, y_}] := {x + y, y - x} / 2
```

```
Apply[Plus, g[{c, d}]²] // Simplify
1
─ (c² + d²)
2
```

```
f[g[{c, d}]] // Simplify
{c, d}
```

This completes the proof of the lemma.   □

Lemma 9.4 means that, in our investigation of $r_2(n)$, we can assume that n is odd. We have already seen that primitive representations play a crucial role, so let $r_2^*(n)$ denote the number of primitive representations of n as a sum of two squares. Formulas for r_2^* and r_2 are closely related to the number-of-divisors function τ introduced in Section 3.3. Recall that $\tau(p^a q^b \cdots) = (a+1)(b+1)\cdots$. Now, we begin by getting a count of the primitive representations. This is easy because our previous work reduces this to counting the number of mod-n square roots of -1.

Lemma 9.5. Suppose that n is odd and $n = p_1^{a_1} \cdots p_t^{a_t}$. Then $r_2^*(n) = 0$ unless each p_i is congruent to 1 (mod 4), in which case $r_2^*(n) = 2^{t+2}$.

Proof. Recall from Theorem 9.3 that the number of primitive representations, where order and sign are ignored, is in one-one correspondence with the number of mod-n square roots of -1 that are less than $n/2$. If n has a prime divisor, q, that is congruent to 3 (mod 4), then because -1 has no mod-q square root, it can have no mod-n square root. This means that there are no primitive representations of n. If no such primes occur in the factorization of n, then we must count the number of square roots of -1 (mod n). Recall from Exercise 6.23 that each power of a prime congruent to 1 (mod 4) admits two square roots of -1; the Chinese Remainder Theorem is used to paste together such roots to get all the square roots modulo n. Therefore there are 2^t square roots in all, and because they come in pairs $\{x, n-x\}$, there are 2^{t-1} of them lying under $n/2$. Because n is odd, each representation yields 8 when sign and order are taken into account. Thus, the total number is $8 \cdot 2^{t-1}$, or 2^{t+2}. □

We can check our results using the routines we have implemented. The example that follows yields 8 basic solutions, which become 64 primitive solutions when sign and order are considered. This agrees with 2^{4+2}.

```
PrimitiveReps[5 13^3 17 101^2]

{{19752, 38921}, {13567, 41484}, {9729, 42548}, {28607, 32964},
 {28201, 33312}, {25047, 35744}, {7551, 42988}, {9207, 42664}}
```

Theorem 9.6. If n is a positive integer, write n as $2^g NM$ where $N = \prod_{i=1}^s p_i^{a_i}$ and $M = \prod_{j=1}^u q_j^{b_j}$, $p_i \equiv 1$ (mod 4), and $q_j \equiv 3$ (mod 4). Then $r_2(n)$ equals 0 if any of the b_j are odd, and equals $4\prod_{i=1}^s (a_i+1)$ otherwise. Alternatively, $r_2(n)$ is 0 unless M is a perfect square, in which case $r_2(n) = 4\tau(N)$.

Proof. Lemma 9.4 tells us that powers of 2 do not affect the left side. Because they leave the right side unchanged too, we can ignore them, and assume $n = MN$. Assume first that all the exponents in N are odd.

Case 1. Each a_i is odd.
Recall that every representation of n arises from the multiplication by d^2 of a primitive representation of n/d^2, where d^2 is a square divisor of n. If d^2 fails to include the entire product M, the number of primitive representations of n/d^2 is 0 by Lemma 9.5. Therefore the number of ways of getting a square divisor that will yield some primitive representations is

$$\left(\frac{a_1+1}{2}\right)\left(\frac{a_2+1}{2}\right)\left(\frac{a_s+1}{2}\right), \quad \text{or} \quad \frac{(a_1+1)(a_2+1)\cdots(a_s+1)}{2^s} \tag{2}$$

and, because each a_i is odd, each such d^2 leaves n/d^2 that is divisible by each of the primes p_i. Therefore in each case $r_2^*(n/d^2) = 4 \cdot 2^s$. But then the power of 2 cancels exactly with the 2^s in (2), and so $r_2(n)$ is $4\prod(a_i+1)$.

In order to motivate the general inductive proof that follows in case 3, consider first the simpler case where $M = 1$ and all primes in N appear to an odd power, except one, which we call r.

Case 2. $n = r^c \prod_{i=1}^s p_i^{a_i}$ where each a_i is odd, r is prime, c is even, and all the primes are congruent to 1 (mod 4).

Note that there are two type of square divisors d^2: those that involve r^c and those that involve a lesser power of r. In the latter case, n/d^2 will involve as many primes as n does, and so $r_2^*(n/d^2)$ equals $4 \cdot 2^{s+1}$. The number of such divisors is

$$\frac{(a_1+1)(a_2+1)\cdots(a_s+1)}{2^s} \cdot \frac{c}{2}$$

and so the number of representations is $4c(a_1+1)(a_2+1)\cdots(a_s+1)$. But if d^2 uses the full power of r, the number of representations is only 2^{s+2}; in this case the possibilities for d number $(a_1+1)(a_2+1)\cdots(a_s+1)/2^s$ and the total number of representations is $4(a_1+1)(a_2+1)\cdots(a_s+1)$. The total number is therefore $4(c+a)(a_1+1)(a_2+1)\cdots(a_s+1)$, as desired.

Now we complete the proof by induction on the number of primes in N whose exponent is even.

Case 3. $n = (\prod_{i=1}^s p_i^{a_i})(\prod_{k=1}^m r_k^{c_k})M$ where each a_i is odd and the primes p_i in N having even exponents are relabeled r_k, with even exponents c_k. In this case we assume that some such primes exist, so $m > 0$.

Use induction on m. The base case, $m = 0$, is covered in case 1, and the $m = 1$ case, which is illustrative of the present case, was just given. Let N_1 denote $(\prod_{i=1}^s p_k^{a_k})(\prod_{k=1}^{m-1} r_k^{c_k})$. The second equality below follows from the fact that if M fails to divide a square divisor, then r_2^* of the number that remains after the square divisor is divided out is 0. For the third inequality we gather the square divisors according to the power of r_m they contain; the case that the full power appears is special and appears at the end. In the last line the inductive hypothesis is used and the 2s appear because of the two primitive representations corresponding to the power of r; the 2 is missing from the last summand because in that case there is no power of r.

$$r_2(n) = \sum_{d^2 \mid n} r_2^*\left(\frac{n}{d^2}\right) = \sum_{d^2 \mid N} r_2^*\left(\frac{N}{d^2}\right)$$

$$= \sum_{d_1^2 \mid N_1} \left(r_2^*\left(\frac{N}{d_1^2}\right) + r_2^*\left(\frac{N}{r_m^2 d_1^2}\right) + r_2^*\left(\frac{N}{r_m^4 d_1^2}\right) + \cdots + r_2^*\left(\frac{N}{r_m^{c_m-2} d_1^2}\right) \right) + \sum_{d_1^2 \mid N_1} r_2^*\left(\frac{N}{r_m^{c_m} d_1^2}\right)$$

$$= \text{(by inductive hypothesis)} \quad \tau(N_1)2 + 4\tau(N_1)2 + \cdots + 4\tau(N_1)2 + r_2(N_1)$$

$$= 4\tau(N_1)2\frac{c_m}{2} + 4\tau(N_1) = 4\tau(N_1)(c_m + 1) = 4\tau(N). \quad \Box$$

Now it is easy to implement a routine to compute $r_2(n)$. The `Cases` phrase picks out the $\{p, a\}$ pairs where $p \equiv 1 \pmod 4$ and replaces each with $a + 1$.

```
r₂[n_] := If[MemberQ[
    fi = FactorInteger[n], {p_ /; Mod[p, 4] == 3, _?OddQ}], 0,
    4 Apply[Times, Cases[fi, {p_ /; Mod[p, 4] == 1, a_} → a + 1]]]
```

There are 804 lattice points on the circle with radius 10^{100}.

```
r₂[10²⁰⁰]
```

```
804
```

There are many results about sums of more than two squares. An encyclopedic reference is [Gro]. Here are a few.

- (Gauss) Every integer is a sum of three squares except those of the form $4^r(8k + 7)$. One direction is an easy exercise.

- It is not known exactly which integers are a sum of three nonzero squares. *Example*: It is conjectured, but not known, that each n congruent to 1 (modulo 8) is a sum of three nonzero squares, except for 1 and 25. It is known that the only squares that are not sums of three nonzero squares are 4^a and $25 \cdot 4^a$.

- (Lagrange) Every integer is a sum of four squares.

- (Descartes and Dubouis [Gro, §6.3]) Every integer is a sum of four nonzero squares except: 1, 3, 5, 9, 11, 17, 29, 41, $2 \cdot 4^r$, $6 \cdot 4^r$, $14 \cdot 4^r$.

- Every integer is a sum of distinct squares, with the following 31 exceptions: {2, 3, 6, 7, 8, 11, 12, 15, 18, 19, 22, 23, 24, 27, 28, 31, 32, 33, 43, 44, 47, 48, 60, 67, 72, 76, 92, 96, 108, 112, 128} (Exercise 9.15).

- [Gro, §2.1] Every integer is a sum of five nonzero squares except 1, 2, 3, 4, 6, 7, 9, 10, 12, 15, 18, 33.

- For each $k \geq 5$, all but a finite set of integers are sums of k nonzero squares. There are two key steps to the proof. The first is to show that 169 is a sum of 1, 2, 3, ..., 156 nonzero squares; proving this is a nice programming exercise (see Exercise 9.16). The second is to appeal to Lagrange's four-square theorem.

- For each n, let $k(n)$ be the least k such that n is not a sum of k nonzero squares. Then $k(n)$ is either 1, 2, 3, or $n - 13$. This somewhat surprising result follows from earlier results in this list (Exercise 9.19).

The proof of Lagrange's theorem is not too difficult, so we present an outline. Call n **good** if n is a sum of four squares, and say that a set A *represents* n if the squares of numbers in A sum to n.

(a) A product of good integers is good. This follows from the identity

$$(a^2 + b^2 + c^2 + d^2)(e^2 + f^2 + g^2 + h^2) = (ae + bf + cg + dh)^2 +$$
$$(af - be + ch - dg)^2 + (ag - bh - ce + df)^2 + (ah + bg - cf - de)^2$$

(b) If p is prime then there is an integer m such that $0 < m < p$ and mp is a sum of three squares. This is proved by defining $S = \{0^2, 1^2, 2^2, ..., (\frac{p-1}{2})^2\}$ and $T = \{-1 - s : s \in S\}$. There must be an entry in S that is congruent to one in T. This means that $x^2 \equiv -1 - y^2 \pmod{p}$, and this gives an appropriate m so that $mp = x^2 + y^2 + 1$.

(c) If p is prime, mp is good, and m is even, then $mp/2$ is good. Let $\{x_i\}$ be a representation for mp. We can assume that x_1 and x_2 have the same parity and x_3 and x_4 have the same parity. It follows that $\{(x_1 + x_2)/2, (x_1 - x_2)/2, (x_3 + x_4)/2, (x_3 - x_4)/2\}$ is a representation of $mp/2$.

(d) Let M be the least integer m for which mp is good. We can use the fact that M is odd to prove that $M = 1$ and so p is good. Let $\{x_i\}$ be a representation of Mp. Then let a_i be the least residue of $x_i \pmod{M}$ in absolute value $-M/2 < a_i < M/2$; let k be $(a_1^2 + a_2^2 + a_3^2 + a_4^2)/M$. If $k = 0$ one can easily get a contradiction to M's minimality. The conditions on a_i imply that $k < M$ and $\{a_i\}$ represents kM. These two representations — of kM and Mp — yield one for kM^2p in which each representing term is divisible by M. This yields a representation for kp, contradicting M's minimality.

These results imply that every prime is a sum of four squares, and then (a) yields the same for any nonnegative integer.   □

A classic problem of number theory is known as Waring's problem: Show that, for every k, there is a g such that every n is a sum of g many kth powers. If we denote the least g for a given k with $g(k)$, then Lagrange's theorem states that $g(2) = 4$. Wieferich proved that $g(3) = 9$. In 1909 Hilbert proved Waring's conjecture that $g(k)$ always exists. Getting exact values of g has proved difficult. We now know that $19 \le g(4) \le 35$, and it has been conjectured that $g(k) = \lfloor (3/2)^k \rfloor + 2^k - 2$.

Lagrange's theorem has an interesting application to logic. Consider the question of whether the natural numbers $\mathbb{N}$ can be defined in the integers $\mathbb{Z}$ using only the terms of basic logic and the two operations of arithmetic, $+$ and $\cdot$. In logical terms, this is asking whether $\mathbb{N}$ is a definable set in the structure $(\mathbb{Z}, +, \cdot, 0, 1)$ using only basic logic. The answer is YES, because of the 4-square theorem. Here is a definition:

$$n \in N \quad \text{if and only if} \quad \exists a\, \exists b\, \exists c\, \exists d\ (n = a \cdot a + b \cdot b + c \cdot c + d \cdot d).$$

It is clear that this works, because a sum of squares cannot be negative, and because of Lagrange's theorem. So, for integers, the notion of being positive is definable; indeed the definition is quite simple. The next question of interest is whether $\mathbb{Z}$ is definable in $\mathbb{Q}$, the rationals. One would like to say that a rational is an integer if and only if its denominator is 1, but there is no obvious way of getting at the denominator, because logic allows us to use only single variables for rationals. In an amazing display of mathematical reasoning, Julia Robinson, in her 1949 doctoral thesis, used some advanced number theory to prove that, indeed, it *can* be done: $\mathbb{Z}$ is definable in $(\mathbb{Q}, +, \cdot, 0, 1)$. The details of her work are presented in [FW].

There are some beautiful formulas that one can use to evaluate r_d, provided that d is not too large. For details see [Gro]. First, some notation: Given n, write it as $2^g N M$ as in the proof of Theorem 9.6. Recall that $\sigma(n)$ is the sum of the divisors of n, $\sigma_k(n)$ is the sum of the kth powers of divisors of n. In Exercises 3.20 and 3.21 the following variations were defined and implemented:

$\sigma_k^{alt}(n)$ is the sum $(-1)^d d^k$ for d dividing n;

$\sigma_k(n, 1)$ is the sum of kth powers of divisors of n that are congruent to 1 (mod 4);

$\sigma_k(n, 3)$ is the sum of kth powers of divisors of n that are congruent to 3 (mod 4).

Here are the formulas for $d = 2$, 4, 6, and 8.

$r_2(n) = 4\tau(N)$, unless M is not a perfect square, in which case it equals 0.

$r_4(n) = 8\sum_{d|n \, \& \, 4\nmid d} d$. This equals $8\sigma(NM)$ if n is odd, or $8\sigma(2NM)$ if n is even.

$r_6(n) = 16n^2[\sigma_{-2}(n, 1) - \sigma_{-2}(n, 3)] - 4[\sigma_2(n, 1) - \sigma_2(n, 3)]$

$r_8(n) = 16(-1)^n \sigma_3^{alt}(n)$

It is possible to use these formulas and recursion to compute values of r_3, r_5, r_7, and r_9, provided that n is not too large (Exercise 9.7).

■ Number Theory and Salt

We can use $\mathbb{Z}^3$, the lattice of integer points in space to visualize salt. Salt has a crystal structure made up of alternating ions of sodium (Na) and chloride (Cl). Imagine a sodium ion placed at the origin, and then chloride and sodium ions placed alternately on the entire infinite lattice (Figure 9.2). An **ion** is an electrically charged atom; sodium is positive, chloride is negative, and the amount of charge in each is the same. Like charges repel and opposite charges attract, with a force inversely proportional to the distance between them. So the basic law is like the law of gravitation, but electricity, unlike gravity, can repel as well as attract. This feature leads to some interesting mathematics, and a surprising application of the sum-of-squares results of this chapter.

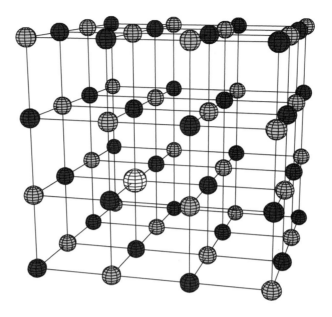

Figure 9.2. Part of an infinite salt lattice. The light spheres represents sodium ions; the dark spheres are the chloride ions. The large sphere represents the salt ion at the origin.

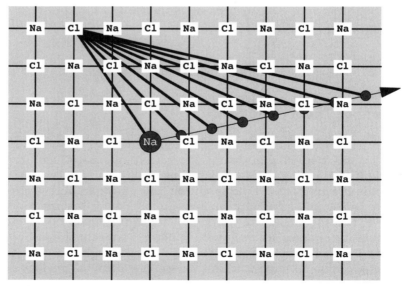

Figure 9.3. As a sodium ion moves to infinity, the attractive force exerted by a single chloride ion leads to an energy demand expressed as an integral of $1/r^2$.

An important constant in a charged lattice is the amount of energy exerted by the lattice to hold a single sodium ion in place. This is the negative of the energy that would be required to move the ion all the way to infinity. We refer here to the energy required to fight the various electrical forces exerted by the other ions: Some of the ions in the lattice will help push the moving ion to infinity, others will work to keep it from moving. The actual path taken to reach the (hypothetical) edge of the lattice turns out not to matter. This constant is called the **Madelung constant**; we will use the term for our salt lattice, though the notion applies to any lattice and leads to different constants for different structures.

To see the subtlety in the mathematics surrounding this constant, it is good to start with a two-dimensional version (Figure 9.3).

As the sodium ion goes to infinity along some line, the amount of energy that must be expended because of the attraction of the single chloride ion at $(-2, 3)$ is given by the following integral, where the 13 arises as $\sqrt{2^2 + 3^2}$. This is just the standard use of integrals in a work $=$ force$\cdot$distance situation.

$$\text{energy} = \int_{\sqrt{13}}^{\infty} \frac{1}{r^2} \, dr$$

This integral is easy to evaluate, even in the general case where the initial distance is r_0 and the force is either attracting or repelling.

$$\int_{r_0}^{\infty} \frac{\pm 1}{r^2} \, dr$$

$$\frac{\pm 1}{r_0}$$

So this means that the total energy exerted by the lattice to hold the sodium ion in place is given by the infinite sum M_2. We use M_2 for the two-dimensional version and M_3 for the true Madelung constant for salt.

$$M_2 = \sum_{\substack{x, y = -\infty \\ xy \neq 0}}^{\infty} \frac{(-1)^{x+y}}{\sqrt{x^2 + y^2}} \qquad M_3 = \sum_{\substack{x, y, z = -\infty \\ xyz \neq 0}}^{\infty} \frac{(-1)^{x+y+z}}{\sqrt{x^2 + y^2 + z^2}}$$

The infinite series in both cases have the property, unusual for series that arise in physical applications, of being conditionally convergent. This means that the series obtained by looking at the absolute values of the terms are divergent. Looking at the values along the x-axis is enough to get divergence of the absolute values. We pause for a second to remind the reader of some basic facts about this notion.

A result of nineteenth-century mathematics that still surprises students of calculus is the fact that the sum of a series can change when the terms are rearranged. This is discussed in any calculus book, so we will just present an example here. The alternating harmonic series converges to log 2. Dividing each term in half reduces the sum to $1/2$ log 2. Spreading out this series and adding it to the first series gives a series that sums to $^3/_2 \log 2$, but is just a rearrangement of the original series.

$$
\begin{array}{l}
1-\tfrac{1}{2}+\tfrac{1}{3}-\tfrac{1}{4}+\tfrac{1}{5}-\tfrac{1}{6}+\tfrac{1}{7}-\tfrac{1}{8}+\tfrac{1}{9}-\tfrac{1}{10}+\tfrac{1}{11}+\cdots \;=\; \log 2 \\[4pt]
+\tfrac{1}{2}\phantom{+\tfrac{1}{3}}-\tfrac{1}{4}\phantom{+\tfrac{1}{5}}+\tfrac{1}{6}\phantom{+\tfrac{1}{7}}-\tfrac{1}{8}\phantom{+\tfrac{1}{9}}+\tfrac{1}{10}-\cdots \;=\; \tfrac{1}{2}\log 2 \\[2pt]
\hline
1\phantom{-\tfrac{1}{2}}+\tfrac{1}{3}-\tfrac{1}{2}+\tfrac{1}{5}+\tfrac{1}{7}-\tfrac{1}{4}+\tfrac{1}{9}+\tfrac{1}{11}+\cdots \;=\; \tfrac{3}{2}\log 2
\end{array}
$$

If a series is such that the absolute values of the terms converge, then the series is called ***absolutely convergent***, and any rearrangement gives the same sum. But if the absolute values fail to converge — as happens for the alternating harmonic series — then the series is called ***conditionally convergent***. A somewhat surprising result is that, for any target T in $[-\infty, \infty]$, the terms of a conditionally convergent series can be rearranged so that the new series sums to T (see [PW]). Indeed, one can program the algorithm to do this explicitly in the case of the alternating harmonic series. The algorithm is sometimes called the "method of repeated U-turns": To get a sum of T, add up positive terms until T is passed; then start with the first unused negative term to go in the other direction, again until T is passed; then add in unused positive terms again until T is passed; and so on ad infinitum. Here is a rearrangement that converges to 2. For more on the general relationship between the target (2 in the next example) and the numbers of positive and negative terms (8, 1, 13, 1, 14, 1, 13, 1, ... in the example) see [PW].

`HarmonicRearrangement[2]`

> The rearranged series for 2 is: $1 + \tfrac{1}{3} + \tfrac{1}{5} + \tfrac{1}{7} + \tfrac{1}{9} + \tfrac{1}{11} + \tfrac{1}{13} +$
>
> $\tfrac{1}{15} - \tfrac{1}{2} + \tfrac{1}{17} + \tfrac{1}{19} + \tfrac{1}{21} + \tfrac{1}{23} + \tfrac{1}{25} + \tfrac{1}{27} + \tfrac{1}{29} + \tfrac{1}{31} + \tfrac{1}{33} + \tfrac{1}{35} +$
>
> $\tfrac{1}{37} + \tfrac{1}{39} + \tfrac{1}{41} - \tfrac{1}{4} + \tfrac{1}{43} + \tfrac{1}{45} + \tfrac{1}{47} + \tfrac{1}{49} + \tfrac{1}{51} + \tfrac{1}{53} + \tfrac{1}{55} +$
>
> $\tfrac{1}{57} + \tfrac{1}{59} + \tfrac{1}{61} + \tfrac{1}{63} + \tfrac{1}{65} + \tfrac{1}{67} + \tfrac{1}{69} - \tfrac{1}{6} + \tfrac{1}{71} + \tfrac{1}{73} + \tfrac{1}{75} +$
>
> $\tfrac{1}{77} + \tfrac{1}{79} + \tfrac{1}{81} + \tfrac{1}{83} + \tfrac{1}{85} + \tfrac{1}{87} + \tfrac{1}{89} + \tfrac{1}{91} + \tfrac{1}{93} + \tfrac{1}{95} - \tfrac{1}{8} +$
>
> $\tfrac{1}{97} + \tfrac{1}{99} + \tfrac{1}{101} + \tfrac{1}{103} + \tfrac{1}{105} + \tfrac{1}{107} + \tfrac{1}{109} + \tfrac{1}{111} + \tfrac{1}{113} + \tfrac{1}{115} +$
>
> $\tfrac{1}{117} + \tfrac{1}{119} + \tfrac{1}{121} + \tfrac{1}{123} - \tfrac{1}{10} + \tfrac{1}{125} + \tfrac{1}{127} + \tfrac{1}{129} + \tfrac{1}{131} + \tfrac{1}{133} +$
>
> $\tfrac{1}{135} + \tfrac{1}{137} + \tfrac{1}{139} + \tfrac{1}{141} + \tfrac{1}{143} + \tfrac{1}{145} + \tfrac{1}{147} + \tfrac{1}{149} + \tfrac{1}{151} - \tfrac{1}{12} \cdots$
>
> The partial sums just before the sign changes:
>
> $\quad$ {2.0218, 1.5218, 2.00406, 1.75406, 2.00945, 1.84278,
>
> $\qquad$ 2.0007, 1.8757, 2.00366, 1.90366, 2.00546, 1.92213, ...}
>
> Numbers of terms that are positive, negative,
>
> $\quad$ positive,...: {8, 1, 13, 1, 14, 1, 13, 1, 14, 1, 14, 1, ...}

Here is what happens, in abbreviated form, for a rearrangement converging to π.

HarmonicRearrangement[π, ShowSeries → False,
  ShowPartialSums → False, MaxTerms → 1000]

> Numbers of terms that are positive, negative, positive, ...:
> {76, 1, 129, 1, 132, 1, 134, 1, 133, 1, 134, 1, 134, 1, 133, 1, ...}

The mathematics of the Madelung constant is of interest because it is a rare example of a conditionally convergent series occurring in nature.

Now, because the series for M_2 and M_3 are conditionally convergent, what happens will be sensitive to the order in which the terms are added. We will focus here on two possibilities, using partial sums that correspond to expanding squares or cubes, or partial sums that correspond to expanding disks or balls. First, squares:

$$M_2^{\text{square}} = \lim_{n \to \infty} \sum_{\substack{x,y=-n \\ (x,y) \neq (0,0)}}^{n} \frac{(-1)^{x+y}}{\sqrt{x^2+y^2}} \;,\; M_3^{\text{square}} = \lim_{n \to \infty} \sum_{\substack{x,y,z=-n \\ (x,y,z) \neq (0,0,0)}}^{n} \frac{(-1)^{x+y+z}}{\sqrt{x^2+y^2+z^2}}$$

One could program these using Sum, but it makes more sense to exploit the symmetry. The M2SquareShell function that follows gets the sum along the border of the square. Then we can add them up and watch the convergence of the partial sums. The computation that follows goes up to a 401×401 square, so 160,800 lattice points are counted.

M2SquareShell[k_] := 4 $\sum_{y=-k}^{k-1} \frac{(-1.)^{k+y}}{\sqrt{k^2+y^2}}$;

termssq = Map[M2SquareShell, Range[200]];
ListPlot[FoldList[Plus, 0, termssq]];

The output (see Figure 9.4) shows that the partial sums appear to converge to a value near -1.6. Expanding disks will lead to the same sum, though the convergence is much slower. For this case one can make a major simplification, and it is here that the sum-of-squares theory comes into play. The disks will pick up their points circle by circle and on any circle there will be only chloride ions, or only sodium ions. An ion at (a, b) lies on the circle of radius $\sqrt{n}$, where $n = a^2 + b^2$, so there will be exactly $r_2(n)$ ions on any such circle. Thus, each term $(-1)^{x+y}/\sqrt{x^2+y^2}$ may be replaced by $(-1)^n r_2(n)/\sqrt{n}$ where we count all the ions at the same distance from the origin at the same time (note that $x+y$ and x^2+y^2 have the same parity). The computation that follows (see Figure 9.4) examines the disk of radius $\sqrt{10,000}$, so about 31,416 points are counted.

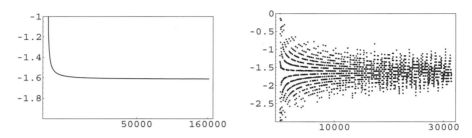

$$\texttt{M2Circle[k\_]} := \frac{(-1.)^k\ \texttt{SumOfSquaresR[2, k]}}{\sqrt{k}}$$

```
terms = Map[M2Circle, Range[10000]];
ListPlot[FoldList[Plus, 0, terms]];
```

Figure 9.4. Expanding squares (left) yield good convergence to the two-dimensional Madelung constant. Expanding disks yield convergence to the same value, but it is very slow. The horizontal labels correspond to the number of lattice points counted.

In fact, it can be shown that M_2, the two-dimensional Madelung constant, equals

$$-4\left(\sum_{n=1}^{\infty} \frac{(-1)^{n+1}}{\sqrt{n}}\right)\left(\sum_{n=0}^{\infty} \frac{(-1)^n}{\sqrt{2n+1}}\right).$$

This in turn can be calculated in terms of zeta functions:

$$\texttt{-2. }\left(1 - \sqrt{2}\right)\texttt{ zeta}\left[\tfrac{1}{2}\right]\left(\texttt{zeta}\left[\tfrac{1}{2}, \tfrac{1}{4}\right] - \texttt{zeta}\left[\tfrac{1}{2}, \tfrac{3}{4}\right]\right)$$

$$-1.61554$$

The case of real three-dimensional salt is more subtle. If we compute approximations as for M_2 we will see that the expanding cubes converge to a constant near -1.74, but the expanding spheres yield highly oscillatory behavior, and indeed one can prove that the spherical partial sums diverge. First we look at the cubic scenario. The following routine uses symmetry and computes terms on the eight vertices, six faces, and twelve edges.

$$\texttt{M3CubeShell[k\_]} :=$$

$$\frac{8\,(-1)^{3k}}{\sqrt{3}\,k^2} + 6 \sum_{y=-k+1}^{k-1}\ \sum_{z=-k+1}^{k-1} \frac{(-1.)^{k+y+z}}{\sqrt{k^2+y^2+z^2}} + 12 \sum_{y=-k+1}^{k-1} \frac{(-1.)^{2k+y}}{\sqrt{2\,k^2+y^2}}$$

Now we build up the sum for a cube of side-length 61 as we did in the two-dimensional case; this counts about 227,000 ions (see Figure 9.5).

```
terms = Map[M3CubeShell, Range[30]];
partialSums = FoldList[Plus, 0, terms];
ListPlot[partialSums, PlotJoined → True];
```

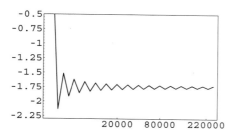
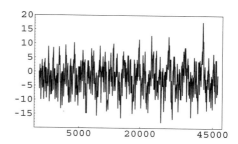

Figure 9.5. Expanding cubes (left) yield convergent to the three-dimensional Madelung constant. Expanding spheres yield oscillatory partial sums that diverge to infinity. The horizontal labels correspond to the number of lattice points counted.

And now we try a sphere of radius $\sqrt{500}$. Its area is $^4/_3\,\pi\,500^{3/2}$, so about 47,000 points are counted. The results, shown in Figure 9.5, look somewhat inconclusive, but it can be proved that the partial sums oscillate to greater and greater heights, and so fail to converge.

$$\texttt{M3Sphere[k\_] :=} \frac{\texttt{(-1.)}^{\texttt{k}}\ \texttt{SumOfSquaresR[3, k]}}{\sqrt{\texttt{k}}}$$

```
terms = Map[M3Sphere, Range[500]];
ListPlot[FoldList[Plus, 0, terms], PlotJoined → True];
```

From a physical perspective the interesting question is: If the sum is dependent on the order of the terms, what order does Nature use to add up the energies? One could use complex numbers and analytic continuation to deduce what the "right" interpretation of the sum is, but does Nature really know about analytic continuation? A more down-to-earth explanation was given recently by physicist Jean Delord (Reed College). He suggested that, because the entire salt crystal is electrically neutral, the partial sums used to approximate the full infinite sum should be corrected to give a neutral charge at each stage. Thus, he suggests looking at the spherical order of terms, but at each stage adding back a charge to the surface of the sphere to render the whole finite ball electrically neutral. Because the total spherical charge, assuming radius $\sqrt{n}$, is the sum of $(-1)^{x+y+z}$ over all points inside or on the sphere of radius $\sqrt{n}$, this requires adding a charge of $\sum_{k \le n}(-1)^k r_3(k)$ to the sphere's surface. This in turn leads to an energy-correction term of $-(1/\sqrt{n})\sum_{k \le n}(-1)^k r_3(k)$. It turns out that the limit of the numbers obtained by correcting in this way converge to the true Madelung constant [BC]; see Exercise 9.22.

There are faster ways to compute sums such as M_3; one such formula is given in Exercise 9.23. The proper mathematical way to define the Madelung constants is to use complex numbers and analytic functions. The

reader who wishes to pursue that should look at [BB]. However, it is noteworthy that the sum-of-squares functions can play a role in Madelung
computations.

Exercises for Section 9.1

9.1. Prove that the following patterns, evident in the table in the middle of
the proof of Theorem 9.2, always hold for the extended Euclidean algorithm
sequence for (p, x) where x is the smaller of the two square roots of -1
modulo p. Make liberal use of the results of Exercise 1.18. Let n denote the
number of quotients, so that $r_n = 1$.

(a) $t_n = -x$. (*Hint*: Show first that t_n must be either x or $p - x$, and then
show that the latter would lead to the contradiction of the t-sequence being
longer than the r-sequence.)

(b) $t_{n+1} = p$.

(c) The t-sequence, in absolute value, is the reverse of the remainder
sequence.

(d) n is even.

(e) The quotient sequence is symmetric about its center.

(f) If m is the index of the first remainder under $\sqrt{p}$, then $m = n/2$.

(g) $|t_m| = r_{m+1}$; therefore the first two remainders under $\sqrt{p}$ are the a and
b of $p = a^2 + b^2$.

9.2. Check that the result in the text for the set of primitive representations of 6565 is correct by doing an exhaustive search of all possibilities.
And perform a similar verification for the complete set of representations of
112,132 as a sum of two squares.

9.3. Find an integer n that has more than 10,000 representations as a sum
of two squares.

9.4. Use the proof of Theorem 9.3 to deduce that if p is prime and congruent to 1 (mod 4) then there is only one way (up to sign and order) to write p
as a sum of two squares.

9.5. Here is an incredibly short proof that every prime p congruent to 1
(mod 4) is a sum of two squares (discovered by D. Zagier [Zag1]).

(a) A function $f : S \to S$ is called an ***involution*** if $f(f(s)) = s$ for each x in S.
Show that if f is an involution on S, then the number of elements in S is
odd if and only if the number of fixed points of f is odd. (A fixed point is a
point y such that $f(y) = y$.)

(b) Suppose that p is prime and S is defined by:

$$S = \{(x, y, z) \in \mathbb{N}^3 : x^2 + 4yz = p\}.$$

Show that f is an involution on S where f is defined by

$$f(x, y, z) = \begin{cases} (x + 2z, z, y - x - z) & \text{if } x < y - z \\ (2y - x, y, x - y + z) & \text{if } y - z < x < 2y \\ (x - 2y, x - y + z, y) & \text{if } x > 2y \end{cases}$$

(c) Show that f has exactly one fixed point; conclude that $|S|$ is odd.

(d) Show that the function $(x, y, z) \mapsto (x, z, y)$ maps S to S. Conclude that this function has a fixed point and show how that yields a representation of p as a sum of two squares.

9.6. Implement procedures to compute r_4, r_6, and r_8 using the formulas given on page 291.

9.7. Implement a procedure to find all the representations of n as a sum of d squares by reducing the problem in a recursive way to the $d = 2$ case, which was handled in this section. For example, for $d = 3$ just subtract all possible squares s^2 from n and compute, in each case, the complete set of representations of $n - s^2$ as a sum of two squares. Such an approach will not work when d or n is very large, but it works reasonably well for d up to 4 and n up to 2000. In the case of $d = 3$ you should check your program for correctness against a brute-force approach, for some values of n.

9.8. Let $\tau_i(n)$ be the number of divisors of n that are congruent to i (mod 4). Show that $r_2(n)$ equals $4[\tau_1(n) - \tau_3(n)]$.

9.9. (I. Vardi) Explain why the concise bit of code that follows correctly computes $r_2(n)$. Im[z] gives the imaginary part of the complex number z.

```
4 Apply[Plus, Im[I^Divisors[n]]]
```

Perform some timing tests to compare this method to the implementation of r_2 given in this section.

9.10. Without appealing to the formulas given in the text, prove the following. (*Hint*: Discover the patterns by looking at the set of representations in each case.)

(a) $r_2(2^n) = 4$

(b) $r_3(2^n) = \begin{cases} 6 & \text{if } n \text{ is even} \\ 12 & \text{if } n \text{ is odd} \end{cases}$

(c) $r_4(2^n) = \begin{cases} 8 & \text{if } n = 0 \\ 24 & \text{if } n > 0 \end{cases}$

9.11. Prove that the long-term average of $r_2(n)$ approaches π; precisely: $\lim_{N \to \infty} \frac{1}{N} \sum_{n=0}^{N} r_2(n) = \pi$. Here is an outline. Let $D(a)$ denote the closed disk of radius a centered at $(0, 0)$. We will examine $D(\sqrt{N})$, the disk given by $x^2 + y^2 \le N$. A *lattice point* is a point of the form (a, b) where a and b are integers.

(a) Prove that $\sum_{n=0}^{N} r_2(n)$ equals the number of lattice points inside $D(\sqrt{N})$.

(b) Take the square of side 1 centered at each lattice point inside $D(\sqrt{N})$ and color them all blue. Then the blue area is $\sum_{n=0}^{N} r_2(n)$.

(c) Step (b) shows that $\sum_{n=0}^{N} r_2(n)$ is near the area of the disk, πN. But the border must be looked at closely. Use geometrical reasoning to show that

$$D\left[\pi\left(\sqrt{N} - \frac{1}{\sqrt{2}}\right)\right] \subseteq \text{Total blue region} \subseteq D\left[\pi\left(\sqrt{N} + \frac{1}{\sqrt{2}}\right)\right]$$

and use this to conclude that $\frac{1}{N} \sum_{n=0}^{N} r_2(n)$ is trapped between two quantities that converge to π.

9.12. Show that the formula for $r_4(n)$ given in the text yields the following formula, which is easy to implement. Here N and M are as defined from n as explained before the formula for r_4.

$$r_4(n) = \begin{cases} 24\,\sigma(NM) & \text{if } n \text{ is even} \\ 8\,\sigma(NM) & \text{if } n \text{ is odd.} \end{cases}$$

9.13. Fill in the details in the proof outline of Lagrange's theorem that every nonnegative integer is a sum of four squares.

9.14. Fill in the proofs and implement the following method, due to Rabin and Shallit, for finding a single representation of n as a sum of four squares. Given $n \equiv 2 \pmod 4$, subtract a pair of randomly chosen squares from n and check if the result is a prime. If it is, it must be 1 (mod 4) and we can represent it as a sum of two squares, using `SumOfSquaresRepresentations`. This gives four squares that sum to n. If the result of the subtraction is not prime, try subtracting a different pair of squares. If $n \equiv 0 \pmod 4$, use recursion on $n/4$, and multiply each entry in the answer by 2. If n is odd, use recursion on $2n$; the result must have two even entries and two odd entries. Let the even ones be r and s and let a and b be the others, with $s > r$ and $b > a$. Then $(s+r)/2$, $(s-r)/2$, $(b+a)/2$, and $(b-a)/2$ solve the problem for n.

9.15. Show that if $n > 128$ then n is a sum of distinct nonzero squares. Here is an outline.

(a) Write a routine `SumNonZeroSquares[n]` that uses recursion to get the set of integers that are a sum of one or more distinct squares of integers in $\{1, 2, ..., n\}$. For example, if $n = 2$ the resulting list should be $\{1, 4, 5\}$.

(b) Prove the auxiliary result: Suppose that n is positive and every integer from $n + 1$ to $4n + 35$ (inclusive) is a sum of distinct positive squares. Then every integer $m > n$ is a sum of nonzero squares. *Hint*: Consider the four cases $m = 4k$, $4k + 1$, $4k + 2$, $4k + 3$.

(c) Write a short program that uses (b) and the code from (a) to get a list of all integers that are not a sum of nonzero squares. The list has length 31 and its largest entry is 128.

9.16. It is not hard to see that 169, or 13^2, is a sum of 1, 2, 3, 4, or 5 squares: $169 = 13^2 = 25 + 144 = 9 + 16 + 144 = 1 + 4 + 64 + 100 = 1 + 4 + 4 + 16 + 144$. Find the first integer k for which 169 is not the sum of k nonzero squares.

9.17. Find the smallest integer that is not a sum of 8 cubes. *Hint*: The number is quite small, so you need consider only small cubes.

9.18. Prove that 79 is not a sum of 18 fourth powers.

9.19. Let $k(n)$ be the least k such that n is not a sum of k nonzero squares. Prove that $k(n)$ is always one of 1, 2, 3, or $n - 13$. *Hint*: You will need the Descartes–Dubouis result cited in the text. Use it to show that if n is a sum of four squares then it is also a sum of j squares for each $j = 5, 6, ..., n - 14$. This leaves the fairly easy assertion that no n is a sum of $n - 13$ nonzero squares.

9.20. (a) Implement the following magic trick, which may or may not impress your friends. Ask a friend to take his or her date of birth and write it as two four-digit numbers. For example, July 9, 1951, becomes $\{0709, 1951\}$. Then ask him or her to square each of the numbers and tell you the sum of the two squares. With the help of `SumOfSquaresRepresentations` you can respond very quickly with the date of birth. Many of the integers generated in this way will have more than one representation as a sum of two squares, but, with rare exceptions, the fact that the answer has so much structure (month between 01 and 12, day between 01 and 31, year between 1900 and 2000) will allow your program to select the correct values. Implement a `BirthDate` routine that does this. Ideally, it will work as follows.

$709^2 + 1951^2$

4309082

<u>BirthDate</u>[4309082]

July 9, 1951

(b) Find some ambiguous birth dates: they should yield the same sum of squares and the years should be within ten years of each other. For example, March 18, 1945, and March 30, 1943, both yield 3,884,149.

9.21. The following variation of the Madelung sum is interesting:

$$\sum_{n=1}^{\infty} \frac{(-1)^n r_2(n)}{n}.$$

Compute some partial sums and try to discover the surprising formula for the infinite sum. *Hint*: The average value of r_2 is π (Exercise 9.11) and the sum of the alternating harmonic series is log 2.

9.22. The Delord interpretation considers the following limit.

$$\lim_{n\to\infty}\left(\sum_{k=1}^{n} \frac{(-1)^k r_3(k)}{\sqrt{k}}\right) + \frac{1}{\sqrt{n}}\sum_{k=1}^{n}(-1)^k r_3(k)$$

Do some computations to gain evidence for the theorem that this limit equals the Madelung constant M_3.

9.23. The following formula due to Crandall and Buhler [CB] seems to be the fastest way to compute the Madelung constant.

$$-2\pi + 2^{7/4}\left[\sum_{k=1}^{\infty} e^{-\pi(k-\frac{1}{2})^2/\sqrt{2}}\right]^2 + 2\sum_{k=1}^{\infty} \frac{(-1)^k r_3(k)}{\sqrt{k}\left(1+e^{4\pi\sqrt{k}}\right)}$$

Compute some approximations and note how few terms in each sum are needed to get accurate results. The true value of M_3 is

$-1.74756459463318219063621203554439740348516143662474175811\cdots$.
5282535076504062353276117989075836269460789

9.2 The Gaussian Integers

■ Complex Number Theory

It makes sense, and it turns out to be quite a useful thing, to define integers and divisibility in the realm of complex numbers. The ***complex numbers***, $\mathbb{C}$, are numbers of the form $a+bi$, where i denotes $\sqrt{-1}$ and a and b

are real. Of course, i suffers from being called an imaginary number, but it can be treated as a concrete object, the important point being that $i^2 = -1$. Each complex number has a real part and an imaginary part, given by a and b in $a + bi$. We can give the complex numbers a geometric interpretation by letting the vertical axis be the imaginary axis. In *Mathematica* I is used for i and Re[ ] and Im[ ] are used for real and imaginary parts, respectively.

We can perform arithmetic in $\mathbb{C}$. Indeed, $\mathbb{C}$ is a ***field***, meaning that we can add, subtract, multiply, or divide any pairs of numbers, except for the case of division by 0. Addition is easy: $(a + bi) + (c + di)$ is $(a + c) + (b + d)i$. Subtraction is similar. Multiplication comes from the identity: $(a + bi) \cdot (c + di) = ac - bd + (ad + bc)i$.

The ***conjugate*** of $z = a + bi \in \mathbb{C}$ comes by replacing the imaginary part of z by its negative and is denoted by $\bar{z}$. So $\overline{a + bi} = a - bi$ (we assume a and b are real). The ***absolute value*** of z, $|z|$, is $\sqrt{a^2 + b^2}$, which is the length of the vector from 0 to z. It is also useful to define the ***norm*** of z, $N(z)$, to be just $a^2 + b^2$.

Now, the integers in $\mathbb{C}$ are called the ***Gaussian integers*** and consist of complex numbers of the form $m + ni$ where m and n are integers; the set is denoted $\mathbb{Z}[i]$. The ***units*** refer to the invertible elements; there are four of them: ± 1 and $\pm i$. If $\alpha = \varepsilon \beta$ where α and β are Gaussian integers and ε is a unit, then we say that α and β are ***associates***. A Gaussian integer α ***divides*** another, β, if there is a Gaussian integer γ such that $\beta = \alpha \gamma$. This definition will allow us to study primes, gcds, and the like in the imaginary realm. Note, for example, that 2 loses its primality because $2 = (1 + i)(1 - i)$. Before investigating the Gaussian primes, we catalog some important properties of $\mathbb{Z}[i]$. The proofs of the nine assertions that follow are straightforward and left as exercises. To avoid confusion, and because the integers are related to the field of rational numbers in the same way that the Gaussian integers are related to the field $\mathbb{C}$, the ordinary integers are called the ***rational integers*** and the ordinary primes are called ***rational primes***.

Proposition 9.7. The following hold for $\alpha, \beta \in \mathbb{Z}[i]$ and $a, b \in \mathbb{Z}$.

(a) $\alpha \bar{\alpha} = N(\alpha)$.

(b) $N(\alpha) = N(\bar{\alpha})$.

(c) $N(\alpha \beta) = N(\alpha) N(\beta)$.

(d) If β divides α in $\mathbb{Z}[i]$, then $N(\beta)$ divides $N(\alpha)$ in $\mathbb{Z}$.

(e) The units are the only elements of $\mathbb{Z}[i]$ that are invertible in $\mathbb{Z}[i]$.

(f) If α divides β and β divides α, then α and β are associates.

(g) If a and b are rational integers and a divides b in $\mathbb{Z}[i]$, then a divides b in $\mathbb{Z}$.

(h) If β divides α, then $\bar{\beta}$ divides $\bar{\alpha}$.

(i) If β divides α and β is neither a unit nor an associate of α, then $1 < N(\beta) < N(\alpha)$.

We can now discuss the fundamental facts about greatest common divisors and unique factorization in $\mathbb{Z}[i]$. First note that gcd is a little tricky to define because there is no notion of "greatest" for complex numbers. Thus, we use the following: γ is the **greatest common divisor** of α and β if γ divides each of α and β and any other common divisor divides γ. And we can define primes in the obvious way: α is a **Gaussian prime** if any divisor of α in $\mathbb{Z}[i]$ is either a unit or an associate of α.

Theorem 9.8. (a) If α and β are in $\mathbb{Z}[i]$ and $\beta \neq 0$, then there exist Gaussian integers q and r such that $\alpha = q\beta + r$ and $N(r) < N(\beta)$.

(b) Every pair of Gaussian integers not both 0 has a greatest common divisor, which is unique up to associates.

(c) If $d = \gcd(\alpha, \beta)$ then there are Gaussian integers s and t such that $d = s\alpha + t\beta$.

(d) Every nonzero Gaussian integer may be written in a unique way (up to order and associates) as a product of Gaussian primes.

Proof. **(a)** Suppose that the true quotient α/β in $\mathbb{C}$ equals $u + iv$. Let m be the nearest integer to u and let n be the nearest integer to v (it will not matter whether rounding is up or down). Then $u + iv = (m + s) + i(n + t)$ where $|s|$ and $|t|$ are at most $1/2$. Take q to be $m + ni$, and let $r = \alpha - q\beta$. Then

$$N(r) = N((u + iv - q)\beta)$$
$$= N(u + iv - q)N(\beta) = N(\beta)N(s + it) \leq N(\beta)\left(\left(\frac{1}{2}\right)^2 + \left(\frac{1}{2}\right)^2\right) < N(\beta)$$

(b) This follows from (a) as for the ordinary integers. Namely, one uses a division algorithm to form a sequence of remainders whose norms are a decreasing sequence of nonnegative integers. The sequence must therefore terminate with a remainder having norm 0, that is, a zero remainder. We leave the details of proving that the final nonzero remainder is a gcd of α and β under the $\mathbb{Z}[i]$-definition as an exercise. Uniqueness follows from Proposition 9.7(f) and the fact that two gcds would divide each other.

(c) See Exercise 9.29.

(d) This is similar in all respects to the proof of the same result in $\mathbb{Z}$, in that one uses induction for existence and the extended gcd ideas to get an analog of Corollary 1.10, which is then used in the proof of uniqueness. In fact, unique factorization for $\mathbb{Z}$ and $\mathbb{Z}[i]$ are special cases of a more general result of modern algebra: Every Euclidean domain is a unique factorization domain. This theorem states that any structure of an appropriate sort (called an **integral domain**) for which the notions of gcd and the central ideas of the Euclidean algorithm hold satisfies a unique factorization theorem. □

Mathematica's PrimeQ, FactorInteger, Divisors, and GCD functions all work in $\mathbb{Z}[i]$ when the GaussianIntegers option is turned on. We present a few examples here, but our understanding of these concepts will be much improved if we implement algorithms for these three operations ourselves, as we will do shortly.

```
Table[p = Prime[i];
  {p, PrimeQ[p, GaussianIntegers → True]}, {i, 5}]
```

{{2, False}, {3, True}, {5, False}, {7, True}, {11, True}}

```
FactorInteger[120] // FactorForm
```

2^3 3 5

```
FactorInteger[120, GaussianIntegers → True] // FactorForm
```

$(1 + I)^6$ (1 + 2 I) (2 + I) 3

Complex gcds are built into GCD, but it is not hard to write a one-line program that does it.

Algorithm 9.2 GCDs in the Gaussian Integers

The code that follows obtains the gcd of two Gaussian integers by a method that is almost identical to the standard Euclidean algorithm. The only difference is that Round[z / u] is used where the classic algorithm uses Floor[z / u]; this works because Round acts on real and imaginary parts separately. The reason for rounding rather than flooring is the proof of Theorem 9.8(a).

```
ComplexGCD[z_, u_] :=
  If[u == 0, z, ComplexGCD[u, z - u Round[z / u]]]
```

```
ComplexGCD[7300 + 12 I, 2700 + 100 I]
```

-4 + 4 I

```
ComplexGCD[324 + 1608 I, -11800 + 7900 I]
```

-52 + 16 I

```
ComplexGCD[Product[a + b I, {a, 1, 8}, {b, 1, 7}],
  Product[a + b I, {a, 12, 16}, {b, 12, 15}]]
```

270015132441600000 + 4044308486400000 I

The CNT package's FullGCD function works on Gaussian integers too. The table that follows shows the extended gcd coefficients *s* and *t*, which can be found in the same way as in $\mathbb{Z}$ (Exercise 9.29).

FullGCD[7300 + 12 I, 2700 + I 100, **ExtendedGCDValues** → **True**]

Remainders	Quotients	s	t
7300 + 12 I		1	0
2700 + 100 I	3	0	1
−800 − 288 I	−3 + I	1	−3
12 + 36 I	−14 + 18 I	3 − I	−8 + 3 I
16	1 + 2 I	25 − 68 I	−61 + 186 I
−4 + 4 I	−2 − 2 I	−158 + 17 I	425 − 61 I
0		−325 − 350 I	911 + 914 I

▣ Gaussian Primes

As mentioned, some rational primes, such as 2 and 5, fail to be prime in $\mathbb{Z}[i]$: 2 factors into $(1+i)(1-i)$; 5 factors as $(1+2i)(1-2i)$. An examination of these and similar factorizations will show that there is a close connection between primality in the Gaussian integers and representations as sums of two squares. Before giving a complete characterization of primes, we show how the connection can be exploited to get another algorithm for writing a prime p congruent to 1 (modulo 4) as a sum of two squares.

By earlier work (either Theorem 9.1 or Exercise 9.5), we know that p can be written as $a^2 + b^2$, and we also know there is an x such that $x^2 \equiv -1 \pmod{p}$. Observe first that $a^2 + b^2 = (a + bi)(a - bi)$ and that each of the factors is a Gaussian prime. For if α divided $a + bi$ and is neither a unit nor an associate of $a + bi$, then, by Proposition 9.7(i), $N(\alpha)$ would be a divisor of $N(a + bi)$ lying strictly between 1 and $N(a + bi)$. But this norm is just p, a prime having no such divisors in $\mathbb{Z}$. Now, because p divides $x^2 + 1$ in $\mathbb{Z}$, $(a + bi)(a - bi)$ divides $(x + i)(x - i)$ in $\mathbb{Z}[i]$. Because each of $a \pm bi$ is prime, each divides one of $x + i$ or $x - i$. But they cannot divide the same factor for that would mean that p divides $x \pm i$, which it does not because it does not divide the imaginary part. It follows that $\gcd(p, x + i)$ is one of $a \pm bi$ (or an associate) and so will yield a solution to the sum-of-two-squares problem.

The ideas of the previous paragraph can also be used to prove that primes congruent to 1 (mod 4) have the form $a^2 + b^2$, and to prove that such representations are unique. The latter is done in Exercise 9.31. Here's an example of the complex algorithm; it is quite fast, but a little bit slower than Smith's method.

```
p = RandomPrime[50, {1, 4}]
x = SqrtNegOne[p]
```

914801667978971291167324586369012689663334430126797
1338332767922066940023029505624457809900005176341778

```
repn = GCD[p, x + I]
```

8616615032376256176870269 + 41513988222916692967037941 I

```
Re[repn]² + Im[repn]² == p
```
```
True
```

We now give a complete characterization of the Gaussian primes.

Theorem 9.9. A Gaussian integer α is prime in $\mathbb{Z}[i]$ if and only if either

(a) α is a rational prime congruent to 3 (mod 4), or

(b) $N(\alpha)$ is a prime integer.

Proof. Part (b) of the reverse direction is easy, for if α is not a prime in $\mathbb{Z}[i]$, then, by Proposition 9.7(i), some β divides α, where $1 < N(\beta) < N(\alpha)$. Then Proposition 9.7(d) implies that $N(\beta)$ divides $N(\alpha)$, contradicting the primality of $N(\alpha)$. For part (a), suppose that p is a rational prime, $p \equiv 3 \pmod 4$, and p fails to be a Gaussian prime. Then $p = \alpha\beta$ where α and β are not units. This implies that $p^2 = N(\alpha\beta) = N(\alpha)N(\beta)$, and this in turn implies that $p = N(\alpha)$, and is therefore a sum of two squares, contradicting the fact that no sum of two squares can be 3 (modulo 4).

For the forward direction, suppose that α is a Gaussian prime. If $\alpha \in \mathbb{Z}$, then α must be a prime integer. It cannot be 2 or a prime congruent to 1 (mod 4) because it would then have the form $a^2 + b^2$ and so be divisible by $a + bi$. Suppose that α is not in $\mathbb{Z}$. Then α divides a product of primes in $\mathbb{Z}$ because $\alpha\,\overline{\alpha} = N(\alpha)$; because of unique factorization in $\mathbb{Z}[i]$, α must divide a prime in $\mathbb{Z}$. But if α divides p, then $N(\alpha)$ divides $N(p)$, which is p^2; therefore $N(\alpha) = p$, as desired. $\square$

Because of the eightfold symmetry to the Gaussian primes — if $a + bi$ is prime so are $\pm a \pm bi$ and $\pm b \pm ai$ — their image in the complex plane makes a pleasing pattern. The CNT package function GaussianPrimePlot produces such images (Figure 9.6), with the primes given as black squares and the units shown in gray.

Exercise 9.32 describes how the methods of this section can be used to obtain the prime factorization of a Gaussian integer; the only tricky part is that one needs a subroutine to write a prime as a sum of two squares, but Smith's algorithm does that in an eyeblink. This is how *Mathematica's* FactorInteger function works on Gaussian integers. In Section 9.1 we showed how one can get all representations of n as a sum of two squares. The full program was a little involved, requiring the Chinese Remainder Theorem to get all the square roots of -1. Another approach (see Exercise 9.33) is to use Gaussian factorization: From a factorization of p one can easily get its set of Gaussian divisors and find the desired representations among them. Because the Gaussian divisor function is built into *Mathematica*, this provides a fast way to get all the representations of an integer as a sum of two squares and is the method used by the NumberTheoryFunctions package function SumOfSquaresRepresentations.

```
GaussianPrimePlot[13];
GaussianPrimePlot[50];
```

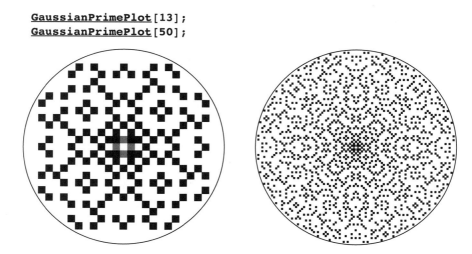

Figure 9.6. The Gaussian primes having norm at most 13 (left) and having norm at most 50. This pleasing pattern is often used to tile walls.

■ The Moat Problem

Recall that it is a simple matter to show that no giant can walk along the real axis from 2 to infinity stepping only on the prime integers. This is because, if the largest step the giant can take has size k, he would eventually run into a string of $k + 1$ composite integers (Exercise 2.22 or see the simpler proof given at the end of Section 4.1). The analogous question in the Gaussian integers is: Can one walk in $\mathbb{Z}[i]$ from $1 + i$ to infinity taking steps of size k or less and stepping only on the Gaussian primes? This is a famous unsolved problem due to Basil Gordon in 1962.

This problem is known as the ***Gaussian moat problem*** because it asks whether, for some k, there is an uncrossable moat of width k: a region of width k that consists only of nonprimes and that surrounds the origin. If there is such a moat, then steps of size less than k will not be adequate to get to infinity. The first computational result was by J. H. Jordan and J. R. Rabung, who showed in 1970 that a moat of width $\sqrt{10}$ exists. In 1998 this was improved by E. Gethner, S. Wagon, and B. Wick [GWW] to show that a moat of width $\sqrt{26}$ exists. This problem leads to the creation of interesting graphics, and we show in this section how to get images of the Gaussian primes that are reachable using steps of size k.

We will consider steps of size 2 first (a simpler case would be steps of size $\sqrt{2}$ or less; that corresponds to squares in the Gaussian prime image that are just touching at the corners, and it is easy to see in Figure 9.6 that such touching primes never leave the disk of radius 12). We can proceed as follows:

Step 1. Identify the Gaussian primes in a given disk (first octant only, because of symmetry).

Step 2. For each identified prime, find all its **2-neighbors**: the Gaussian primes at distance two or less.

Step 3. Form the road network consisting of all lines between Gaussian primes and their 2-neighbors.

Step 4. Determine the connected component of $1 + i$ in this network.

Step 1 is easy. We just select the Gaussian primes from all possibilities in the disk; because we want a graphic image we use pairs {a, b} rather than Gaussian integers a + b I. Because we want to be sure to include edges that cross the lines of symmetry, we fatten the first octant wedge (via the argument k) to include primes that are near the lines of symmetry but not in the first octant.

```
GaussianPrimesInExpandedFirstOctant[rad_, k_] :=
 Select[Flatten[Table[{a, b}, {a, -Ceiling[k], rad},
    {b, -Ceiling[k], Min[a + Floor[k √2], √(rad² - a²)]}], 1],
  GaussianPrimeQ[#[[1]] + #[[2]] I] &]
```

Figure 9.7 shows how we have captured the Gaussian primes in the fattened wedge.

```
Show[Graphics[{Line[{{0, 0}, {11, 11}}],
   Line[{{0, 0}, {15, 0}}], PointSize[0.02],
   Map[Point, GaussianPrimesInExpandedFirstOctant[15, 2]]}],
 AspectRatio → Automatic, Frame → True,
 PlotRange → {{-3, 15.5}, {-3, 11.5}}, FrameTicks →
  {Range[-2, 15, 2], Range[-2, 11, 2], None, None}];
```

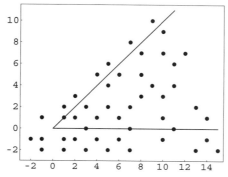

Figure 9.7. The Gaussian primes in the first octant, slightly fattened.

For step 2, the possibilities are limited. With some small exceptions, the primes within two units of a Gaussian prime p must be exactly two steps away in any direction, where a step refers to a horizontal or vertical step; this is because a and b must have opposite parity for $a + bi$ to be a Gaussian prime. There is an exception at $1 + i$, for which a and b are odd, but $a^2 + b^2$ is the prime 2. The following two cases take care of it, where the second argument, 2, refers to the bound on the step size.

```
neighbors[{1, 1}, 2] := {{2, 1}, {1, 2}, {-1, 1}, {1, -1}}
```

```
neighbors[pt_, 2] := Select[Map[pt + # &, {{1, 1},
    {1, -1}, {-1, 1}, {-1, -1}, {2, 0}, {0, 2}, {0, -2}, {-2, 0}}],
    GaussianPrimeQ[#[[1]] + #[[2]] I] &]
```

Step 3 is easy too: Just agglomerate all the neighbors of all the Gaussian primes in a disk of radius 50. We use `Sort` so that `Union`, which we use in `roads`, will eliminate duplicates. Note that `edges` and `roads` consist of pairs of primes, to be viewed as edges connecting two primes that are sufficiently close.

```
primes = GaussianPrimesInExpandedFirstOctant[50, 2];
edges[p_, d_] := Map[Sort[{p, #}] &, neighbors[p, d]]
roads = Union[Flatten[Map[edges[#, 2] &, primes], 1]];
Length /@ {primes, roads}
```

{235, 341}

So we have 235 primes and 341 edges in the road network. Time for a reality check to see if we have what we want. The result, shown in Figure 9.8, is correct; note that it contains some edges lying outside the first octant. This will lead to some duplicates eventually, but they will be eliminated; they arise because we want to be certain to get all the edges that cross the lines of symmetry.

```
Show[Graphics[{Line /@ roads,
    {GrayLevel[0.4], Line[{{0, 0}, {40, 40}}]}}],
  AspectRatio → Automatic, Frame → True, FrameTicks →
   {Range[0, 50, 10], Range[0, 50, 10], None, None}, GridLines →
   {{{0, {GrayLevel[0.4]}}}, {{0, {GrayLevel[0.4]}}}}];
```

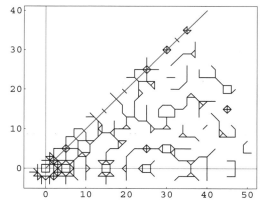

Figure 9.8. The road network in the Gaussian primes using steps of size 2 or less, first octant only.

For step 4 we start with the single point and build the set by repeatedly adding neighbors. To avoid redundancy, at each step we look only at the neighbors of the points that were just added. And we restrict the neighbor search to points not already in the component. These important efficiencies are introduced by keeping track of the pair (component, new) and stopping when new is empty, and using compl to keep track of the set of points not yet in the component.

```
FindComponent[s_, {m_Integer, n_}] := Module[
  {new, component, newneighbors, compl}, newneighbors[{1, 1}] :=
    Intersection[{{2, 1}, {1, 2}, {-1, 1}, {1, -1}}, compl];
  newneighbors[pt_] := Intersection[
    Map[pt + # &, {{1, 1}, {1, -1}, {-1, 1}, {-1, -1},
      {2, 0}, {0, 2}, {0, -2}, {-2, 0}}], compl];
  component = new = {{m, n}};
  While[new ≠ {}, compl = Complement[s, component];
   new = Union @@ (newneighbors /@ new);
   component = Join[component, new]];
  component]

component = FindComponent[primes, {1, 1}];
Length[component]
```

```
121
```

This shows that 121 primes are reachable from $1 + i$. To see the result in its symmetric form, we define symmetrize to apply the eight transformations, and then symmetrize the road network consisting of reachable Gaussian primes and use just dots for the unreachable ones. The following code generates the image of the **2-component** (the set of reachable primes using steps of size 2) shown in Figure 9.9. We make liberal use of Union to eliminate duplicate edges or points.

```
symmetrize[{a_Integer, b_}] := {{a, b}, {-a, b},
  {a, -b}, {-a, -b}, {b, a}, {-b, a}, {b, -a}, {-b, -a}};
symmetrize[{p_List, q_List}] :=
  Transpose[{symmetrize[p], symmetrize[q]}];
symmetrize[pts_List] :=
  Flatten[symmetrize /@ pts, 1] /; Depth[pts] == 3;

firstOctant = Select[roads, MemberQ[component, #[[1]]] &];
roadNetwork = Union[Flatten[symmetrize /@ firstOctant, 1]];

unreachablePrimes = Union[
  symmetrize[Complement[primes, component]]];

Show[Graphics[{Circle[{0, 0}, 50],
   {AbsolutePointSize[1], Point /@ unreachablePrimes},
   {AbsoluteThickness[0.4], Line /@ roadNetwork}}],
 AspectRatio → Automatic];
```

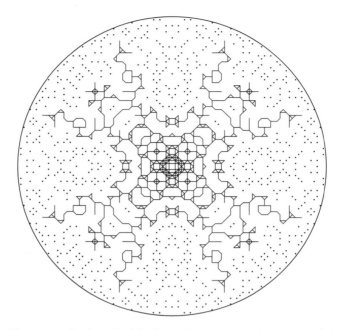

Figure 9.9. The network of reachable Gaussian primes using steps of size 2 or smaller.

The preceding computation shows that steps of size 2 or smaller will not get one beyond the circle of radius 46. Note that because of parity considerations the differences between Gaussian primes (excluding $1 + i$ and ignoring signs) must have one of the forms $1 + i$, 2, $2i$, $2 + 2i$, $1 + 3i$, $3 + i$, 4, $4i$, $3 + 3i$, $2 + 4i$, $4i + 2$, $1 + 5i$, $5i + 1$, and so on. Thus, the step-size bounds that one considers in the moat problem should be: $\sqrt{2}$, 2, $\sqrt{8}$, $\sqrt{10}$, 4, $\sqrt{18}$, $\sqrt{20}$, $\sqrt{26}$, ... Similar computations show that steps of size $\sqrt{8}$ will not get escape the disk of radius 94 (Figure 9.10). And the complete Jordan–Rabung collection of Gaussian primes is shown in Figure 9.11; that shows that steps of size $\sqrt{10}$ will not get one past the circle of radius 1025. Slightly different techniques were used in [GWW] to show that a walk using steps of size $\sqrt{26}$ or less will never get farther from the origin than distance 5,656,855.

For heuristic arguments that no step size will lead to a Gaussian prime walk to infinity, see [Var].

▦ The Gaussian Zoo

In the familiar world of rational primes, a primality pattern is called ***admissible*** if it might occur infinitely often. For example, the pattern $(n, n + 1)$ is not admissible because no matter what n is, one of the pair is divisible by 2; therefore there cannot be infinitely many prime pairs of the form $(n, n + 1)$.

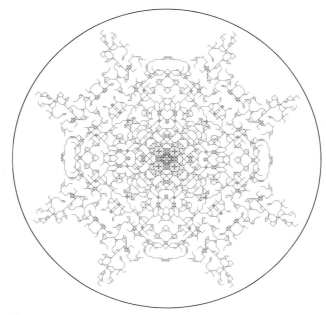

Figure 9.10. The network of reachable Gaussian primes using steps of size $\sqrt{8}$ or smaller; the greatest distance from the origin is 93.47.

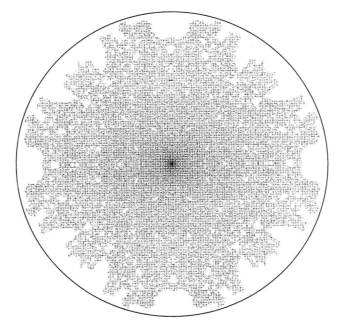

Figure 9.11. The collection of reachable Gaussian primes using steps of size $\sqrt{10}$ or smaller. The greatest distance from the origin is 1024.35; there are 249,508 reachable primes.

Of course, (2, 3) is the only prime pair that fits this pattern. But the pattern $(n, n + 2)$ is admissible because, for any prime p and any n, the pair, reduced modulo p, will miss one of the residue classes. If $p \geq 3$ this is obvious, and if $p = 2$, the pair becomes (0, 0) or (1, 1) (mod p). This means that arbitrarily large pairs can be found such that no entry is divisible by p, and so p does not block the primality of that pair. The assertion that there are infinitely many n such that the pattern $(n, n + 2)$ yields two primes is just the Twin Prime Conjecture. Note that $(n, n + 2, n + 4)$ is not admissible because $p = 3$ turns this into $(n, n + 2, n + 1)$, which hits all three classes. But $(n, n + 2, n + 6)$ is admissible. Note that to check whether a k-term pattern (sometimes called a ***constellation***) is admissible it suffices to check it modulo the primes up to k.

The main conjecture — called the ***prime k-tuples conjecture*** and due to Hardy and Littlewood — is that every admissible pattern occurs infinitely often among the primes, but not even the simplest case (the notorious Twin Prime Conjecture) has been proved (see [Rie] and [RWW] for more information, including heuristic arguments in favor of the conjecture).

Here is a short routine for checking admissibility, where we assume the input is just a list of integers.

```
AdmissibleQ[c_] := And @@ Table[p = Prime[i];
   Length[Union[Mod[c, p]]] ≠ p, {i, π[Length[c]]}]

AdmissibleQ[{0, 2, 6, 8}]
```

```
True
```

We can easily find all the examples below 10,000 that fit the {0, 2, 6, 8} pattern. The next output contains the first members of the sequences.

```
Select[Prime[Range[π[10000]]], And @@ PrimeQ[# + {2, 6, 8}] &]
```

```
{5, 11, 101, 191, 821, 1481, 1871, 2081, 3251, 3461, 5651, 9431}
```

And here is an inadmissible constellation.

```
AdmissibleQ[{1, 7, 11, 13, 17, 19, 23, 29}]
```

```
False
```

This pattern fails because all the mod-7 classes are represented. Therefore any sequence that fits this pattern will have an entry that is divisible by 7.

```
Mod[{1, 7, 11, 13, 17, 19, 23, 29}, 7]
```

```
{1, 0, 4, 6, 3, 5, 2, 1}
```

Here is the pattern determined by the primes between 11 and 29. It is admissible and Exercise 9.38 asks you to find some occurrences of it in the primes.

```
largepattern = Prime[Range[5, 10]] - 11
```

{0, 2, 6, 8, 12, 18}

<u>AdmissibleQ</u>[largepattern]

True

One might think that an admissible constellation could not be denser than the initial segment of primes. But D. Gordon and G. Rodemich [GR] found an admissible constellation spanning an interval of size 4893 and having 655 numbers; $\pi(4894)$ is only 654 (see Exercise 9.39). Such a constellation is called a ***superdense constellation***. If the prime k-tuples conjecture is true, then there would be an interval of primes that follows this pattern. Such an interval would have a greater prime density than the corresponding interval starting at 2, which would contradict an earlier conjecture of Hardy and Littlewood that such an interval cannot exist!

Now, in the hope of understanding what a Twin Prime Conjecture for the Gaussian primes might be, we can look at admissible patterns in the Gaussian primes. Let us use the term ***animal*** for a connected pattern of squares in the Gaussian lattice; an ***n-animal*** is an animal with n squares. Animals related by any of the eight symmetries of Gaussian primes are considered the same. An animal is ***admissible*** if for every Gaussian prime p, the animal, when reduced modulo p, does not cover all classes. Here is an extension of the routine just given that uses an option to handle both the rational and Gaussian cases. If the input has a complex number in it, then it will automatically work in the Gaussian integers.

```
Options[AdmissibleQ] = {GaussianIntegers → False};

AdmissibleQ[c_, opts___] := Module[{primes, gi, l = Length[c]},
  gi = GaussianIntegers /. {opts} /. Options[AdmissibleQ];
  If[!FreeQ[c, Complex], gi = True];
  If[gi, primes =
    Select[Flatten[Table[a + b I, {a, 0, Floor[√l]},
      {b, 0, Floor[√(1 - a²)]}]], GaussianPrimeQ];
  And @@ (Length[Union[Mod[c, #]]] ≠ GaussianNorm[#] &) /@
    primes,
  And @@ Table[p = Prime[i]; Length[Union[Mod[c, p]]] ≠ p,
  {i, π[l]}]]]
```

Of course, by parity considerations any admissible animal cannot have two squares that touch along an edge (in particular, one of two such touching squares would be divisible by $1 + i$); an admissible animal can have only squares touching at a corner. Simple enumeration of possibilities (using the fact that the longest 45° row must have length 4, 3, or 2) shows that the following list contains all the 4-animals.

```
animals[4] = {{0, 1 + I, 2 + 2 I, 3 + 3 I}, {0, 1 + I, 2 + 2 I, 3 + I},
  {0, 1 + I, 2 + 2 I, 2}, {0, 1 + I, 2, 3 + I}, {I, 1 + 2 I, 2 + I, 1}};
```

```
Show[GraphicsArray[(Graphics[#, AspectRatio → Automatic,
        PlotRange → {{-1, 5}, {-1, 5}}]] &) /@
    Map[Rectangle[{Re[#], Im[#]}] &, animals[4], {2}]]];
```

Figure 9.12. The five admissible patterns of four Gaussian primes.

And all five are admissible.

```
AdmissibleQ /@ animals[4]

{True, True, True, True, True}
```

However, one should focus on maximal admissible animals: ones that become inadmissible if a square is added anywhere. Let us call these *lions*. Then the admissible animals are just the connected subsets of any of the lions.

Here are several points that can be proved with hand or light computer computations (see Exercise 9.40).

- Of the five admissible 4-animals, only the diamond is a lion.
- There are twelve 5-animals.
- There are seven admissible 5-animals.
- There are no 5-lions.

Bringing more complicated computer programs to bear ([JR], [Var1], [RWW]) leads to:

- The largest lion has size 48, and there are seven of them.
- There are 52 lions in all, having size profile as follows, where an exponent indicates the number of lions of a given size: 4 12 17 19 20 23 24 25^2 27^4 28 29^3 30 31 32^5 33^3 34^2 36^2 39^3 41^4 42^2 43 44 45 46 47 48^7.

Using a combination of computer techniques it is possible to capture all the lions. Renze, Wagon, and Wick [RWW] accomplished this by combining some ideas from computer graphics with the approach given by Vardi [Var1]. Vardi showed that every admissible animal is contained in one of 506 animals, the largest of which have sizes 37, 50, 51, and 71 (details in Exercise 9.41). One can then prune the animals in all possible ways to yield connected subsets and check for admissibility; such connected sets, when they first occur, will be lions, and all lions arise in this way. Some care is necessary to check for maximality and eliminate the duplicates under symmetry. The complete collection is shown in Figure 9.13.

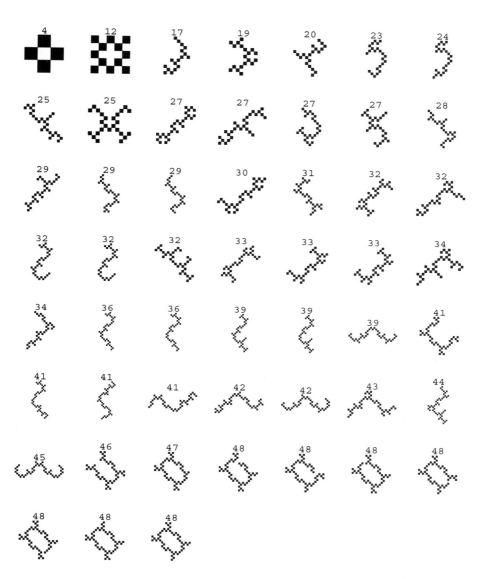

Figure 9.13. The complete catalog of lions, labelled by size. There are 52 of them.

The CNT package contains a function that we will describe only briefly here. For more information on its derivation, implementation, and use, see [RWW]. Given an admisible constellation, such as {0, 2}, one can make a heuristic estimate of how many times the pattern is expected to appear among the primes under x. The ConstellationEstimate function does it, leaving the answer in symbolic form and li(x) is the logarithmic integral function (see Section 4.1). The symbolic form of the output involves an infinite product over primes p and an integral of a negative power of log.

ConstellationEstimate[{0, 2}, EvaluateIntegral → False][x]

$$\left(2 \prod_{p \geq 3} \frac{(p-2)\,p}{(p-1)^2}\right) \int_0^x \frac{1}{\text{Log}[t]^2}\,dt$$

One can simplify the integrals that arise by omitting the option of the preceding input.

ConstellationEstimate[{0, 2}][x]

$$\left(2 \prod_{p \geq 3} \frac{(p-2)\,p}{(p-1)^2}\right) \left(\text{li}(x) - \frac{x}{\log(x)}\right)$$

And one can get the numerical value of the infinite product.

N[ConstellationEstimate[{0, 2}][x]]

$$1.32032 \left(\text{li}(x) - \frac{x}{\log(x)}\right)$$

And so we can find out the expected number of twin primes less than, say, 10^{14}.

Round[N[ConstellationEstimate[{0, 2}][10^{14}]]]

135780282954

In fact, it is known that there are 135,780,321,665 twin primes beneath 10^{14}, so the error in the estimate is quite small. The good agreement here provides some evidence that the Twin Prime Conjecture is true.

The estimates also work on constellations in the Gaussian integers, where the count is of the expected number of times the pattern occurs inside the first octant and inside the disk of radius x.

ConstellationEstimate[{0, 2, 1 + I, 1 - I}][x]

$$\frac{4}{\pi^3}\left(2^3 \prod_{N(p) \geq 5} \frac{(N(p)-4)\,(N(p))^3}{(N(p)-1)^4}\right) \cdot$$
$$\left(-\frac{2\,x^2}{3\,\log(x)} - \frac{x^2}{3\,\log^2(x)} - \frac{x^2}{3\,\log^3(x)} + \frac{4\,\text{li}(x^2)}{3}\right)$$

Rationalize[N[ConstellationEstimate[{0, 2, 1 + I, 1 - I}][x]]]

$$1.37959 \,\frac{4}{\pi^3}\left(-\frac{2\,x^2}{3\,\log(x)} - \frac{x^2}{3\,\log^2(x)} - \frac{x^2}{3\,\log^3(x)} + \frac{4\,\text{li}(x^2)}{3}\right)$$

N[ConstellationEstimate[{0, 2, 1 + I, 1 - I}][200000]]

193922.

And, as shown in [RWW], there are 193,628 diamonds in the first octant and within distance 200,000 of the origin.

Exercises for Section 9.2

9.24. Prove all parts of Proposition 9.7.

9.25. Explain why, in the division algorithm for Gaussian integers, we cannot just let $q = \lfloor u \rfloor + \lfloor v \rfloor i$ (where $\alpha / \beta = u + iv$), which would be the exact analog to what is done in the ordinary integers.

9.26. Explain why, if $a + bi$ is a Gaussian prime, then the four nearest neighbors $a + (b \pm 1)i$ and $(a \pm 1) + bi$ are not Gaussian primes, except for a few cases surrounding $1 + i$.

9.27. Prove that if p is a Gaussian prime that divides $\alpha \beta$ where α and β are Gaussian integers, then p must divide at least one of α, β.

9.28. Find a characterization of the Gaussian integers that are squares of Gaussian integers.

9.29. Explain why the extended gcd coefficients s and t can be defined by a recursive formula identical to that formula for the integers (given in Section 1.2).

9.30. When ordinary integers are used in `PrimeQ`, `Divisors`, or `Factor-Integer`, the `GaussianIntegers` option must be turned on if one wants the integer to be considered as a Gaussian integer. Explain why `GCD` can handle both domains without having a `GaussianIntegers` option. *Hint*: Use Proposition 9.7(h).

> **Options[PrimeQ]**
>
> {GaussianIntegers → False}

> **Options[GCD]**
>
> {}

9.31. Use the Gaussian integers and Gaussian primes to give an alternative proof that if a prime integer is a sum of two squares, then its representation is unique, up to order and sign. *Hint*: Assume that p has two representations $a^2 + b^2$ and $c^2 + d^2$ and use the fact that $a \pm bi$ and $c \pm di$ are all Gaussian primes.

9.32. Formulate and implement an algorithm to factor a Gaussian integer α into powers of prime Gaussian integers. Outline:

(a) Factor $N(\alpha)$ in $\mathbb{Z}$.

(b) If $N(\alpha)$ is even use $2 = (1 + i)^2$ to get the appropriate power of $1 + i$.

(c) Deal with the primes in $N(\alpha)$ that are congruent to 3 (modulo 4).

(d) Deal with the remaining primes in $N(\alpha)$, by factoring such a p into the primes $(a + bi)(a - bi)$ and determining how many of each divide α.

Check your program by using the `Unfactor` command from the `CNT` package.

9.33. Show how the built-in `Divisors` function for Gaussian integers can be used to obtain all representations of an integer n as a sum of two squares.

9.34. (a) Show that if n is an integer and $\alpha = a + bi$ is a Gaussian integer, then $\gcd(\alpha, n) = 1$ if and only if $\gcd(n - \alpha, n) = 1$ iff $\gcd(a - bi, n) = 1$ iff $\gcd(b + ai, n) = 1$.

(b) Show that if $n = 2m$ is an even integer and α is a Gaussian integer then $\gcd(\alpha, n) = 1$ if and only if $\gcd(m + mi - \alpha, n) = 1$.

(c) To study the set of Gaussian integers coprime to an integer n, it suffices to look at integers whose real and imaginary parts lie in $[0, n)$. Use parts (a) and (b) to show that, if $n = 2m$ is even, there are 16 symmetry functions for the coprime data within this square. In fact, all the coprime information is given by the triangular region $\{a + bi : 0 \le a \le m,\ b \le a,\ b \le m - a\}$. Figure 9.14 shows the symmetries in the case $n = 48$.

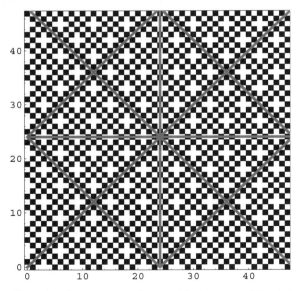

Figure 9.14. The Gaussian integers coprime to 48. Any one of the triangles can be used to generate the entire figure.

9.35. The Gaussian integers are related to some important formulas for π. For example, there is the relatively simple

$$\frac{\pi}{2} = \arctan 1 + \arctan \frac{1}{2} + \arctan \frac{1}{3}$$

which is equivalent to

$$\pi = 4 \arctan \frac{1}{2} + 4 \arctan \frac{1}{3}$$

But this formula, when expressed as an infinite series using the Maclaurin series for arctan x, is not too efficient in computing many digits of π because the powers of 2 and 3 do not grow very quickly. Much more impressive is Machin's formula

$$\pi = 16 \arctan \frac{1}{5} - 4 \arctan \frac{1}{239}$$

or Gauss's

$$\pi = 48 \arctan \frac{1}{18} + 32 \arctan \frac{1}{57} - 20 \arctan \frac{1}{239}.$$

These formulas, expressed as infinite series, were used for several centuries to compute hundreds, thousands, and then millions of digits of π. Here is how these formulas can be understood using the Gaussian integers.

The following factorizations are the key to Machin's formula.

FactorInteger[239 + I] // FactorForm

$I (1 + I) (2 + 3 I)^4$

FactorInteger[(5 + I)^4] // FactorForm

$(1 + I)^4 (2 + 3 I)^4$

FactorInteger[2, GaussianIntegers → True] // FactorForm

$-I (1 + I)^2$

These factorizations tell us that $(5 + i)^4 = 2(239 + i)(1 + i)$. Therefore $z = (5 + i)^4 / (239 + i)$ is a complex number whose angle (with the positive real axis) is $\pi/4$. But $5 + i = r_1 e^{i\theta_1}$ and $239 + i = r_2 e^{i\theta_2}$, where $\theta_1 = \arctan \frac{1}{5}$ and $\theta_2 = \arctan \frac{1}{239}$; it follows that the angle of z is $4\theta_1 - \theta_2$.

Use similar techniques to prove Gauss's formula. Can you discover (and prove) Störmer's formula, which is a combination of arctan $\frac{1}{8}$, arctan $\frac{1}{57}$, and arctan $\frac{1}{239}$?

9.36. Find the longest possible sequence of adjacent Gaussian primes on a line of slope 1. More precisely, find a long sequence of the form p_1, p_2,

p_3, ... such that $p_{i+1} = p_i + 1 + i$. Prove that your sequence is the longest one that exists.

9.37. Modify the code in the text to generate an image of the Gaussian primes that are reachable from $1 + i$ in steps of size bounded by $\sqrt{8}$. If you have a computer with a lot of memory, you can try for an image of the $\sqrt{10}$-reachable primes. For that you should arrange to show only points at the reachable primes (don't bother with the network of lines), as there are over 249,000 reachable points. If memory is tight, just generate the image in the first octant (under the 45° line in the first quadrant), because, by symmetry, that contains the essence of the component. Also look at the $\sqrt{2}$-component; this consists of the Gaussian primes that are "connected" to $1 + i$. Exercise 9.41 looks at connected sets in more detail.

9.38. Find a sequence of primes not starting with 5 or 11 that matches the pattern $\{n, n+2, n+6, n+8, n+12, n+18\}$

9.39. (Gordon and Rodemich [GR])   Let A consist of those integers n between 1 and 4893 that satisfy each of the following 45 incongruences

$n \not\equiv 0 \,(\mathrm{mod}\,2)$	$n \not\equiv 26 \,(\mathrm{mod}\,29)$	$n \not\equiv 22 \,(\mathrm{mod}\,67)$	$n \not\equiv 42 \,(\mathrm{mod}\,107)$	$n \not\equiv 150 \,(\mathrm{mod}\,157)$
$n \not\equiv 2 \,(\mathrm{mod}\,3)$	$n \not\equiv 4 \,(\mathrm{mod}\,31)$	$n \not\equiv 24 \,(\mathrm{mod}\,71)$	$n \not\equiv 12 \,(\mathrm{mod}\,109)$	$n \not\equiv 70 \,(\mathrm{mod}\,163)$
$n \not\equiv 4 \,(\mathrm{mod}\,5)$	$n \not\equiv 24 \,(\mathrm{mod}\,37)$	$n \not\equiv 70 \,(\mathrm{mod}\,73)$	$n \not\equiv 106 \,(\mathrm{mod}\,113)$	$n \not\equiv 72 \,(\mathrm{mod}\,167)$
$n \not\equiv 6 \,(\mathrm{mod}\,7)$	$n \not\equiv 32 \,(\mathrm{mod}\,41)$	$n \not\equiv 28 \,(\mathrm{mod}\,79)$	$n \not\equiv 52 \,(\mathrm{mod}\,127)$	$n \not\equiv 77 \,(\mathrm{mod}\,173)$
$n \not\equiv 5 \,(\mathrm{mod}\,11)$	$n \not\equiv 10 \,(\mathrm{mod}\,43)$	$n \not\equiv 30 \,(\mathrm{mod}\,83)$	$n \not\equiv 54 \,(\mathrm{mod}\,131)$	$n \not\equiv 78 \,(\mathrm{mod}\,179)$
$n \not\equiv 0 \,(\mathrm{mod}\,13)$	$n \not\equiv 12 \,(\mathrm{mod}\,47)$	$n \not\equiv 84 \,(\mathrm{mod}\,89)$	$n \not\equiv 125 \,(\mathrm{mod}\,137)$	$n \not\equiv 61 \,(\mathrm{mod}\,181)$
$n \not\equiv 6 \,(\mathrm{mod}\,17)$	$n \not\equiv 30 \,(\mathrm{mod}\,53)$	$n \not\equiv 9 \,(\mathrm{mod}\,97)$	$n \not\equiv 58 \,(\mathrm{mod}\,139)$	$n \not\equiv 84 \,(\mathrm{mod}\,191)$
$n \not\equiv 17 \,(\mathrm{mod}\,19)$	$n \not\equiv 18 \,(\mathrm{mod}\,59)$	$n \not\equiv 71 \,(\mathrm{mod}\,101)$	$n \not\equiv 47 \,(\mathrm{mod}\,149)$	$n \not\equiv 98 \,(\mathrm{mod}\,193)$
$n \not\equiv 0 \,(\mathrm{mod}\,23)$	$n \not\equiv 17 \,(\mathrm{mod}\,61)$	$n \not\equiv 40 \,(\mathrm{mod}\,103)$	$n \not\equiv 64 \,(\mathrm{mod}\,151)$	$n \not\equiv 188 \,(\mathrm{mod}\,197)$

Show that A is admissible, and that A has more elements than the set of primes between 2 and 4894.

9.40. Find all 5-animals, all admissible 5-animals, and all maximal admissible 5-animals.

9.41. (J. H. Jordan, J. R. Rabung, I. Vardi) The goal of this exercise is to show that there is no admissible m-animal where $m > 71$. The reduction to the correct maximum of 48 takes a little more work along the lines of Exercise 9.42.

(a) Let $n = 130 = 2 \cdot 5 \cdot 13$. Consider the Gaussian integers having real and imaginary parts between 0 and $n - 1$ and mark the ones that are relatively prime to n. Referring to Exercise 9.34, consider the triangular region — $\{a + bi : 0 \le a \le m, b \le a, b \le m - a\}$ — that generates the full coprime data, and let T be the set of the Gaussian integers in that triangle that are coprime to n. Generate an image of T, as in Figure 9.15.

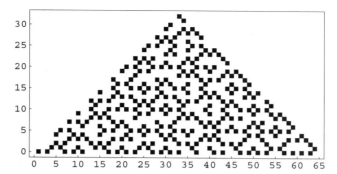

Figure 9.15. The Gaussian integers coprime to 130.

(b) Examine Figure 9.15 and conclude that there is no path in T that touches all three sides. Locate the largest path in T that touches two sides; it has 75 points.

(c) Show that the 75-animal of (b), when extended through the symmetric copies of the triangle, leads to one of size 580 (this comes from $8 \cdot 70 + 4 \cdot 5$). This is the largest animal in the gcd-lattice in the 130×130 square.

(d) Show that any admissible animal must be contained in the gcd-lattice in the 130×130 square. This proves that the largest admissible animal has size no greater than 580. In particular, there is no infinite animal, and so no prime walk to infinity using steps bounded by $\sqrt{2}$, no matter where one starts (see [GS]).

(e) Repeat parts (a)–(d) with $n = 390 = 2 \cdot 3 \cdot 5 \cdot 13$ to deduce that the largest admissible animal has size no larger than 71. For more information on the types of admissible animals see [RWW]. The largest admissible animal has size 48.

9.42. Consider the pattern in Figure 9.16, which we call a ***protolion*** and which has 50 elements; the source for this is discussed in Exercise 9.41. The coordinates are as follows:

```
protolion50 = {0, 2, 1 + I, 2 + 2 I, 1 + 3 I, 2 + 4 I, 3 + 5 I, 4 + 6 I,
    5 + 5 I, 6 + 6 I, 7 + 7 I, 6 + 8 I, 7 + 9 I, 8 + 10 I, 9 + 11 I, 10 + 10 I,
    11 + 11 I, 12 + 12 I, 12 + 10 I, 8 + 12 I, 9 + 13 I, 8 + 14 I,
    7 + 15 I, 6 + 16 I, 5 + 15 I, 4 + 16 I, 3 + 17 I, 4 + 18 I, 5 + 19 I,
    3 + 19 I, 4 I, -1 + 3 I, -2 + 4 I, -3 + 5 I, -4 + 6 I, -3 + 7 I,
    -4 + 8 I, -5 + 9 I, -6 + 8 I, -7 + 7 I, -7 + 9 I, -3 + 9 I, -2 + 10 I,
    -1 + 11 I, -2 + 12 I, -1 + 13 I, 14 I, 1 + 13 I, 2 + 14 I, 3 + 15 I};
```

(a) Show that the only prime moduli for which this pattern fails to omit a residue are $p_1 = 2 + 5i$ and $p_2 = 2 - 5i$. This shows that the 50-element set is inadmissible.

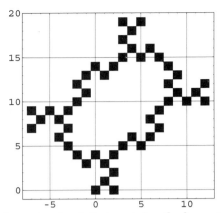

Figure 9.16. A 50-element animal that contains the largest lions.

(b) Show that omitting one square from the set cannot lead to a 49-element admissible animal. Use Figure 9.17, which shows the residues modulo p_1 and p_2 (where we are using a lemma, to be proved in Section 9.3, that residues mod $2 \pm 5i$ can be taken to be one of {0, 1, …, 28}).

(c) Find all ways of deleting two squares from the main pattern so that the remainder is connected and a residue is omitted in each case, thus making the 48-element set admissible. For example, deleting the two squares corresponding to 25 (mod p_1) and 0 (mod p_2) (the boldface entries in Figure 9.17) works and leaves an admissible set. Up to symmetry there are seven ways to perform such double deletions, which leads to the seven 48-lions. A similar analysis on the 51-component and 71-component of Exercise 9.41 shows that the largest lions arising from those have sizes 45 and 44, respectively, thus proving that 48 is the largest lion size.

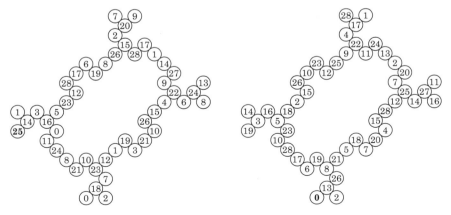

Figure 9.17. The reduction of the protolion modulo $2 + 5i$ (left) and $2 - 5i$. Removal of the two boldface entries (25 and 0) yields a 48-animal that is admissible.

9.43. Use the `ConstellationEstimate` function to predict how many times the pattern $\{n, n+2, n+6, n+8\}$ should be expected to appear beneath 10,000, and then count how many times the pattern does appear.

9.3 Higher Reciprocity

The Legendre symbol is an example of a ***character***: a mapping χ from $\mathbb{Z}_p^*$ into $\mathbb{C}$ such that $\chi(a)\chi(b) = \chi(ab)$. There are similar characters that reveal when a number is a perfect cube or fourth or higher power in $\mathbb{Z}_p^*$. We shall restrict attention to ***quartic residues***, perfect fourth powers, and the corresponding quartic residue symbol $\left(\frac{a}{p}\right)_4$. We would like this symbol to be $+1$ if and only if a is a quartic residue modulo p. But if $1 = \chi(a^4) = \chi(a)^4$, this means that the symbol needs to take on the values ± 1 and $\pm i$; therefore we will work over the Gaussian integers; the primes we use as a modulus will be Gaussian primes.

This makes life a little more complicated. For the purposes of the quartic residue symbol, 5 is not prime. We will have to work either modulo $1 + 2i$ or $1 - 2i$. Moreover, if we are working modulo 7, say, then 3 is a quartic residue because $(1 + i)^4 = -4$; but there is no rational integer whose fourth power is 3 (mod 7). Let π be a Gaussian prime. Two Gaussian integers are ***congruent modulo*** π, written $a + bi \equiv c + di \pmod{\pi}$, if and only if π divides their difference.

Lemma 9.10. Given a Gaussian prime π, there are exactly $N(\pi)$ distinct residue classes modulo π, and each nonzero residue class has an inverse modulo π. Moreover, if $\pi = a + bi$ and neither a nor b is 0, then any Gaussian integer μ is congruent modulo π to one of $0, 1, ..., N(\pi) - 1$.

Proof. The existence of the inverse follows from the fact that we have an extended Euclidean algorithm (Theorem 9.8(c)). If π does not divide α, then we can find Gaussian integers s and t so that $s\alpha + t\pi = 1$; s is the inverse, modulo π, of α.

If $\pi = q$, a rational prime congruent to 3 (mod 4), then each Gaussian integer is congruent to a Gaussian integer of the form $a + bi$, $0 \le a < q$, $0 \le b < q$, modulo q, and no two Gaussian integers of this form can be congruent modulo q (Exercise 9.44). The number of distinct residue classes is $q^2 = N(q)$.

If $\pi = a + bi$, where $ab \ne 0$, let $p = a^2 + b^2$; p is a rational prime, π divides p, and p does not divide b. We claim that each Gaussian integer is congruent to exactly one of the integers $0, 1, ..., p - 1 \pmod{\pi}$. Given any Gaussian integer $\mu = m + ni$, we can find a rational integer c such that $cb \equiv n \pmod{p}$. It follows that $\mu - c\pi \equiv m - ca \pmod{p}$, and so $\mu \equiv m - ca \pmod{\pi}$. This shows that each Gaussian integer is congruent to a rational integer modulo π, and so must be congruent to one of the integers from 0 through $p - 1 \pmod{\pi}$. Because no two of these rational integers can be congruent modulo π (use Proposition 9.7(d)), there are exactly $p = N(\pi)$ residue classes modulo π. $\square$

Mathematica's `Mod[a, z]` does not always give residues in the two classes of the preceding lemma (rather, it finds a residue whose absolute value is bounded by $\sqrt{N[z]}$); therefore we will want our own version for the Gaussian integers that makes the conditions of Lemma 9.10 the highest priority. Here are several utilities for $\mathbb{Z}[i]$: `GaussianIntegerQ` detects the Gaussian integers, `GaussianPrimeQ` detects the Gaussian primes, `GaussianNorm` gives the norm, and `GaussianMod` gives the residue lying in the complete systems specified in Lemma 9.10 (when the modulus is prime). These are included in the CNT package. We cannot use π, and so often use p for Gaussian primes. The `_Complex` in the last case ensures that that case does not apply to primes of the form p or pi, where p is a rational prime. Note that any w is congruent to one of $\{0, 1, ..., N[z] - 1\}$ modulo z so long as $\gcd(a, b) = 1$, where $z = a + bi$. This case is included in the code that follows.

```
GaussianIntegerQ[a_Integer] := True
GaussianIntegerQ[Complex[a_Integer, b_Integer]] := True
GaussianIntegerQ[a_] := False

GaussianPrimeQ[p_ ?GaussianIntegerQ] :=
  PrimeQ[p, GaussianIntegers → True]

GaussianNorm[a_ ?GaussianIntegerQ] := Abs[a]²

GaussianMod[a_ ?GaussianIntegerQ, p_ ?GaussianIntegerQ] :=
  Module[{pp = GaussianNorm[p]},
    Mod[Re[a] - Im[a] Re[p] PowerMod[Im[p], -1, pp], pp]] /;
  !IntegerQ[p] && !MatchQ[p, Complex[0, _Integer]] &&
    GCD[Re[p], Im[p]] == 1

GaussianMod[a_ ?GaussianIntegerQ, p_ ?GaussianIntegerQ] :=
  Mod[a, p]

SetAttributes[#, Listable] & /@
  {GaussianIntegerQ, GaussianPrimeQ, GaussianMod};

GaussianPrimeQ[{7, 1 + 2 I, 3 + 8 I}]
```

{True, True, True}

```
GaussianMod[2 + Range[30] I, 3 + 8 I]
```

{29, 56, 10, 37, 64, 18, 45, 72, 26, 53, 7, 34, 61, 15, 42,
  69, 23, 50, 4, 31, 58, 12, 39, 66, 20, 47, 1, 28, 55, 9}

```
GaussianMod[34 + 12 I, 14 + 9 I]
```

Theorem 9.11. Fermat's Little Theorem for Gaussian Integers If π is a Gaussian prime and α is a Gaussian integer that is not divisible by π, then $\alpha^{N(\pi)-1} \equiv 1 \pmod{\pi}$.

The proof is essentially identical to the classical Fermat Little Theorem, and is left to Exercise 9.45. Here are some examples.

```
GaussianMod[PowerMod[{19, 3 + 2 I, 11, 10^10 + 1 + 10 I},
    GaussianNorm[#] - 1, #], #] & /@ {1 + 2 I, 7, 3 + 8 I}
```

$$\{\{1, 1, 1, 1\}, \{1, 1, 1, 1\}, \{1, 1, 1, 1\}\}$$

Now, if π is a Gaussian prime and $N(\pi) \neq 2$, then $N(\pi) \equiv 1 \pmod 4$. Therefore, if π does not divide α and $N(\pi) \neq 2$, then $\alpha^{(N(\pi)-1)/4}$ is a Gaussian integer and it satisfies $x^4 \equiv 1 \pmod{\pi}$. But this congruence has the four incongruent solutions ± 1 and $\pm i$; it has no other solutions because if π divides $x^4 - 1$, then π divides the product $(x+1)(x-1)(x+i)(x-i)$, and hence divides one of the factors. Therefore we have that

$$\alpha^{\frac{N(\pi)-1}{4}} \equiv \pm 1 \quad \text{or} \quad \pm i \pmod{\pi}.$$

We define the ***quartic residue symbol***, $\left(\frac{\alpha}{\pi}\right)_4$, to be ± 1 or $\pm i$, so that

$$\alpha^{\frac{N(\pi)-1}{4}} \equiv \left(\frac{\alpha}{\pi}\right)_4 \pmod{\pi}.$$

Corollary 9.12. If π is a Gaussian prime and α is a Gaussian integer that is not divisible by π, then α is a fourth power modulo π if and only if $\left(\frac{\alpha}{\pi}\right)_4 = 1$.

Proof. If α is a fourth power modulo π, say $\alpha \equiv \beta^4 \pmod{\pi}$, then $\alpha^{(N(\pi)-1)/4} \equiv \beta^{N(\pi)-1} \equiv 1 \pmod{\pi}$, so $\left(\frac{\alpha}{\pi}\right)_4 = 1$. For the converse, let g be a primitive root modulo π (this requires generalizing the work in Chapter 4 on primitive roots for primes to the Gaussian prime context, but in fact this uses no new ideas; see Exercise 9.51). Then $\{g^e : e = 1, \ldots, N(\pi) - 1\}$ is a complete residue system for π. If e is a multiple of 4 then g^e is of course a fourth power and the quartic residue is correct. If e is not a multiple of 4, then $g^{e(N(\pi)-1)/4}$ cannot be 1 $\pmod{\pi}$ because the exponent is not a multiple of $N(\pi) - 1$, which is the order of $g \pmod{\pi}$. And these powers of g are not the fourth powers of any other power of g, so they are not fourth powers at all, because powers of g exhaust the possible fourth roots. $\square$

The quartic residue symbol is not defined for $\pi = 1 + i$. Various aspects of the quartic residue symbol are discussed in the exercises, the highlight being the law of quartic reciprocity (Exercise 9.53) which was discovered by Gauss, and proved later by F. G. Eisenstein.

Here is code to compute the quartic residue symbol.

```
Attributes[QuarticResidueSymbol] = Listable;

QuarticResidueSymbol[α_ ?GaussianIntegerQ, p_ ] :=
  (GaussianMod[PowerMod[α, (GaussianNorm[p] - 1) / 4, p], p] /.
    {If[IntegerQ[p], p - 1, GaussianNorm[p] - 1] → -1,
      GaussianMod[I, p] → I, GaussianMod[-I, p] → -I}) /;
  GaussianIntegerQ[p] && GaussianNorm[p] ≠ 2

p = 8 + 3 I;
powers = PowerMod[13 + {1, 2, 3} I, 4, p]

{7 + 5 I, 1 + 8 I, 3 + 4 I}
```

```
QuarticResidueSymbol[powers, p]

{1, 1, 1}
```

We next show how working in the Gaussian integers allows us to determine which elements of $\mathbb{Z}_p$ are fourth powers mod p, when p is a rational prime. If $p \equiv 3 \pmod 4$ then it is not hard to see that the fourth powers coincide with squares, which correspond to integers x for which $\left(\frac{x}{p}\right) \neq -1$. Every fourth power is obviously a square. And if $x \equiv y^2 \pmod p$ then also $x \equiv (-y)^2 \pmod p$. But we know that -1 is a nonsquare mod p, so it follows that one of $\pm y$ is a square mod p; this means that x is a fourth power. Now, if $p \equiv 1 \pmod 4$, then p splits into $(a + bi) \cdot (a - bi)$. If x is a fourth power mod p it will be a fourth power mod each of these factors and so $\left(\frac{x}{a+bi}\right)_4 = \left(\frac{x}{a-bi}\right)_4 = 1$. Conversely, if these two quartic symbols are both 1, then x is a fourth power modulo each of $a \pm bi$ and, by Lemma 9.10, the two fourth roots can be taken to be integers. But if $x \equiv y^4 \pmod{a + bi}$ then $x \equiv y^4 \pmod{a - bi}$: just take conjugates in the equation $x = k(a + bi) + y^4$. Because $a - bi$ and $a + bi$ are relatively prime, this means that x is a fourth power modulo $(a + bi)(a - bi)$, or modulo p.

Here is a routine that detects the fourth powers for all rational primes.

```
FourthPowerQ[m_, p_ ?PrimeQ] :=
  JacobiSymbol[m, p] ≠ -1 /; Mod[p, 4] == 3;

FourthPowerQ[m_, 2] := True

FourthPowerQ[m_, p_ ?PrimeQ] := (
    {a, b} = SumOfSquaresRepresentations[2, p][[1]];
    MemberQ[{0, 1}, QuarticResidueSymbol[m, a + b I]]) /;
  Mod[p, 4] == 1
```

And here we check by comparing it to an exhaustive listing.

```
n = 13;
Select[Range[0, n - 1], FourthPowerQ[#, n] &]
Union[PowerMod[Range[0, n - 1], 4, n]]

{0, 1, 3, 9}

{0, 1, 3, 9}
```

```
n = 19;
Select[Range[0, n - 1], FourthPowerQ[#, n] &]
Union[PowerMod[Range[0, n - 1], 4, n]]
```

{0, 1, 4, 5, 6, 7, 9, 11, 16, 17}

{0, 1, 4, 5, 6, 7, 9, 11, 16, 17}

Recalling the power grids of Chapter 2, the present work tells us that, when p is prime, the number of distinct integers in the fourth column is either $(p-1)/2$ or $(p-1)/4$ according as p is or is not a Gaussian prime.

Other quadratic extensions of the integers and rationals are possible. If d, called the **discriminant**, is squarefree, we can consider the field $\mathbb{Q}(\sqrt{d})$, which consists of all reals of the form $r + s\sqrt{d}$, where r and s are rational. The **integers** of this field refer to the elements of the field that satisfy a monic polynomial with integer coefficients.

If $d \equiv 1 \pmod 4$, then the integers are of the form $(m + n\sqrt{d})/2$, where m and n are rational integers of the same parity. Otherwise the integers are of the form $m + n\sqrt{d}$ where m and n are arbitrary rational integers. The norm of $m + n\sqrt{d}$ is always $m^2 - dn^2$.

There are relatively few quadratic extensions for which the norm can be used to create a Euclidean algorithm. The complete list of cases for which the norm does yield a Euclidean algorithm is: $d = -1, -2, -3, -7, -11, 2, 3,$ 5, 6, 7, 11, 13, 17, 19, 21, 29, 33, 37, 41, 57, and 73. And in these cases, because there is a Euclidean algorithm there is unique factorization for the integers of $\mathbb{Q}(\sqrt{d})$.

But there are other quadratic extensions that have unique factorization even though the norm does not admit a Euclidean algorithm. When d is negative they are $d = -19, -43, -67, -163$. In the 1960s Harold Stark proved that there are no other quadratic extensions with negative discriminant (d) that have unique factorization. For positive discriminants it has been conjectured but not proven that there are infinitely many cases with unique factorization. Among those that have unique factorization but for which one cannot use the norm to get a Euclidean algorithm are $d = 14, 22,$ 23, 31, 38, 43, 46, 47, 53, 59, 61, 62, 67, 69, 71, 77, 83, 86, 89, 93, 94, and 97. All other square-free discriminants between -100 and 100 yield quadratic extensions with nonunique factorization.

These distinctions can help explain many mysteries involving rational integers, and also real numbers. For instance, the fact that the integers of $\mathbb{Q}(\sqrt{41})$ admit a Euclidean algorithm is related to the amazing property of $x^2 + x + 41$ discovered by Euler: This polynomial is prime for every value of x between -40 and 40. And the fact that $\mathbb{Q}(\sqrt{-163})$ has unique factorization is related to the curiosity, observed by Ramanujan, among others, that $e^{\pi\sqrt{163}}$ is very close to an integer.

```
AccountingForm[N[E^(π √163), 35]]
```

262537412640768743.99999999999925007

For more information in these directions an excellent place to start is the book by by D. Flath [Fla]; see also [Els], [EM], [Mol], or [Rib].

Exercises for Section 9.3

9.44. Prove that if q is a rational prime congruent to 3 (mod 4), then q divides $a + bi$ if and only if q divides a and q divides b. Show that this implies that the q^2 values $a + bi$, $0 \le a$, $b < q$, are distinct modulo q.

9.45. Let π be any Gaussian prime. Use the fact that every residue class other than 0 has an inverse modulo π to prove that if π does not divide α, then $\alpha^{N(\pi)-1} \equiv 1 \pmod{\pi}$.

9.46. Show that, for any Gaussian integer $z = a + bi$ with $d = \gcd(a, b)$, the rectangle $\{r + si : 0 \le s < d,\ 0 \le r \le d((a/d)^2 + (b/d)^2)\}$ is a complete residue system modulo z. Implement a routine that, given x and z, returns the entry in this rectangle that is congruent to x (mod z).

9.47. Show that if q is a rational prime congruent to 3 (mod 4), and a is a rational integer not divisible by q, then $\left(\frac{a}{q}\right)_4 = 1$.

9.48. Show that if q is a rational prime congruent to 3 (mod 4), then $\left(\frac{i}{q}\right)_4 = (-1)^{(q+1)/4}$. If π is a Gaussian prime that is not rational with $N(\pi) = p$, then $\left(\frac{i}{\pi}\right)_4 = i^{(p-1)/4}$.

9.49. Prove that a polynomial of degree n has at most n incongruent solutions over the Gaussian integers modulo π, where π is any Gaussian prime. *Hint*: This is similar to the rational case (Theorems 3.17 and 3.18), using Exercise 9.27 as necessary.

9.50. Prove that if π is a Gaussian prime, then the order of any integer in the reduced residue system modulo π must divide $N(\pi) - 1$.

9.51. Prove that if π is a Gaussian prime and d divides $N(\pi) - 1$, then in the reduced residue system modulo π there are exactly $\phi(d)$ elements of order d. In particular, there is at least one element of order $N(\pi) - 1$, which is called a ***primitive root*** for π.

9.52. Each Gaussian prime π (other than $1 + i$ and its associates) has four representations: $\pm \pi$, $\pm \pi i$. Prove that exactly one of them is either congruent to 1 (mod 4) or congruent to $3 + 2i$ (mod 4). This is called the ***primary representation*** of π. For example, $-1 - 2i$ is the primary representation of $1 + 2i$, -7 is the primary representation of 7, and $-5 + 2i$ is the primary representation of $2 + 5i$. Implement a routine that finds primary representations (the built-in Mod can be used to do this).

9.53. There is a law of quartic reciprocity for Gaussian primes. Experiment with pairs of Gaussian primes written in their primary representation until you can state it with confidence. Gauss stated this law in 1832, but he never published a proof. He claimed that the proof "belongs to the mysteries of higher arithmetic." The first proof was published by F. G. Eisenstein in 1844 (see [IR]).

Mathematica Basics

A.0 Introduction

Mathematica is a somewhat unusual programming language. While it has some features in common with BASIC, Pascal, and C, it is really a radical departure from anything that has come before. In fact, the breadth and depth of the program are phenomenal and we can only touch the surface in this brief appendix. For a fuller treatment of the capabilities of *Mathematica* see [Wag, Wol].

A central point is that it is an interpreted language, which means that one can execute commands in real-time, without having to compile them into a program. Just start typing and an Input cell will form (to create other types of cells, type some text, select the cell via its cell bracket on the right, and try some of the entries in the Format:Style menu). Hit the *enter* or *shift-return* keys and your command will be evaluated.

```
2^314
```

```
333747974362642200374222141588992517906672581618226995304225
25122222183215322508594108782608384
```

Here are a few examples that are particularly relevant to the study of number theory. First, a modest factorization.

```
FactorInteger[1000000001]
```

```
{{7, 1}, {11, 1}, {13, 1}, {19, 1}, {52579, 1}}
```

This means that $1000000001 = 7 \cdot 11 \cdot 13 \cdot 19 \cdot 52579$. Here is the thousandth prime.

```
Prime[1000]
```

```
7919
```

Here is how to tell (with a reasonable amount of certainty — see Chapter 4) whether an integer is prime.

```
PrimeQ[10^12 + 1]
```

False

```
PrimeQ[10^30 + 57]
```

True

Here are all the divisors of 10000.

```
Divisors[10000]
```

{1, 2, 4, 5, 8, 10, 16, 20, 25, 40, 50, 80, 100, 125, 200,
  250, 400, 500, 625, 1000, 1250, 2000, 2500, 5000, 10000}

The `PrimePi` function gives the number of primes under x.

```
PrimePi[1000000]
```

78498

Here is the 100th Fibonacci number.

```
Fibonacci[100]
```

354224848179261915075

Of course you can write programs. It is tempting to use `Do`-loops, much as one would do in a traditional languages. Often more advanced techniques will be better, but `Do`-loops have their place, and the next code shows how to implement a loop to compute the sum of the first 10 integers. The semicolons at the end of the lines suppress the output from that line. The phrase {i, 1, 10} is called an *iterator* in this context. The 1 can be omitted. Or a step can be added, as in {i, 1, 10, 2}.

```
sum = 0;
Do[sum = sum + i, {i, 1, 10}];

sum
```

55

There is a built-in `Sum` function, though in fact the `Do`-loop is a good approach when adding thousands of numbers.

```
Sum[i, {i, 1, 10}]
```

55

There are several important notational conventions.

- All built-in *Mathematica* functions begin with a capital letter (functions and variables you define have no such restriction).

- A space will, in most cases, be interpreted as multiplication. thus a b is the same as a*b (but ab denotes a single variable called "ab").
- Round brackets, ( ), are used for grouping, as in 2(3+5).
- Square brackets, [ ], are used for arguments to functions.
- Curly braces, { }, are used to denote lists, such as {1, 2, 3}.

One can use the `Table` command to make a list.

```
Table[i^2, {i, 1, 10}]
```
{1, 4, 9, 16, 25, 36, 49, 64, 81, 100, 121}

One can add a step-size to an iterator as a fourth entry.

```
Table[i^2, {i, 2, 10, 2}]
```
{4, 16, 36, 64, 100}

But some lists, like the integers from a to b, can be obtained more easily using `Range`.

```
Range[10, 20]
```
{10, 11, 12, 13, 14, 15, 16, 17, 18, 19, 20}

And because numerical operations are "listable" — meaning that they work on the elements of a list — this can be a quick way of avoiding programming. Here is an example that squares the even integers, as we just did using `Table`.

```
Range[2, 10, 2]^2
```
{4, 16, 36, 64, 100}

A handy abbreviation is the use of % to refer to the last-computed output.

```
Sqrt[1234.]
```
35.1283

```
%^2
```
1234.

A.1 Plotting

A powerful feature of *Mathematica* is its sophisticated graphics capabilities. Here is a standard plot.

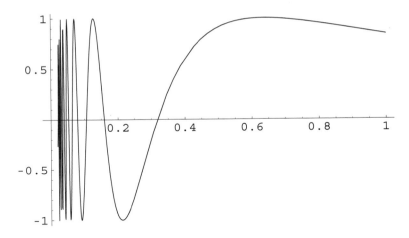

Next we add some options to improve the look.

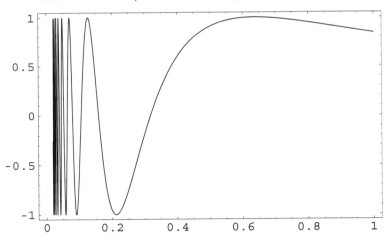

To get information about a function one can first ask for its usage message.

```
? Plot
```

```
Plot[f, {x, xmin, xmax}] generates a plot of f
    as a function of x from xmin to xmax. Plot[{f1,
    f2, ... }, {x, xmin, xmax}] plots several functions fi.
```

```
Options[Plot]
```

$$\left\{ \text{AspectRatio} \to \frac{1}{\text{GoldenRatio}} \text{, Axes} \to \text{Automatic, AxesLabel} \to \text{None,} \right.$$

AxesOrigin → Automatic, AxesStyle → Automatic,
Background → Automatic, ColorOutput → Automatic,
Compiled → True, DefaultColor → Automatic, Epilog → {},
Frame → False, FrameLabel → None, FrameStyle → Automatic,
FrameTicks → Automatic, GridLines → None, ImageSize → Automatic,
MaxBend → 10., PlotDivision → 30., PlotLabel → None,
PlotPoints → 25, PlotRange → Automatic, PlotRegion → Automatic,
PlotStyle → Automatic, Prolog → {}, RotateLabel → True,
Ticks → Automatic, DefaultFont :→ $DefaultFont,
DisplayFunction :→ $DisplayFunction,
FormatType :→ $FormatType, TextStyle :→ $TextStyle}

? MaxBend

MaxBend is an option for Plot which measures the maximum
  bend angle between successive line segments on a curve.

One can also use the Help Browser in the **Help** menu, which provides
examples and more detailed information about each *Mathematica* function.

Often one has data points that need plotting. This is done with List-
Plot. Here is some data that counts the number of divisors of *n*.

data = Table[{n, Length[Divisors[n]]}, {n, 30}]

{{1, 1}, {2, 2}, {3, 2}, {4, 3}, {5, 2}, {6, 4},
 {7, 2}, {8, 4}, {9, 3}, {10, 4}, {11, 2}, {12, 6},
 {13, 2}, {14, 4}, {15, 4}, {16, 5}, {17, 2}, {18, 6},
 {19, 2}, {20, 6}, {21, 4}, {22, 4}, {23, 2}, {24, 8},
 {25, 3}, {26, 4}, {27, 4}, {28, 6}, {29, 2}, {30, 8}}

ListPlot[data];

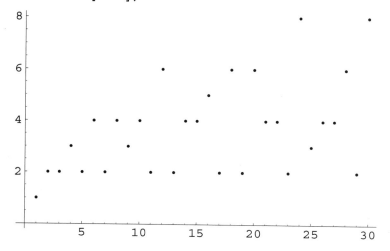

`ListPlot[data, PlotJoined -> True, Frame -> True, Axes -> None];`

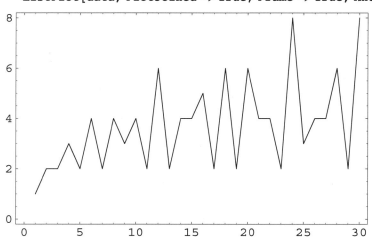

A.2 Typesetting

Mathematica provides an environment that allows simple typesetting of complicated mathematical expressions. And equally important, one can e-mail these expressions, or entire notebooks, to users who are on different platforms. Here are the basic ideas.

Mathematica input and output comes in three flavors: `InputForm`, `StandardForm`, and `TraditionalForm`. The first two are the most important to a beginner. `InputForm` is straightforward inline computer-code style, such as

`Sin[x] Pi / Sqrt[13] * Sum[1 / n ^ (3 / 2), {n, 1, Infinity}]`

`StandardForm` is very close to traditional mathematical notation, but is perfectly precise and unambiguous as input. Here is the `StandardForm` version of the preceding expression.

$$\frac{\mathtt{Sin[x]}\,\pi}{\sqrt{13}} \sum_{n=1}^{\infty} \frac{1}{n^{3/2}}$$

It is easy to convert between the two forms by using the items in the Cell: Convert To menu. One can arrange that inputs and outputs be in either of the two forms by using the Cell:DefaultInputFormatType and Cell:DefaultOutputFormatType menu items. We recommend that beginners use `InputForm` for input and `StandardForm` for output. When you learn more about the typesetting system you might prefer to use `StandardForm` for input too. `StandardForm` is fine for short programs, but longer programs should probably be written in `InputForm`. The following table shows some of the differences in these two forms.

InputForm	StandardForm
`Integrate[f[x], {x, 0, 1}]`	$\int_0^1 f[x]\,dx$
`Sum[1 / n^2, {n, 1, Infinity}]`	$\sum_{n=1}^\infty \frac{1}{n^2}$
`Sqrt[1 + (x / Sqrt[y]) ^ (1 / 3)]`	$\sqrt{1 + \left(\frac{x}{\sqrt{y}}\right)^{1/3}}$
`Product[x + i, {i, 1, 4}]`	$\prod_{i=1}^4 (x + i)$
`LogIntegral[Log[x] / Sqrt[x]]`	$\text{LogIntegral}\left[\frac{Log[x]}{\sqrt{x}}\right]$
`Floor[x / 5]`	$\text{Floor}\left[\frac{x}{5}\right]$

When working in a `StandardForm` cell there are many shortcuts. For example, one can enter an expression in `InputForm`, select the cell and ConvertToStandardForm. This will turn

```
Integrate[1 / Log[x] ^15, {x, 0, Infinity}]
```

into

$$\int_0^\infty \frac{1}{\mathtt{Log[x]}^{15}}\,dx$$

Alternatively, one can use the palettes (use the File:Palettes menu item to make the basic palettes operative). But there are many easy-to-remember escape-key and control-key sequences that make typesetting simple. Table A.1 shows some of the keyboard shortcuts that we have found most useful. The most important ones are CTRL 6 for superscript, CTRL - for subscripts, CTRL 2 for square roots, and CTRL / to create a fraction. These can be found in the Edit:Expression Input menu.

Here's an example to show how simple keystrokes can be used to create the formula

$$\phi(n) \geq n \prod_{i=1}^r \left(1 - \frac{1}{p_i}\right) \geq \frac{n}{\log_2(2n)}$$

CTRL 9	To start an inline cell.
ESC f ESC (n)	f yields ϕ
ESC >= ESC n	OPTION >] or ALT >] also works for this
`Product[1 - 1/p_i, {i, 1, r}]`	Then select this phrase and convert it StandardForm.
ESC >= ESC n CTRL /	
log CTRL -] 2 right arrow(2n)	

Key sequence	Result	Full name
⎡ESC⎤ a ⎡ESC⎤	α	\ [Alpha]
⎡ESC⎤ D ⎡ESC⎤	Δ	\ [CapitalDelta]
⎡ESC⎤ inf ⎡ESC⎤	∞	\ [Infinity]
⎡ESC⎤ p ⎡ESC⎤	π	\ [Pi]
⎡ESC⎤elem⎡ESC⎤	$\in$	\ [Element]
⎡ESC⎤e⎡ESC⎤	$\in$	\ [Epsilon]
⎡ESC⎤ deg⎡ESC⎤	°	\ [Degree]
⎡ESC⎤ -> ⎡ESC⎤	$\rightarrow$	\ [Rule]
⎡ESC⎤ lf ⎡ESC⎤	$\lfloor$	\ [LeftFloor]
⎡ESC⎤ rf ⎡ESC⎤	$\rfloor$	\ [RightFloor]
⎡ESC⎤ * ⎡ESC⎤	$\times$	\ [Times]
⎡ESC⎤ . ⎡ESC⎤	·	\ [CenterDot]
⎡ESC⎤ === ⎡ESC⎤	$\equiv$	\ [Congruent]
⎡ESC⎤ ~ ⎡ESC⎤	~	\ [Tilde]
⎡ESC⎤ ' ⎡ESC⎤	$\prime$	\ [Prime]
⎡ESC⎤ un ⎡ESC⎤	$\cup$	\ [Union]
⎡ESC⎤ int ⎡ESC⎤	$\cap$	\ [Intersection]
⎡ESC⎤ dsZ ⎡ESC⎤	$\mathbb{Z}$	\ [DoubleStruckCapitalZ]
⎡ESC⎤ scP ⎡ESC⎤	$\wp$	\ [DoubleStruckCapitalZ]
x ⎡CTRL⎤6 2	x^2	SuperscriptBox["x", "2"]
x ⎡CTRL⎤– 1	x_1	SubscriptBox["x", "1"]
⎡CTRL⎤2 x	$\sqrt{x}$	SqrtBox["x"]
1 ⎡CTRL⎤ / 23	$\frac{1}{23}$	FractionBox["1", "23"]
⎡CTRL⎤9	begin an inline cell	

Table A.1. Keyboard shortcuts for useful typesetting commands.

For something like $\sum_{d|n} \phi(d) = n$ one would create a normally indexed sum, convert it, and then delete the upper index and edit the lower one into $d\,|\,n$.

▤ Sending Files By E-Mail

If you have a complete file you would like to send via e-mail, then that can be done with the standard *send file* or *attach file* feature of your e-mail software. The user will receive a text-only version of the notebook, but when it is opened in *Mathematica* it will look like a formatted notebook.

If you just wish to send a single cell that is mostly text, use the Edit: Copy As:Text menu item and paste the text into an e-mail message. If the cell has typesetting in it then you can copy it as a Cell object and e-mail that. The recipient will be able to paste it and get it to look like a cell. Note that the internal form of any cell can be viewed by selecting the cell and choosing the Format:Show Expression menu item.

A.3 Types of Functions

A dummy variable is represented by a blank (_). Here is how to define a function f that squares its input. The := indicates a delayed assignment, meaning that the right-side is not evaluated until f is called.

```
f[x_] := x²
```

```
f[15]
```
225

```
f[w]
```
w²

Using `Select` is a convenient way to select from a list numbers having a specified quality. Here are the evens.

```
Select[Range[10], EvenQ]
```
{2, 4, 6, 8, 10}

Next we define a function called good that gives True only if its input is prime and of the form $100k + 1$. Then we can easily find all good numbers below 1000. Note that a check for equality is done via ==, as opposed to just =.

```
good[x_] := PrimeQ[x] && Mod[x, 100] == 1
Select[Range[1000], good]
```
{101, 401, 601, 701}

The preceding discussion concerns the familiar way of applying functions to arguments. But there are slightly less familiar ways, and these are built-into *Mathematica*. For example, we might wish to apply a function to each member of a list. Map accomplishes that.

```
Map[func, {a, b, c, d}]
```
{func[a], func[b], func[c], func[d]}

Or sometimes we have a list but we wish to treat its entries as arguments to a function. This is done by Apply.

```
Apply[func, {1, 2, 3}]
```
func[1, 2, 3]

Here is an example: The greatest common divisor function wants its arguments in sequence, not as a list.

```
GCD[100, 642]
```

```
2
```

When we use a list, nothing happens.

```
GCD[{100, 642}]
```

```
{100, 642}
```

But `Apply` turns the list into a sequence of arguments.

```
Apply[GCD, {100, 642}]
```

```
2
```

Iteration is an important concept. The `Nest` and `NestList` functions apply a given function n times. Here is an example where we apply the squaring function 5 times, starting with 2.

```
sq[x_] := x^2;
Nest[sq, 2, 5]
```

```
4294967296
```

`NestList` causes the entire sequence of iterates to be shown.

```
NestList[sq, 2, 5]
```

```
{2, 4, 16, 256, 65536, 4294967296}
```

Having to define functions such as `sq` and `good` in the examples above can be a bother. For example, if we wish to iterate the cosine function, it is very simple.

```
NestList[Cos, 0.2, 15]
```

```
{0.2, 0.980067, 0.556967, 0.848862, 0.660838, 0.789478,
  0.704216, 0.76212, 0.723374, 0.749577, 0.731977,
  0.743854, 0.735864, 0.741251, 0.737624, 0.740068}
```

There is a construction, called a ***pure function***, that allows us to specify a function without defining it with a separate name. The idea is that # is the generic variable, and & is used to mark the end of the pure function. Thus the squaring function is #^2 &. This can be used as a function.

```
#^2 &[5]
```

```
25
```

But the more important use of this construction is as follows, where we repeat the computation given earlier using `Select`.

```
Select[Range[1000], (PrimeQ[#] && Mod[#, 100] == 1) &]
```

$\{101, 401, 601, 701\}$

This is a very powerful mechanism. For another example, here is how to find all the primes under 5000 having the form $n^2 + 1$.

```
Select[Range[5000], (PrimeQ[#] && IntegerQ[Sqrt[# - 1]]) &]
```

$\{2, 5, 17, 37, 101, 197, 257, 401,$
$\quad 577, 677, 1297, 1601, 2917, 3137, 4357\}$

Fold and FoldList are useful in some special situations.

```
FoldList[f, 0, {a, b, c}]
```

$\{0, f[0, a], f[f[0, a], b], f[f[f[0, a], b], c]\}$

To see this in action, note that pure functions of two or more arguments use #1, #2,... for the variables. The first argument to Fold must be a function of two arguments.

```
Fold[f, 1, {3, 4, 5}]
```

$f[f[f[1, 3], 4], 5]$

We can use this idea to quickly code the left-to-right method of modular exponentiation (see Section 1.4) as follows. Rest[IntegerDigits[n, 2]] gives all the base-two digits except the leading 1.

```
powermodLR1[a_, n_, m_] :=
  Fold[Mod[Mod[#1 #1, m] * If[#2 == 1, a, 1], m] &,
    a, Rest[IntegerDigits[n, 2]]]

powermodLR1List[a_, n_, m_] :=
  FoldList[ Mod[Mod[#1 #1, m] * If[#2 == 1, a, 1], m] &,
    a, Rest[IntegerDigits[n, 2]]]

powermodLR1List[a, 41, Infinity] //. Mod[x_, ∞] :> x
```

$\{a, a^2, a^5, a^{10}, a^{20}, a^{41}\}$

```
{powermodLR[3, 41, 7], powermodLR1[3, 41, 7]}
```

$\{5, 5\}$

A.4 Lists

Lists in *Mathematica* are delineated by braces.

```
s = {1, 2, 3, 9, 8, 7}
```

$\{1, 2, 3, 9, 8, 7\}$

Here are some examples of operations on lists.

Sort[s]

{1, 2, 3, 7, 8, 9}

Reverse[s]

{7, 8, 9, 3, 2, 1}

s[[4]]

9

In StandardForm one can use a special character for double brackets: ESC [[ESC yields ⟦.

s⟦-1⟧

7

Take[s, 3]

{1, 2, 3}

A matrix is just a list of lists

{{1, 2}, {3, 4}}

{{1, 2}, {3, 4}}

MatrixForm[{{1, 2}, {3, 4}}]

$\begin{pmatrix} 1 & 2 \\ 3 & 4 \end{pmatrix}$

IdentityMatrix[3]

{{1, 0, 0}, {0, 1, 0}, {0, 0, 1}}

Matrix multiplication is carried out by a dot: $\begin{pmatrix} 1 & 2 \\ 3 & 4 \end{pmatrix}\begin{pmatrix} 2 \\ 2 \end{pmatrix} = \begin{pmatrix} 6 \\ 14 \end{pmatrix}$.

A = {{1, 2}, {3, 4}};
A.{2, 2}

{6, 14}

Here are three other common operations on matrices.

Transpose[A]

{{1, 3}, {2, 4}}

```
Det[A]
```

-2

```
Eigenvalues[A]
```

$$\left\{ \frac{1}{2} \, (5 - \sqrt{33}\,), \; \frac{1}{2} \, (5 + \sqrt{33}\,) \right\}$$

As a final example, here is how the `IntegerDigits` function, which returns the list of digits of an integer, can be used to decide if an integer is a palindrome (reads the same forward or backward, like 1991). Then we can very easily find all the palindromic primes between, say, 10,000 and 11,000.

```
palindromeQ[x_] :=
 IntegerDigits[x] == Reverse[IntegerDigits[x]]
```

```
palindromeQ[121]
```

```
True
```

```
palindromeQ[1211]
```

```
False
```

```
Select[Range[10000, 11000], PrimeQ[#] && palindromeQ[#] &]
```

$\{10301, 10501, 10601\}$

A.5 Programs

Here is a simple example to show how a `While` loop can be used to add up 1, $1/2^2$, $1/3^2$, until the terms are less than $1/1000$. Note the decimal point after the 1, which turns everything into approximate reals. Try it without the decimal point to see the perfectly precise rational answer.

```
sum = 0; n = 1;
While[ 1/n^2 > 1/1000, sum = sum + 1./n^2; n = n + 1];
sum
```

```
1.61319
```

Note the comma in the `While` phrase that separates the condition from what is being done. And what is being done consists of two statements — change the sum; change the term — separated by a semicolon.

Often in a program one wishes to see the progress of a computation. Here is how to use a `Print` statement to print a tracing message every tenth step.

```
sum = 0; n = 1;
While[ 1/n² > 1/1000 ,
  If[Mod[n, 10] == 0, Print[{n, "terms. The sum is", sum}]];
  sum = sum + 1./n² ; n = n + 1];
sum
```

{10, terms. The sum is, 1.53977}

{20, terms. The sum is, 1.59366}

{30, terms. The sum is, 1.61104}

1.61319

The preceding ideas can be turned into a general program as follows.

```
sumSquareReciprocals[tolerance_] := (sum = 0; n = 1;
  While[ 1/n² > tolerance, sum = sum + 1./n² ; n = n + 1]; sum)
```

```
sumSquareReciprocals[1 / 1000]
```

1.61319

Here is another simple example of a program, one that finds the first prime greater than or equal to the input.

```
nextprime[z_] := (n = z; While[!PrimeQ[n], n = n + 1]; n)
nextprime[100]
```

101

```
nextprime[10¹²]
```

1000000000039

An important point in writing programs is make the auxiliary variables local, so that they are unaffected by any prior definitions. For example, if n was already set to something the value would be lost by the n = z part of the nextprime program. This can be avoided by using Module, whose first argument is a list of symbols that are to be local only. The second argument is a compound statement comprising the main program.

```
nextprime1[z_] := Module[{n},
  n = z; While[!PrimeQ[n], n = n + 1]; n]
```

```
n = 159;
nextprime1[100]
```

101

And n is still 159.

```
n

159
```

A cautionary note: Code such as the following will fail, where the z is both the dummy variable and a variable being incremented. The reason this fails is that when `nextprime2[100]` is called, the $z = z + 1$ part of the program becomes $100 = 100 + 1$, which is of course not allowed and yields error messages. And the loop runs forever because the z in `PrimeQ[z]` never departs from 100. If you run this code you will have to abort the program, which is accomplished by either control-period or ⌘-period.

```
nextprime2[z_] := (While[! PrimeQ[z], z = z + 1]; z)
nextprime2[100]

Set::setraw : Cannot assign to raw object 100.
```

It is often useful to restrict the inputs to, say, odd integers or prime integers, and that can be done with the `n_?OddQ` construction. Here is an example of a function that works only on odd integers n and returns `True` if and only if $n = 5$ or $n > 1$ and if $n \equiv 1$ or 4 (mod 5) then n divides F_{n-1} while if $n \equiv 2$ or 3 (mod 5), then n divides F_{n+1}. (Note: If $n \equiv 0$ (mod 5) then it is not hard to see than n does not divide F_{n+1}, and this is why the 0 case is ignored in the code.)

```
FibonacciPseudoprimeTest[n_?OddQ] := n == 5 ||
    (n > 1 && 0 == Mod[
        Fibonacci[n - If[MemberQ[{1, 4}, Mod[n, 5]], 1, -1]], n])
```

And here is how one would find composite numbers that pass the Fibonacci pseudoprime test (see Exercise 1.38).

```
Select[Range[1000], !PrimeQ[#] && FibonacciPseudoprimeQ[#] &]

{323, 377}
```

A.6 Solving Equations

Mathematica uses substitution rules, such as `x -> 2` (same as $x \to 2$), to replace every occurrence of `x` in an expression by 2. One reason this is important is that the global value of `x` is unchanged. The replacement is carried out by the slash-dot operator, `/.`.

```
x^2 + 2 /. x -> 2

6


x

x
```

And this explains why the solutions to equations are given as substitution rules.

```
soln = Solve[x³ + 7 x² - 5 x - 35 == 0, x]
```

$\{\{x \to -7\}, \{x \to -\sqrt{5}\}, \{x \to \sqrt{5}\}\}$

```
x /. soln[[2]]
```

$-\sqrt{5}$

Solve works only for equations having a solution that can be expressed in terms of standard functions. In general, fifth-or-higher degree polynomials do not have such solutions. One can always try to use Solve. Be aware that solutions often involve imaginary numbers.

```
soln = Solve[Cos[2 x] Sin[x] == 1/2, x]
```

$$\left\{\left\{x \to -\text{ArcSin}\left[\frac{(9-\sqrt{57})^{1/3}}{2\ 3^{2/3}} + \frac{1}{(3\ (9-\sqrt{57}))^{1/3}}\right]\right\},\right.$$

$$\left\{x \to \text{ArcSin}\left[\frac{(1+I\sqrt{3})\ (9-\sqrt{57})^{1/3}}{4\ 3^{2/3}} + \frac{1-I\sqrt{3}}{2\ (3\ (9-\sqrt{57}))^{1/3}}\right]\right\},$$

$$\left.\left\{x \to \text{ArcSin}\left[\frac{(1-I\sqrt{3})\ (9-\sqrt{57})^{1/3}}{4\ 3^{2/3}} + \frac{1+I\sqrt{3}}{2\ (3\ (9-\sqrt{57}))^{1/3}}\right]\right\}\right\}$$

Here are the numerical values of the three solutions.

```
nSoln = N[x /. soln]
```

$\{-1.08573, 0.434174 - 0.319561\ I, 0.434174 + 0.319561\ I\}$

We can check them using the listability of sine and cosine.

```
Cos[2 nSoln] Sin[nSoln]
```

$\{0.5, 0.5 - 2.77556 \times 10^{-17}\ I, 0.5 + 2.77556 \times 10^{-17}\ I\}$

Small imaginaries are caused by roundoff error; Chop gets rid of small quantities.

```
Chop[%]
```

$\{0.5, 0.5, 0.5\}$

Sometimes there are built-in functions that can help solve equations that you might think are unsolvable. The ProductLog function is useful for some simple equations involving powers and logarithms.

```
Solve[x 10^x == 2, x]
```

$$\left\{\left\{x \to \frac{\text{ProductLog}[2\,\text{Log}[10]]}{\text{Log}[10]}\right\}\right\}$$

Of course, many equations cannot be solved symbolically but have to be approached numerically using Newton's method or the like. This is accomplished by `FindRoot`, but we must provide a starting value.

```
FindRoot[(x^2 + 1) 10^x == 2, {x, 1}]
```

$$\{x \to 0.270386\}$$

A.7 Symbolic Algebra

Here are a few examples to show the power of symbolic computation. First, an easy geometric series.

```
Sum[1 / 2^k, {k, 1, ∞}]
```

$$1$$

Here is the sum of a finite geometric series. We switch to `StandardForm`.

$$\sum_{k=1}^{n} \frac{1}{2^k}$$

$$2^{-n}\,(-1 + 2^n)$$

Here is an elementary sum.

$$\sum_{i=1}^{n} i$$

$$\frac{1}{2}\,n\,(1 + n)$$

Here is one that is harder to remember.

$$\sum_{i=1}^{n} i^2$$

$$\frac{1}{6}\,n\,(1 + n)\,(1 + 2\,n)$$

And here is an infinite geometric series with x as the ratio.

$$\sum_{n=0}^{\infty} x^n$$

$$\frac{1}{1 - x}$$

Here is a more complicated infinite series.

$$\sum_{n=0}^{\infty} \frac{x^n}{(n+2)!}$$

$$\frac{-1+E^x-x}{x^2}$$

And of course derivatives and (many) integrals can be done effortlessly.

$$D\left[\text{Sin}[x]\,\text{Cos}[x^2]^{17},\, x\right]$$

$$\text{Cos}[x]\,\text{Cos}[x^2]^{17} - 34\, x\, \text{Cos}[x^2]^{16}\,\text{Sin}[x]\,\text{Sin}[x^2]$$

$$\int \frac{1}{1+x^4}\, dx$$

$$\frac{\text{ArcTan}\left[\frac{-\sqrt{2}+2\,x}{\sqrt{2}}\right]}{2\sqrt{2}} + \frac{\text{ArcTan}\left[\frac{\sqrt{2}+2\,x}{\sqrt{2}}\right]}{2\sqrt{2}} - \frac{\text{Log}\left[-1+\sqrt{2}\,x-x^2\right]}{4\sqrt{2}} + \frac{\text{Log}\left[1+\sqrt{2}\,x+x^2\right]}{4\sqrt{2}}$$

Lucas Certificates Exist

In Section 8.2 we discussed how to obtain a Lucas certificate that proves an integer n is prime. However, it remains to prove that such certificates always exist for the method as described in that section. We present such a proof in this appendix.

Let p be an odd prime and d a quadratic nonresidue modulo p. We shall prove that among the $p^2 - p$ pairs of integers a, b for which $1 \le a < p$ and $0 \le b < p$, exactly $(p-1)\phi(p+1)$ of them are such that the Lucas sequence defined by $P = 1$ and $Q \equiv (1 - a^{-2}b^2 d)/4 \pmod{p}$ will certify the primality of p. In particular, a Lucas certificate of primality (as defined in Section 8.2) exists.

We will be working with numbers of the form $a + b\sqrt{d}$ where a and b are integers. We first need a divisibility result for these generalized integers. To say that p divides $a + b\sqrt{d}$ means that $a + b\sqrt{d} = p(u + v\sqrt{d})$ for some pair of integers u and v, and therefore (see Exercise 7.19) p divides a and p divides b.

Lemma. Suppose that p is an odd prime and d is a quadratic nonresidue modulo p. Then if p divides $(a + b\sqrt{d})(f + g\sqrt{d})$, then p divides $a + b\sqrt{d}$ or p divides $f + g\sqrt{d}$.

Proof. If p divides $(a + b\sqrt{d})(f + g\sqrt{d}) = af + bgd + (bf + ag)\sqrt{d}$, then p divides $(a - b\sqrt{d})(f - g\sqrt{d}) = af + bgd - (bf + ag)\sqrt{d}$. It follows that p^2 divides $(a^2 - db^2)(f^2 - dg^2)$, and so p divides $a^2 - db^2$ or $f^2 - dg^2$. We can assume it divides $a^2 - db^2$. If p does not divide b, then b has an inverse modulo p and $(ab^{-1})^2 \equiv d \pmod{p}$, which contradicts the fact that d is a nonresidue modulo p. Therefore p divides b and so it must divide a. $\quad\square$

We now do our arithmetic on these ***extended residues*** modulo p. To say that

$$a + b\sqrt{d} \equiv f + g\sqrt{d} \pmod{p}$$

means that p divides the difference of the left and right sides, so $a \equiv f$ and $b \equiv g$. This residue system has p^2 residues, $p^2 - 1$ of which are relatively prime to p.

The lemma implies that any monic polynomial of degree n has at most n incongruent solutions modulo p (as in Theorem 3.17; see Exercise B.3). In a manner exactly analogous to what we did in Section 3.5 where we proved that there are $\phi(p - 1)$ primitive roots modulo p, it is possible to use this lemma to prove that there are exactly $\phi(p^2 - 1)$ residues of order $p^2 - 1$ in the set of extended residues modulo p. We leave this for the exercises where you are led through the proof that the order of each extended residue that is relatively prime to p must divide $p^2 - 1$, and if m divides $p^2 - 1$, then there are $\phi(m)$ extended residues of order m.

Among the $p^2 - 1$ extended residues that are not divisible by p, $p - 1$ of them are rational integers. These satisfy the congruence $x^{p-1} \equiv 1 \pmod{p}$. Since this congruence has at most $p - 1$ solutions in the extended residues (Exercise B.3), we know that the order of $\beta = a + b\sqrt{d}$ divides $p - 1$ if and only if p divides b (if and only if β is a rational integer).

Now consider the Lucas sequence defined by parameters P and Q with $d = P^2 - 4Q$ and $\alpha = (P + \sqrt{d})/2$. Suppose that p is a prime such that d is a nonresidue modulo p. Then the Lucas sequence will certify the primality of p if and only $p + 1$ is the smallest power of α that is congruent to a rational integer modulo p. This is true if and only if the order of α is $(p^2 - 1)/t$ where t is relatively prime to $p + 1$. The divisors of $p^2 - 1$ that are relatively prime to $p + 1$ are precisely the odd divisors of $p - 1$.

If $a + b\sqrt{d}$ has order $(p^2 - 1)/t$ where t is an odd divisor of $p - 1$, then any Lucas sequence with $P \equiv 2a$ and $Q \equiv a^2 - b^2 d \pmod{p}$ will certify the primality of p.

The certification protocol we presented restricts the search to Lucas sequences with $P = 1$. This is not a significant restriction. Let $\beta = a + b\sqrt{d}$ be a residue for which $p + 1$ is the smallest power that is congruent to a rational integer. If p does not divide a, then this will still be true of $(2a)^{-1}\beta$. The corresponding Lucas sequence is $P = 1$, $Q \equiv (1 - a^{-2}b^2 d)/4 \pmod{p}$.

Exercises for Appendix B

B.1. Imitate the proof of Fermat's Little Theorem to prove that for any extended residue, β, modulo p that is not divisible by p, $\beta^{p^2-1} \equiv 1 \pmod{p}$.

B.2. Prove that the order of β modulo p must divide $p^2 - 1$.

B.3. Prove that $x^n \equiv 1 \pmod{p}$ can have at most n incongruent roots among the extended residues modulo p.

B.4. In Section 3.5, we used the fact that $x^n \equiv 1 \pmod{p}$ has at most n solutions modulo p to prove that for each m that divides evenly into $p - 1$, there are exactly $\phi(m)$ integers modulo p with order m. In a precisely analogous manner, prove that if m divides $p^2 - 1$, then exactly $\phi(m)$ extended residues have order m.

B.5. Let $p - 1 = 2^e m$ where m is odd. Show that

$$\sum_{t \mid m} \phi\left(\frac{p^2 - 1}{t}\right) = (p - 1)\phi(p + 1).$$

B.6. Show that if $P = 1$ and Q is chosen at random from among those integers for which $1 - 4Q$ is a nonresidue modulo p, then the probability that this sequence will certify the primality of p is $\phi(p + 1)/p$.

B.7. Check computationally the main claim of this appendix: Given a prime p and a quadratic nonresidue d for p, then among all the pairs a, b with $1 \leq a < p$, $0 \leq b < p$, exactly $(p - 1)\phi(p + 1)$ of them are such that the Lucas sequence defined by $P = 1$ and $Q \equiv (1 - a^{-2}b^2 d)/4 \pmod{p}$ has the property that $\omega(p) = p + 1$.

References

[AS]   W. Adams and D. Shanks, Strong primality tests that are not sufficient, *Mathematics of Computation*, 39, 255–300 (1982).

[AGP] W. R. Alford, A. Granville, and C. Pomerance, There are infinitely many Carmichael numbers, *Annals of Mathematics*, 140, 703–722 (1994).

[BS]   E. Bach and J. Shallit, *Algorithmic Number Theory*, MIT Press, Cambridge, MA (1996).

[BW]   R. Baillie and S. S. Wagstaff, Jr., Lucas pseudoprimes, *Mathematics of Computation*, 35, 1391–1417 (1980).

[Ble] D. Bleichenbacher, *Efficiency and Security of Cryptosystems Based on Number Theory,* ETH Ph.D. dissertation 11404, Swiss Federal Instititute of Technology, Zurich (1996).

[BB]   J. Borwein and P. Borwein, *Pi and the AGM: A Study in Analytic Number Theory and Computational Complexity*, Wiley, New York (1987).

[BLS] J. Brillhart, J. L. Selfridge, and D. H. Lehmer, New primality criteria and factorizations of $2^m \pm 1$, *Mathematics of Computation*, 39, 620–647 (1975).

[BC]   J. P. Buhler and R. E. Crandall, On the convergence problem for lattice sums, *Journal of Physics, Series A: Mathematical and General*, 23, 2523–2528 (1990).

[BW1] J. Buhler and S. Wagon, Secrets of the Madelung constant, *Mathematica in Education and Research*, 5:2, 49–55 (1996).

[CELV] F. W. Clarke, W. N. Everitt, L. L. Littlejohn, and S. J. R. Vorster, H. J. S. Smith and the Fermat two squares theorem, *American Mathematical Monthly*, 106, 652–665 (1999).

[Coh] H. Cohen, *A Course in Computational Algebraic Number Theory*, Springer-Verlag, Berlin (1993).

[CB]   R. E. Crandall and J. P. Buhler, Elementary function expansions for Madelung constants, *Journal of Physics, Series A: Mathematical and General*, 20, 5497–5510 (1987).

[CP]   R. Crandall and C. Pomerance, *Prime Numbers: A Computational Perspective,* Springer-Verlag, New York (2000).

[DR]   M. Déléglise and J. Rivat, Computing $\pi(x)$: The Meissel, Lehmer, Lagarias, Miller, Odlyzko method, *Mathematics of Computation*, 65, 235–45 (1996).

[DM]   E. Dunne and M. McConnell, Pianos and continued fractions, *Mathematics Magazine*, 72, 104–115 (1999).

[ELS]   R. B. Eggleton, C. B. Lacampagne, and J. L. Selfridge, Euclidean quadratic fields, *American Mathematical Monthly*, 99, 829–837 (1992).

[EM]   J. Esmonde and M . R. Murty, *Problems in Algebraic Number Theory*, Springer-Verlag, New York (1999).

[Fla]   D. Flath, *Introduction to Number Theory*, Wiley, New York (1989).

[FW]   D. Flath and S. Wagon, How to pick out the integers in the rationals: An application of number theory to logic, *American Mathematical Monthly*, 98, 812–823 (1991).

[For]   K. Ford, The distribution of totients, *Electronic Research Announcements of the American Mathematical Society*, 4, 27–34 (1998).

[GW]   J. Gallian and S. Winters, Modular arithmetic in the marketplace, *American Mathematical Monthly*, 95, 548–551 (1988).

[Gal]   J. Gallian, Error detection methods, *ACM Computing Surveys*, 28:3, 504–516 (Sept. 1996).

[GS]   E. Gethner and H. M. Stark, Periodic Gaussian moats, *Experimental Mathematics*, 6, 289–292 (1997).

[GWW]   E. Gethner, S. Wagon, and B. Wick, A stroll through the Gaussian primes, *American Mathematical Monthly*, 104, 327–337 (1998).

[GR]   D. M. Gordon and G. Rodemich, Dense admissible sets, *Proceedings of Algorithmic Number Theory Symposium III*, Lecture Notes in Computer Science 1423, Springer-Verlag, New York, 216–225 (1998).

[Gro]   E. Grosswald, *Representations of Integers as Sums of Squares*, Springer-Verlag, New York (1985).

[Guy]   R. K. Guy, *Unsolved Problems in Number Theory*, 2nd ed., Springer-Verlag, New York (1994).

[HW]   G. H. Hardy and E. M. Wright, *An Introduction to the Theory of Numbers*, 5th ed., Oxford Univ. Pr., Oxford (1979).

[Hod]   A. Hodges, *Alan Turing: The Enigma*, Simon and Schuster, New York (1983).

[HB]   R. H. Hudson and A. Brauer, On the exact number of primes in the arithmetic progressions $4n \pm 1$ and $6n \pm 1$, *Journal für die Reine und Angewandte Mathematik*, 291, 23–29 (1977).

[IR]   K. Ireland and M. Rosen, *A Classical Introduction to Modern Number Theory*, Springer-Verlag, New York (1982).

[Jon]   J. Jones, Diophantine representation of the Fibonacci numbers, *Fibonacci Quarterly*, 13, 84–88 (1975).

[JSWW]   J. P. Jones, D. Sato, H. Wada, and D. Wiens, Diophantine representation of the set of prime numbers, *American Mathematical Monthly*, 83, 449–464 (1976).

[JR]   J. H. Jordan and J. R. Rabung, Local distribution of Gaussian primes, *Journal of Number Theory*, 8, 43–51 (1976).

[Kah]   D. Kahn, *The Codebreakers: The Story of Secret Writing*, Macmillan, New York (1983).

[KW]   V. Klee and S. Wagon, *Unsolved Problems in Plane Geometry and Number Theory*, Mathematical Association of America, Washington, DC (1991).

[Knu]   D. E. Knuth, *The Art of Computer Programming*, vol. 2, Addison-Wesley, Reading, MA (1971).

[KSW]   G. C. Kurtz, D. Shanks, and H. C. Williams, Fast primality tests for numbers less than $50 \cdot 10^9$, *Mathematics of Computation*, 46, 691–701 (1986).

[Leh]   E. Lehmer, On the infinitude of Fibonacci pseudo-primes, *Fibonacci Quarterly*, 2, 229–230 (1964).

[Mar]   G. Martin, The smallest solution of $\phi(30n + 1) < \phi(30n)$ is ..., *American Mathematical Monthly*, 106, 449–451 (1999).

[Mat]   Y. Matijasevich, *Hilbert's Tenth Problem*, MIT Press, Cambridge, MA (1993).

[Mau]   U. Maurer, Fast generation of prime numbers and secure public-key cryptographic parameters, *Journal of Cryptology*, 9, 123–155 (1995).

[Mor]   S. B. Morris, *Magic Tricks, Card Shuffling, and Dynamic Computer Memories*, Mathematical Association of America, Washington, DC (1998).

[NZM]   I. Niven, H. S. Zuckerman, and H. L. Montgomery, *An Introduction to the Theory of Numbers*, 5th ed., Wiley, New York (1991).

[PW]   E. Packel and S. Wagon, *Animating Calculus*, Springer/TELOS, New York (1997).

[PSW]   C. Pomerance, J. L. Selfridge, and S. S. Wagstaff, Jr., The pseudoprimes to 25×10^9, *Mathematics of Computation*, 35, 1003–1026 (1980).

[RWW]   J. Renze, S. Wagon, and B. Wick, The Gaussian zoo, forthcoming.

[Rib]   P. Ribenboim, *The New Book of Prime Number Records*, 3rd ed., Springer-Verlag, New York (1996).

[Rie]   H. Riesel, *Prime Numbers and Computer Methods for Factorization*, Birkhäuser, Boston (1985).

[Ros]   K. H. Rosen, *Elementary Number Theory and Its Applications*, 3rd ed., Addison-Wesley, Reading, MA (1993).

[RS]   M. Rubenstein and P. Sarnak, Chebyshev's bias, *Experimental Mathematics*, 3, 173–197 (1994).

[SW]   A. Schlafly and S. Wagon, Carmichael's conjecture is valid below $10^{10,000,000}$, *Mathematics of Computation*, 63, 415–419 (1994).

[Schn] B. Schneier, *Applied Cryptography*, 2nd ed., Wiley, New York (1996).

[Schr] M. Schroeder, *Number Theory in Science and Communication*, Springer-Verlag, Berlin (1984).

[SRK]   A. S. Sethi, V. Rajaraman, and P. S. Kenjale, An error-correcting coding scheme for alphanumeric data, *Information Processing Letters*, 7, 72–77 (1978).

[Tuc]   B. Tuchman, *The Zimmermann Telegram*, Macmillan, New York (1966).

[Var]   I. Vardi, Prime percolation, *Experimental Mathematics*, 7, 275–289 (1998).

[Var1] I. Vardi, Archimedes' cattle problem, *American Mathematical Monthly*, 105, 305–319 (1998).

[Wag]   S. Wagon, *Mathematica in Action*, 2nd ed., Springer/TELOS, New York (1999).

[Wil]   H. Wilf, *Generatingfunctionology*, Academic Press, San Diego (1994).

[Wil1] H. C. Williams, A numerical investigation into the length of the period of the continued fraction expansion of $\sqrt{d}$, *Mathematics of Computation*, 36, 593–601 (1981).

[Wil2] H. C. Williams, *Édouard Lucas and Primality Testing*, Wiley, New York (1998).

[Wol]   S. Wolfram, *The Mathematica Book*, 4th ed., Cambridge Univ. Press, New York (1996).

[Zag]   D. Zagier, The first 50 million prime numbers, *The Mathematical Intelligencer*, 0, 7–19 (1977).

[Zag1] D. Zagier, A one-sentence proof that every prime $\equiv 1 \pmod 4$ is a sum of two squares, *American Mathematical Monthly*, 97, 144 (1990).

Mathematica Index

Items in boldface are from the CNT package; items in italics denote symbols defined in the book, but not in the CNT package.

@@, Apply, 51
// (function application), xviii
/., ReplaceAll, 19–20, 347–348
_, Blank, xviii
__, BlankSequence, xviii
___, BlankNullSequence, xviii
==, Equal, xviii
===, SameQ, xviii
#, Slot, 62, 342–343
#1, 33
#2, 33
&, Function, 342, 343
&&, And, 343
:>, RuleDelayed, 73
% (preceding output), 335
? (show information), 336
:=, SetDelayed, 341

AbundantQ, 76
AccountingForm, 329
AdditionChain, 32, 37–38
AdmissibleQ, 312, 315–316
AliquotSequence, 77–78
AmicableQ, 76
And, 312
animals, 315–316
Apply, 22, 44, 51, 299, 341–342
AspectRatio, 311, 316
AssumedPrimeBound, 126
Assumptions, 44

Attributes, 123
Average, 71
AverageList, 71
Axes, 336
AxesLabel, 36

BinaryLeftToRight, 32
BinaryRightToLeft, 37
Binomial, 29
BirthDate, 301
BLSPrimeProof, 121–122

CarmichaelLambda, 94
CarmichaelQ, 61, 98
Cell, 340
CentralRemainder, 281, 284
Certificate, 132
CertifiedPrime, 100, 127–128
CFFactor, 243–245
CFPeriodLength, 220
CFSqrt, 222
ChineseRemainder, 53
Chop, 348
Circle, 311
CNT package, xii, 6
Collect, 214, 230
ColumnForm, 29
Columns, 66
ComplexGCD, 305
Conductor, 47

ConstellationEstimate, 317–318, 325
ContinuantQ, 10–12
ContinuedFraction, 207, 217
ContinuedFractionForm, 202
Convergents, 203, 207, 209
CorrectPrime, 134
CRT, 51–52
CRTGrid, 50

DeficientQ, 76
DeleteCases, 241
Denominator, 79, 214
Det, 345
DiophantineSolve, 42–43
Divisors, 72, 299, 305, 319, 334
DivisorSigma, 72
Do, 33, 334
Drop, 10

EALength, 23
Eigenvalues, 345
Epilog, 36
Eratosthenes, 107
EratosthenesTable, 106
Euclid, 111, 120, 122
EvaluateIntegral, 317
EvenQ, 341
Expand, 43, 224, 230
ExpandDenominator, 214
ExponentForm, 62, 68
exponentsBinary, 241
ExtendedGCD, 15, 28, 42
ExtendedGCDValues, 15, 204, 282, 306

Factor, 225
FactorForm, 6, 17, 305
FactorInteger, 6, 17, 305, 319, 321, 333
FactorMethod, 37
FermatFactor, 170
FermatNumber, 183
Fibonacci, 3, 12, 334
FindComponent, 311
FindRoot, 349
FirstSPSPWitness, 117
Fit, 36
FixedPointList, 71
Flatten, 78, 187
Floor, 7, 176
Fold, 33, 190, 343
FoldList, 33, 297, 343
FourthPowerQ, 328–329
Frame, 310, 336

FrameTicks, 310
FromCFSquared, 224
FromContinuedFraction, 203, 217
FromDigits, 88
FromSRKInteger, 167
FullContinuant, 204
FullContinuedFraction, 203–204, 217
FullGCD, 8–9, 12, 15, 204, 282, 305–306
FullPrattCertificate, 126
FullSimplify, 223
FunctionHeading, 70, 171
FunctionTable, 66, 70, 171

GaussianIntegerQ, 326–328
GaussianIntegers, 305, 315, 319, 321
GaussianMod, 326–328
GaussianNorm, 326–328
GaussianPrimePlot, 308
GaussianPrimeQ, 309–310, 326
GCD, 9, 61, 305–306, 319
GeneralSum2Squares, 285
Get, 11
GoldenRatio, 3–4, 207

HarmonicRearrangement, 294–295
HoldForm, 60, 173–174

I, 303
IdentityMatrix, 248–249, 344
If, 346
Im, 299, 303, 307
InputForm, 338
IntegerDigits, 31, 33
IntegerExponent, 116
IntegerToString, 149, 152

jacobi, 189
jacobiIter, 191
JacobiSymbol, 189, 250

λ, 94–95
Labels, 12, 15
LCM, 9, 51
Legendreφ, 135–136
LegendrePi, 137
LegendrePiFull, 137
LegendreSymbol, 185
li, 103–104
LinearCongruenceSolve, 45
LinearDiophantineSolve, 43–44
LinearRecurrence, 249, 255
LinearSolve, 45

Listable, 123
ListPlot, 220, 297, 337
Log, 36
Log*, 95
LogIntegral, 103
LogListPlot, 77
LogStar, 95
LucasΛ, 258
LucasΦ, 257–258
LucasBoth, 254–255
LucasCertificate, 260–261
LucasParameters, 265–266
LucasPseudoprimeQ, 264–265, 269
LucasPseudoprimesAllTypes, 270–271
LucasPseudoprimeTest, 264, 270
LucasRank, 258
LucasUFast, 255
LucasUMat, 255
LucasUPair, 254–255
LucasVFast, 255
LucasVMat, 255
LucasVPseudoprimeTest, 271

M2Circle, 296
M2SquareShell, 295
M3CubeShell, 296–297
M3Sphere, 297
Map, 3, 36, 341
MatrixForm, 3, 230, 344
MatrixPower, 3, 230
MatrixPowerMod, 34, 249
MeisselPi, 139
MemberQ, 116
Mersenne, 62
MersennePrimeQ, 264
Message, 173–174
MethodA, 269–270
MethodAStar, 269, 271
MillerRabinPrimeQ, 118
Mod, 7
Module, 33
Modulus, 45, 87, 197

N, 61, 237, 318
Needs, 6
Nest, 264, 342
NestList, 60, 95, 116, 342
NextPrime, 194, 244
Nonresidue, 196, 280
NullSpace, 242–243
NumberTheory`ContinuedFractions`
    package, 202

NumberTheory`NumberTheoryFunctions`
    package, 285–286, 307
NumberTheory`PrimeQ` package, 132
Numerator, 214

OddQ, 120, 347
Options, 315, 319, 336
Or, 116
OrderMod, 89
Outer, 187

π, 52, 61, 101, 103, 138
φ, 67–68
PairwiseCoprimeQ, 51
Partition, 241–243
PascalTriangle, 29
PellSolutions, 231
PellSolve, 231, 236
PhiInverse, 69
PhiMultiplicity, 69–70, 72
Plot, 36, 336
PlotPoints, 336
Plus, 44
PollardpMinus1, 174–175
PollardRho, 173
Position, 78
PowerAlgorithm, 31–32, 37
PowerGrid, 56
PowerMod, 33
powermodLR, 190–191
PrattCertificate, 122–124, 126, 130
PrattCertificateBasic, 123, 130
Prime, 52, 99, 134, 333
PrimeAndCertificate, 129
PrimePi, 101, 334
PrimeQ, 99, 101, 305, 319, 333
PrimeQWithLucasProof, 260
PrimeQWithProof, 120, 122
PrimitiveReps, 284, 287
PrimitiveRoots, 85, 90
Print, 345
Product, 305
ProductLog, 348–349
Protect, 95
ProvablePrimeQ, 132
PseudoprimeQ, 69, 271
PseudoprimeSearch, 265–266
PseudoprimeTable, 60
PythagoreanTriples, 20

QuadraticConjugate, 213–214, 250
QuadraticMod, 250

QuadraticModuli, 181
QuadraticNormalForm, 214
QuadraticPowerMod, 250, 267
QuadraticPseudoprimeQ, 265, 269
QuadraticPseudoprimeTest, 265, 270
QuadraticResidueQ, 181, 185
QuarticResidueSymbol, 328
QuickPrimeQ, 127
Quotient, 6, 22, 176

r_2, 289
Random, 9, 35
RandomPrime, 104–105, 147, 149, 183,
            281–282, 306
Range, 335
Rationalize, 318
Re, 303, 307
RealDigits, 88–89
ReducedResiduesOnly, 81
ReplacePart, 248–249
Repunit, 64
Rest, 10
Reverse, 344
RhoTrials, 177
RightSide, 231
RootReduce, 214, 224
RotateRight, 248–249
Round, 209, 318
RSADecode, 151
RSAEncode, 150, 154

σ, 72
σ^-, 72, 74–75
Scan, 242–243
Select, 54–55, 59, 341, 343
SetAttributes, 326
Short, 78
Show2s, 123–124, 126
ShowCertificate, 127–128
ShowData, 243–245
ShowPrimitiveRoots, 57
Simplify, 24
SmallSqrtsNegOne, 284
Solve, 45, 87, 197, 349
Sort, 82, 344
SpspSequence, 116
Sqrt, 335
SqrtFloor, 170
SqrtModAll, 197, 234

SqrtModPrimeTonelli, 196–197
SqrtNegOne, 280, 306
SqrtSquareDivisors, 284
SRKCorrect, 165
SRKProtect, 165
StandardForm, 338–339
StringForm, 120–121, 260
StringToInteger, 147, 149, 152
StrongPseudoprimeData package, 115
StrongPseudoprimeQ, 116–117
StrongPseudoprimes, 115, 275
StrongQuadraticPseudoprimeQ, 275
StrongQuadraticPseudoprimeTest, 275
Subtract, 187
Sum, 334, 349
Sum2Squares, 281
SumOfSquaresR, 285–286
SumOfSquaresRepresentations, 285, 307,
            328

Table, 9–12, 335
Take, 10
Times, 51
Timing, 9, 35
ToCharacterCode, 147
Together, 110
ToSRKInteger, 166
TowerMod, 96–97
TraceIntervalFactor, 245
TraditionalForm, 14, 338
Transpose, 242, 344
Tree, 124, 126, 260–261
TrialDivide, 168
TwoPseudoprimeQ, 59
TwoPseudoprimeTest, 59, 62
Type, 265–266

Unprotect, 95
UpArrow, 95
UseGridBox, 274–275
UVSequence, 274–275

VisualPowerGrid, 57, 81

While, 22, 345–346

$\mathbf{Z}_n^*$, 82, 93

$RecursionLimit, 135

Subject Index

absolute value, 303
absolutely convergent, 294
abundant number, 76
Adams, W., 355
addition chain, 31
Adleman, L., 131, 149, 176
admissible animal, 315
admissible pattern, 312
Alford, W. R., 61, 355
algorithm, 2
aliquot cycle, 80
aliquot sequence, 77
    problem, 77
alternating harmonic series, 294
American Chemical Society, 162
amicable pair, 75
Anderson, C. W., 78
animal, 315
Archimedes, 227, 232–233, 238
    sun god cattle problem, 232
associate, 303
asymptotic, 177
Atkin, A. O. L., 132
average, 286

b-sequence, 114
b-strong pseudoprime, 114
Bach, E., 355
Bailey, D., 53
Baillie, R., 64, 269, 271, 355
Belgian house number problem, 232
Bertrand, J. L. F., 110
Bertrand's Postulate, 110, 112
    extension, 142
big-O notation, 34
billiard balls, 232
Binet's formula, 4
binomial coefficient, 29

birthday paradox, 171
birthday trick, 301
Bleichenbacher, D., 117, 128, 355
Borwein, J., 355
Borwein, P., 355
Brauer, A., 356
Brent, R. P., 172, 175
Brillhart, J., 121, 131, 175–176, 355
Brizolis, D., 89
Buhler, J. P., 302, 355

Cabal, The, 176
Carmichael, R. D., 69
Carmichael number, 61, 97, 132
Carmichael's conjecture, 69
Carmichael's lambda, 93
Catalan's equation, 41
certificate of primality, 118, 120–121
    cyclotomy test, 131
certified prime, 126
CF, *see* continued fraction
CFRAC, 245
character, 325
Chebyshev, P., 109–110
Chebyshev's theorem, 109
check digit, 158
Chicken McNugget problem, 46
Chinese Remainder Theorem, 49, 51
Cipolla, M. 64
Clark, F. W., 355
coconut problem, 48
Cohen, H., 118, 131, 355
coin tossing, 146, 153
complete reduced residue system, 82
complete residue system, 56
complex number, 302
composite, 6
conditionally convergent, 294

conductor, 46
congruent, 26, 325
conjugate, 213, 303
constellation, 314
    superdense, 315
continuant, 10
continuant reversal, 12
continued fraction, 201
    convergent, 203
    existence and uniquenes, 208
    factoring with, 238
    period length, 220
    periodicity for quadratic irrational, 215
    purely periodic, 221
Costa Pereira, N., 109
Crandall, R. E., 176, 302, 355
Cunningham Project, 176
Curry, C., 176
cyclotomy test, 131

Davenport, J. H., 129
de la Vallée-Poussin, C.-J., 103
deficient number, 76
Deléglise, M., 137, 355
Delord, J., 297
dense, 25
Descartes, R., 75, 79, 289
Diffie, W., 149
dihedral group of pentagon, 160
Diophantus, 41
direct product, 50
Dirichlet's theorem, 112
discriminant, 247, 329
divide, 6, 303
divisible, 6
divisor, 6
    sum of, 72
Dubner, H., 64
Dubouis, E., 289
Dunne, E., 355

Eggleton, R. B., 356
Eisenstein, F. G., 327, 331
Eratosthenes, 105
Erdős, P., 29, 108, 110
Esmonde, J., 356
Euclid, 6, 79
Euclid number, 111
Euclidean algorithm, 8
Euler, L., 182, 329
Euler phi function, 66
Euler pseudoprime, 193
Euler–Euclid formula, 74
Euler's criterion, 180
Euler's theorem, 82
Everitt, W. N., 355

extended Euclidian algorithm, 14
extended integer, 17
extended residue, 352

factor base, 238
factor method, 38
fast powers, 33
Fermat, P., 227, 280
Fermat number, 63, 180
Fermat's challenge, 227
Fermat's factorization algorithm, 170
Fermat's Little Theorem, 56
    for Gaussian integers, 327
    in quadratic field, 249
Fibonacci number, 2
Fibonacci pseudoprime, 264
Fibonacci pseudoprime challenge, 273
field, 303
finite simple continued fraction, 202
first category algorithm, 168
Flath, D., 330, 356
Ford, K., 69–70, 356
Fundamental Theorem of Arithmetic, 16

Gallian, J., 356
Garcia, M., 80
Gardner, M., 176
Gauss, C. F., 86, 182, 191, 289, 327, 330
Gauss's criterion, 186
Gauss's formula, 321
Gaussian integer, 303
Gaussian moat problem, 308
Gaussian prime, 304
gcd, 7, 304–305
Germain prime, 142
German currency, 162
Gethner, E., 308, 356
Gilbreath's conjecture, 142
GIMPS, 261
Giuga's conjecture, 142
Goldbach's conjecture, 142
golden ratio, 3
Goldwasser, S., 132
good, 289
googol, 34
Gordon, D., 315, 322, 356
Granville, A., 61, 355
Great Internet Mersenne Prime Search, 261
Grosswald, E., 356
group, 50, 160
Guy, R. K., 356

Hadamard, J., 103
Hajratwala, N., 261
Hardy, G. H., 315, 356
Heilbronn, H., 23

Hellman, M., 149
Hermite, C., 280
Hilbert, D., 290
Hilbert's tenth problem, 5
Hodges, A., 356
Holy Grail, 276
homogeneous equation, 42
Hudson, R. H., 356
Hurwitz, A., 211

IBM check digit scheme, 159
industrial-grade prime, 118
infinite simple continued fraction, 207
integer of quadratic extension, 329
integral domain, 304
involution, 298
ion, 291
Ireland, K., 356
irrational, 18
ISBN numbers, 165
isomorphic, 50
iterator, 334

Jacobi symbol, 189–190
Jones, J., 356
Jordan, J. H., 308, 322, 356
Julius Ceasar, 146

Kahn, D. 356
Kenjale, P. S., 162, 357
Kilian, J., 132
Klee, V., 356
Knuth, D. E., 356
Kraitchik, M., 169, 176, 238
Kraitchik family, 169
Kurtz, G. C., 356

Lacampagne, C. B., 356
Lagarias, J., 137
Lagrange, J. L., 289
Lagrange's theorem, 86
Lamé's theorem, 12
lattice point, 300
lcm, 7
least universal exponent, 93
Legendre, A.-M., 134, 182, 185
Legendre sum, 134
Legendre symbol, 185
Lehmer, D. H., 63, 67, 121, 131, 137, 139,
        240, 355
Lehmer, E., 30, 356
Lenstra, A. K., 131, 169
Lenstra, H. W., 131
li(), 103
Li, S., 129
liar, 133

Lichtblau, D., 54
linear Diophantine equation, 42
linear time, 34
lions, 316, 317
Littlejohn, L. L., 355
Littlewood, J. E., 112, 315
Littlewood's theorem, 112
$\log_f$, 71
logarithmic integral, 103
love potion, 76
Lucas, E., 131, 247
Lucas certification, 259
Lucas pseudoprime, 251, 264
Lucas pseudoprime challenges, 271
Lucas sequences, 247, 254
        computing 254
        divisibility of, 252
        prime testing with, 259
Lucas–Lehmer algorithm, 261
LucasV pseudoprime, 266

Machin's formula, 321
Madelung constant, 293
Manasse, M. S., 169
Martin, G., 70
Martin's number, 70
Matijasevic, Y., 227, 357
Maurer, U., 126, 131, 357
McConnell, M., 355
Meissel, E. D. F., 137–138
Mersenne primes, 262
Mertens, F., 108
Mertens's theorem, 108
method A, 269
method A^+, 277
method A^*, 269
method A^{**}, 269
Miller, G., 137
Miller–Rabin test, 118
modular square root, 195
modular towers, 95
modulus, 26
Monier, L.,, 64
Monte Carlo method, 170
Montgomery, H. L., 357
Montgomery, P., 176
Morris, S. B., 357
Morrison, M., 176
multiplicative function, 67
multiplicative inverse, 28
Murty, M. R., 356

n-animal, 315
Newman, D. J., 70
Niven, I., 357
nondeterministic polynomial time, 125

nonresidue (quadratic), 180
norm, 24, 303
normal form, 213
$\mathcal{NP}$, 125

$O()$, 35
odd perfect number problem, 75
odd unit fraction, 77
Odlyzko, A., 137
order, 84
order-of-powers theorem, 85

$\mathcal{P}$, 125
Packel, E., 357
Pascal's triangle, 29
peasant multiplication, 39
Pell's equation, 227
pentagon, group of symmetries, 160
Pépin, J. F. T., 180
Pépin's test, 183
perfect number, 72
    odd, 75
perfect shuffles, 39
Perrin pseudoprime, 278
Perrin's sequence, 278
Perrin's test, 278
phi function, 66
    Carmichael's conjecture, 69
    divisibility of $n-1$ by $\phi(n)$, 67
    Newman's theorem, 70
ϕ-multiplicity, 69
π, formulas for, 320
Pocklington, H. C., 126
Pollard, J. M., 170
Pollard $p-1$, 170, 174
Pollard rho, 170, 173
polynomial time, 35, 117, 125
Pomerance, C., 61, 117, 131, 176, 355, 357
postage stamp problem, 25
power grid, 56, 57
power tower, 94
Powers, R. E., 240
powers of matrices, 34
Pratt, V., 122, 125
Pratt certificate, 123
preperiod, 88
primary representation, 330
prime factorization, 6
prime k-tuples conjecture, 314
prime number, 6
    certification, 118, 120, 121, 259
    certified, 126
    cyclotomy test, 131
    finding the nth, 134
    industrial-grade, 118
    mysteries, 141

Prime Number Theorem, 103
    Erdős–Selberg proof, 108
primitive root, 57, 84, 330
primitive sum, 283
primitive triple, 18
proper unit fraction, 79
protolion, 323
pseudoperfect number, 79
pseudoperfect test, 271
pseudoprime, 59
    b-strong, 114
    comparison of psp tests, 272–273
    Euler, 193
    Fibonacci, 264
    Fibonacci challenge, 273
    Lucas, 251, 264
    Lucas challenges, 271
    LucasV, 266
    Perrin, 278
    quadratic, 265, 267
    strong psp test, 115
    strong quadratic psp, 273
    symmetric Perrin, 278
pseudoprime witness, 64
PSW challenge, 271
public key, 148
pure function, 342
Pythagorean triple, 18

quadratic congruences, 194, 199
quadratic irrational, 213
    periodicity of CF, 215
    purely periodic, 221
quadratic pseudoprime, 265
quadratic reciprocity, 182, 185
quadratic residue, 180
quartic residue, 325
quartic residue symbol, 327
quotient, 8

Rabin, M., 300
Rabung, J. R., 308, 322, 356
Rajaraman, V., 162, 357
Ramanujan, S., 329
rank, 251
rational, 18
rational integer, 303
rational point, 20
rational prime, 303
rectangular bulls, 233
relatively prime, 7
remainder, 8
Renze, J., 316, 357
repeating decimals, 87
represent, 289
repunit, 64

residue class, 26
Ribenboim, P., 357
Riemann, B., 105
Riemann hypothesis, 105
    extended, 117
Riemann zeta function, 105
Riesel, H., 357
Rivat, J., 137, 355
Rivest, R., 149, 176
Robin, G., 134
Robinson, J., 291
Rodemich, G., 315, 322, 356
Rosen, K. H., 357
Rosen, M., 356
Rosser, J. B., 134
RSA encryption, 149
Rubenstein, M., 112, 357
Rumely, R., 131

salt, 291
Sarnak, P., 112, 357
Sato, D., 356
Schlafly, A., 70, 357
Schneier, B., 357
Schroeder, M., 357
Schur's theorem, 47
second category algorithm, 169
Selberg, A., 108
Selfridge, J., 117, 121, 131, 175, 269,
    355–357
series rearrangement, 294
Serret, J.-A., 280
Sethi, A. S., 162, 357
Shallit, J., 300, 355
Shamir, A., 149, 176
Shanks, D., 355–356
Shanks's algorithm, 200
Sierpiński, W., 69
Sierpiński's conjecture, 69
sieve of Eratosthenes, 106
sigma function, 72
signature, digital, 151
Skewes number, 112
Smith, H. J. S., 280, 355
Smith's algorithm, 282
Solovay–Strassen algorithm, 193
square root, modular, 195
square triangular numbers, 231
SRK scheme, 162
Stark, H., 329, 356
Störmer's formula, 321
strong liar, 118, 132
strong pseudoprime, 114
strong pseudoprime test, 115
sum of divisors, 72

sums of squares, 285, 289
sun god cattle problem, 232
superdense constellation, 315
Sylvester, J. J., 75
symmetric Perrin pseudoprime, 278
Szekeres, G., 29

3-weight method, 159
time
    linear, 34
    nondeterministic polynomial, 125
    polynomial, 35, 117, 125
Tonelli's algorithm, 195
trial division, 167
Tuchman, B., 357
Turing, A., 149, 356
12-tone scale, 211
twin prime conjecture, 142
2-component, 311
2-pseudoprime, 59

U. S. Postal Service money orders, 158
unit, 7, 303
unit fraction, 79
universal exponent, 93

Vardi, I., 232–233, 238, 299, 316, 322, 357
Verhoeff, J., 160
Vorster, S. J. R., 355

Wada, H., 356
Wagon, S., 70, 308, 316, 355–357
Wagstaff, S. S., Jr, 64, 117, 269, 271, 355,
    357
Wallis, J., 149
Waring's problem, 290
Weiferich prime, 58
Wick, B., 308, 316, 356–357
Wiens, D., 356
Wiles, A., 41
Wilf, H., 80, 357
Williams, B. A., 48
Williams, H. C., 220, 356–357
Wilson's theorem, 49
Winters, S., 356
witness, 116, 119
Wolfram, S., 357
Wright, E. M., 356
Wurm, J. F., 233

Yao millionaire problem, 155

Zagier, D., 298, 357
Zimmermann telegram, 150
Zuckerman, H. S., 357